Advances in Bridge Design and Construction: Technologies and Applications

Contents

About the Editors

Munzer Hassan

Munzer Hassan is an associate professor in the Department of Construction Engineering at the École de Technologie Supérieure of the University of Quebec. He is dedicated to research in the Accelerated Bridge Construction (ABC) field, with a particular focus on concrete encased-steel (CES) piers. His primary interests also encompass bridge design, bridge rehabilitation, structure strengthening using FRP materials, soil–structure interaction, deep foundations and FE modeling.

Khaled Sennah

Khaled Sennah is a Professor of Structural Engineering in the Department of Civil Engineering at Toronto Metropolitan University, Canada. His interests focus on the design, evaluation and rehabilitation of bridges, with extensive projects advancing economic design and sustainable construction adopted in practice and standards. He has published over 370 papers, supervised more than 100 graduate students, received major ASCE and CSCE awards and is a Fellow of multiple international engineering academies.

Preface

Bridges are among the most critical components of modern infrastructure, serving as indispensable links that support economic development, social connectivity and regional resilience. Rapid urbanization, increasing traffic demands, aging infrastructure and heightened exposure to natural hazards have collectively placed unprecedented performance expectations on bridge systems worldwide. In this context, continued innovation in bridge design and construction is essential to ensure safety, durability, sustainability and cost-effectiveness throughout the life cycle of these structures.

This Special Issue brings together recent research developments and practical applications that address contemporary challenges in bridge engineering. The contributions reflect the growing integration of advanced analytical methods, novel materials and innovative construction techniques into both new bridge projects and the rehabilitation of existing structures.

Topics covered in this collection include material innovation, seismic resilience and structural optimization, digital design and construction technologies, construction monitoring and adaptive control, retrofitting and lifecycle management. Collectively, these studies contribute to a deeper understanding of bridge behavior and provide actionable insights for improving design reliability and construction efficiency.

This Special Issue is intended to serve as a valuable reference for researchers, practicing engineers and decision-makers involved in bridge engineering. By showcasing state-of-the-art technologies and their practical applications, it aims to foster knowledge exchange, inspire further innovation and support the development of safer and more resilient bridge infrastructure. The editors sincerely thank all authors and reviewers for their contributions and dedication, which have made this Special Issue possible.

Munzer Hassan and Khaled Sennah
Guest Editors

paper presents a comprehensive numerical study of their seismic performance under varying geometries and loading conditions. A design-oriented framework for assessing ductility and failure modes was proposed, offering guidelines for seismic-resistant bridge design using steel columns. The study contributes to the performance-based seismic design of bridge substructures.

Contribution (6) is titled Efficient Design Optimization of Cable-Stayed Bridges: A Two-Layer Framework with Surrogate-Model-Assisted Prediction of Optimum Cable Forces. Cable-stayed bridges are complex structures where optimization of cable forces is critical. This study proposes a two-layer optimization framework combining finite element modeling with surrogate modeling and machine learning to expedite the search for optimum cable forces. The framework reduces computational demand while maintaining accuracy, making it applicable to large-scale, practical bridge projects. The integration of AI into structural optimization exemplifies the future of computational bridge engineering.

Contribution (7) is titled 3D-Printed Concrete Bridges: Material, Design, Construction, and Reinforcement. Additive manufacturing is making its way into civil infrastructure. This review article provides an in-depth synthesis of 3D printing technologies applied to concrete bridges, covering material properties, design considerations, structural performance, and reinforcement strategies. The paper identifies current limitations, such as weak interlayer bonding and reinforcement integration, and outlines potential directions for future research. It serves as a roadmap for the gradual adoption of 3D printing in bridge construction.

Contribution (8) is titled Transition Effects in Bridge Structures and Their Possible Reduction Using Recycled Materials. Bridge transitions (e.g., between approach slabs and superstructures) often suffer from differential settlement and durability issues. This research explores the use of recycled materials, particularly rubber-modified backfill, to improve long-term performance. Laboratory testing and numerical analysis showed enhanced energy absorption and reduced stiffness contrast at transitions. The use of recycled materials contributes to both sustainability and functional performance, reflecting a growing trend in eco-conscious engineering.

Contribution (9) is titled The Seismic Behavior of Rectangular Concrete-Encased Steel Bridge Piers: A Review. Concrete-encased steel (CES) members combine the advantages of both materials and have demonstrated excellent seismic resilience. This paper provides a comprehensive literature review on CES bridge piers, focusing on their hysteretic behavior, failure mechanisms, and analytical modeling. The authors identify research gaps in cyclic loading behavior, confinement effects, and design codes. This review highlights the potential of CES piers in seismic bridge design and recommends further experimental and numerical research.

Contribution (10) is titled Retrofitting of Steel Structures with CFRP: Literature Review and Research Needs. Fiber-reinforced polymer (FRP) composites are widely used for retrofitting concrete structures, but their application to steel bridges is less mature. This review summarizes existing studies on CFRP retrofitting of steel members, focusing on adhesion mechanisms, fatigue performance, and environmental durability. The authors propose a research agenda addressing interface modeling, field validation, and long-term behavior. The article offers a foundation for expanding CFRP applications in steel bridge rehabilitation.

3. Conclusions

The contributions in this Special Issue represent a robust cross-section of the latest developments in bridge engineering, combining material innovation, computational advancements, construction techniques, and sustainability considerations. While each study targets a specific aspect of bridge design and construction, together they form a comprehensive response to the multifaceted challenges facing modern bridge infrastructure.

3.1. Key Outcomes and Advances

1. **Material Innovation**: Several articles emphasize the shift toward non-traditional materials. GFRP and CFRP (Contributions 1 and 10) offer promising alternatives to traditional steel, especially in corrosive environments and retrofitting applications. The incorporation of recycled and sustainable materials (Contribution 8) further reinforces the environmental responsibility now central to infrastructure design.
2. **Seismic Resilience and Structural Optimization**: Contributions 5 and 9 present advances in understanding seismic behavior and design of bridge substructures. These studies, combined with machine learning-based geotechnical characterization (Contribution 3), offer a pathway toward more resilient and site-specific seismic design practices.
3. **Digital Design and Construction Technologies**: Digital tools are transforming bridge design and analysis. The surrogate modeling and optimization techniques for cable-stayed bridges (Contribution 6) and the application of neural networks for geotechnical modeling (Contribution 3) point to the future of AI-assisted engineering. Moreover, the article on 3D printing (Contribution 7) pioneers the fusion of digital design with physical construction.
4. **Construction Monitoring and Adaptive Control**: Contributions 4 and 2 emphasize the role of construction technology and monitoring in ensuring build quality. As construction becomes more complex, the integration of real-time data and intelligent control methods will be vital in reducing risk and improving precision.
5. **Retrofitting and Lifecycle Management**: With many existing bridges aging beyond their design life, the importance of retrofitting technologies cannot be overstated. Contributions 9 and 10 address these needs by reviewing and proposing advancements in seismic retrofitting and CFRP reinforcement.

3.2. Bridging the Gap and Future Directions

Despite the progress highlighted in this Special Issue, several research gaps persist. Field validation of novel materials and methods remains a pressing challenge. Long-term performance data, especially in aggressive environments, is critical for mainstream adoption [4]. Additionally, while digital tools offer promising capabilities, their integration into daily engineering practice requires user-friendly interfaces, standardized workflows, and training. The bridge engineering community must also continue advancing performance-based design frameworks that consider not only strength but also durability, serviceability, and sustainability throughout the lifecycle [5]. There is a growing need for integrated approaches where structural, geotechnical, material, and construction perspectives are unified in decision-making.

This Special Issue provides a timely and insightful contribution to the field of bridge engineering. It demonstrates that the convergence of new materials, advanced analytics, and innovative construction methods is no longer aspirational, as it is actively shaping the present and future of bridge design and construction. We hope that the insights offered here will inspire further research, collaboration, and practical implementation that advance the state-of-the-art and state-of-practice in bridge engineering worldwide.

Conflicts of Interest: The authors declare no conflicts of interest.

List of Contributions:

1. Dervishhasani, G.; Sennah, K.; Afefy, H.M.; Diab, A. Ultimate Capacity of a GFRP-Reinforced Concrete Bridge Barrier–Deck Anchorage Subjected to Transverse Loading. *Appl. Sci.* **2025**, *15*, 7771. https://doi.org/10.3390/app15147771.
2. Hraib, F.; Tabsh, S.W. Influence of Girder Flaring on Load Effect in Girders of Composite Steel Bridges. *Appl. Sci.* **2025**, *15*, 4674. https://doi.org/10.3390/app15094674.
3. Boudghene Stambouli, A.; Guizani, L. Neural Network-Based Prediction of Amplification Factors for Nonlinear Soil Behaviour: Insights into Site Proxies. *Appl. Sci.* **2025**, *15*, 3618. https://doi.org/10.3390/app15073618.
4. Huang, H.; Xue, X.; Ma, H.; Zhou, Z.; Dou, J.; Yang, H.; Guo, E.; Ma, J. Construction Control Technology and Monitoring Analysis of Walking Incremental Launching Construction of Small-Curvature Steel Box Girder Bridges Across Expressways. *Appl. Sci.* **2025**, *15*, 585. https://doi.org/10.3390/app15020585.
5. Njiru, M.; Mamaghani, I.H.P. Seismic Design and Ductility Evaluation of Thin-Walled Stiffened Steel Square Box Columns. *Appl. Sci.* **2024**, *14*, 8554. https://doi.org/10.3390/app14188554.
6. Ma, Y.; Song, C.; Wang, Z.; Jiang, Z.; Sun, B.; Xiao, R. Efficient Design Optimization of Cable-Stayed Bridges: A Two-Layer Framework with Surrogate-Model-Assisted Prediction of Optimum Cable Forces. *Appl. Sci.* **2024**, *14*, 2007. https://doi.org/10.3390/app14052007.
7. Miri, Z.S.; Baaj, H.; Polak, M.A. 3D-Printed Concrete Bridges: Material, Design, Construction, and Reinforcement. *Appl. Sci.* **2025**, *15*, 3054. https://doi.org/10.3390/app15063054.
8. Spyrowski, M.; Ostrowski, K.A.; Furtak, K. Transition Effects in Bridge Structures and Their Possible Reduction Using Recycled Materials. *Appl. Sci.* **2024**, *14*, 11305. https://doi.org/10.3390/app142311305.
9. Moradian, M.; Hassan, M. The Seismic Behavior of Rectangular Concrete-Encased Steel Bridge Piers: A Review. *Appl. Sci.* **2024**, *14*, 6627. https://doi.org/10.3390/app14156627.
10. Delzendeh Moghadam, M.; Fathi, A.; Chaallal, O. Retrofitting of Steel Structures with CFRP: Literature Review and Research Needs. *Appl. Sci.* **2024**, *14*, 5958. https://doi.org/10.3390/app14135958.

References

1. Statistics Canada. Inventory of Publicly Owned Bridge and Tunnel Assets, Infrastructure Canada. Government of Canada, 2018. Available online: https://www150.statcan.gc.ca/t1/tbl1/en/tv.action?pid=3410007801 (accessed on 31 July 2025).
2. Qiu, Z.; Prabhakaran, A.; Su, L.; Zheng, Y. Performance-based seismic resilience and sustainability assessment of coastal RC bridges in aggressive marine environments. *Ocean. Eng.* **2023**, *279*, 114547. [CrossRef]
3. Tan, X.; Mahjoubi, S.; Zhang, Q.; Dong, D.; Bao, Y. A framework for improving bridge resilience and sustainability through optimizing high-performance fiber-reinforced cementitious composites. *J. Infrastruct. Preserv. Resil.* **2022**, *3*, 18. [CrossRef]
4. Lamei, T.; Abdel-Rahim, A.; Ibrahim, A. Development of performance-based deterioration models for bridge components. *Innov. Infrastruct. Solut.* **2025**, *10*, 230. [CrossRef]
5. Zhao, X.L.; Li, W.; Stanbrook, J. A Framework for the Integration of Performance Based Design and Life Cycle Assessment to Design Sustainable Structures. *Adv. Struct. Eng.* **2014**, *17*, 461–470. [CrossRef]

applied sciences

MDPI

Article

Ultimate Capacity of a GFRP-Reinforced Concrete Bridge Barrier–Deck Anchorage Subjected to Transverse Loading

Gledis Dervishhasani [1,†], Khaled Sennah [1], Hamdy M. Afefy [2,3,*] and Ahmed Diab [1]

1 Civil Engineering Department, Toronto Metropolitan University, Toronto, ON M5B 2K3, Canada; glediscenderv@gmail.com (G.D.); ksennah@torontomu.ca (K.S.); adiab@torontomu.ca (A.D.)
2 Construction Engineering and Management Department, Faculty of Engineering, Pharos University, Alexandria 21648, Egypt
3 Faculty of Engineering, Tanta University, Tanta 31527, Egypt
* Correspondence: hamdy.afefy@pua.edu.eg or hamdy.afefy@f-eng.tanta.edu.eg; Tel.: +20-106-177-3174
† Current address: Skampis Ltd., Toronto, ON M6B 4J3, Canada.

Abstract: This paper outlines a structural qualification process to assess the use of newly developed high-modulus (HM) glass fiber-reinforced polymer (GFRP) bars with headed ends in the joint between concrete bridge barriers and decks. The main goals of the study are to evaluate the structural performance of GFRP-reinforced TL-5 barrier–deck systems under transverse loading and to determine the pullout capacity of GFRP anchorage systems for both new construction and retrofit applications. The research is divided into two phases. In the first phase, six full-scale Test-Level 5 (TL-5) barrier wall–deck specimens, divided into three systems, were constructed and tested up to failure. The first system used headed-end GFRP bars to connect the barrier wall to a non-deformable thick deck slab. The second system was similar to the first but had a deck slab overhang for improved anchorage. The third system utilized postinstalled GFRP bars in a non-deformable thick deck slab, bonded with a commercial epoxy adhesive as a solution for deteriorated barrier replacement. The second phase involves an experimental program to evaluate the pullout strength of the GFRP bar anchorage in normal-strength concrete. The experimental results from the tested specimens were then compared to the factored applied moments in existing literature based on traffic loads in the Canadian Highway Bridge Design Code. Experimental results confirmed that GFRP-reinforced TL-5 barrier–deck systems exceeded factored design moments, with capacity-to-demand ratios above 1.38 (above 1.17 with the inclusion of an environmental reduction factor of 0.85). A 195 mm embedment length proved sufficient for both pre- and postinstalled bars. Headed-end GFRP bars improved pullout strength compared to straight-end bars, especially when bonded. Failure modes occurred at high loads, demonstrating structural integrity. Postinstalled bars bonded with epoxy performed comparably to preinstalled bars. A design equation for the barrier resistance due to a diagonal concrete crack at the barrier–deck corner was developed and validated using experimental findings. This equation offers a conservative and safe design approach for evaluating barrier–deck anchorage.

Keywords: bridge barriers; Test-Level 5 (TL-5); barrier–deck junction; anchorage resistance; experimental capacity; GFRP bars; pullout test

1. Introduction

The development of epoxy-coated reinforcement as a corrosion protection method was started in the early 1970s. Following successful demonstration projects in the mid-1970s, the adoption of epoxy-coated bars in highway bridges increased significantly, establishing it as the preferred corrosion protection technique. However, the first signs of inadequate field performance were noted in 1986 in Florida bridges, with additional examples reported in the United States and Canada during the 1990s, particularly in bridge barriers [1,2]. In environments subjected to de-icing, severe corrosion was observed on the outer side of the vertical bars on the traffic-facing side of the barrier wall, accompanied by section loss in the horizontal bars after 10 to 20 years of service. An investigation by the Ontario Ministry of Transportation (MTO) [3] indicated that the epoxy coating on steel bars deteriorates after a certain period, resulting in corrosion that leads to concrete cracking and spalling. Consequently, the MTO prohibited using epoxy-coated bars in bridge decks, while permitting the use of non-corrosive glass fiber-reinforced polymer (GFRP) bars to enhance bridge designs, thereby improving life expectancy and significantly lowering maintenance costs.

Traditionally, GFRP bars are anchored to concrete through mechanisms such as friction, mechanical adhesion, and mechanical interlock provided by the profile of the bar's surface. The necessary embedment length of the bar within the concrete is determined by its bond strength with the surrounding concrete and the specific end condition [4,5], which may include straight ends, 90-degree bends, or 180-degree hooks. In contrast to steel bars, which can be manually bent, GFRP bar bends and hooks are produced in a factory setting and subsequently transported to the construction site for installation. Due to the significant reduction in pullout strength at the curved sections of the bar, attributed to the reorientation of fibers [6,7], GFRP manufacturers have introduced GFRP bars with cast anchorage. This type of headed anchor bar is anticipated to alleviate reinforcement congestion at cast joints.

A limited number of researchers have examined the pullout capacity and bond strength of fiber-reinforced polymer (FRP) bars embedded in normal-strength concrete [8,9], as well as in high-strength concrete [4]. Additional research has focused on the pullout capacity of glass fiber-reinforced polymer (GFRP) bars within fiber-reinforced concrete [10,11], in geopolymer normal concrete [12,13], and in ultra-high-performance fiber-reinforced concrete (UHPFRC) [14–18]. Furthermore, Firas et al. [19] conducted a study on the bond strength of carbon fiber-reinforced polymer (CFRP) bars in UHPFRC.

Recently, some manufacturers specializing in glass fiber-reinforced polymer (GFRP) have launched GFRP bars that incorporate cast-headed ends, aimed at improving their anchorage with concrete. Several researchers have examined the pullout capacity and bond strength of these headed-end GFRP bars in various concrete types, including normal-strength concrete [6,20–23], high-strength concrete [4,5,24], and geopolymer normal concrete [13]. Johnson and Sheikh [25] conducted experimental studies to assess the viability of employing double-headed GFRP bars as shear reinforcement in concrete beams. Additionally, Sennah and Khederzadeh [26] and Sennah and Hedjazi [27] verified the use of GFRP bars with headed ends in TL-5 barrier design using vehicle crash tests, while verifying the design by loading the barrier–deck overhang to collapse [28–30]. Others verified the use of GFRP bar with 180° hook in Test-Level 5 (TL-5) barrier design using a crash test and static tests to collapse [31,32]. Hasaballa and El-Salakawy [33] investigated the behavior of beam–column joints reinforced with GFRP-headed bars under seismic conditions, concluding that the headed bars demonstrated superior seismic performance compared to bent bars. Furtehermore, El-Badry et al. [34] developed and tested a hybrid GFRP-reinforced bridge truss girder system, which incorporated headed-end GFRP bars for improved anchorage at truss joints under both fatigue and static loading conditions.

Bridge barriers, especially in Canada, are subjected to significant quantities of de-icing salts throughout the winter months, which serve as catalysts for the corrosion of conventional steel reinforcement. Figure 1 illustrates a photograph of a deteriorated bridge barrier resulting from the corrosion of steel bars and concrete spalling, highlighting the necessity for an alternative to steel reinforcing bars for the barrier wall. Given that GFRP bars do not possess steel properties, corrosion is effectively mitigated, making them an appropriate substitute [35–38]. While GFRP bars are resistant to corrosion, their long-term performance may vary based on exposure conditions, and design considerations should account for potential reductions due to environmental effects.

Figure 1. Photo of a deteriorated reinforced steel bridge barrier in Canada.

The use of GFRP bars in bridge barriers has drawn significant attention due to their corrosion resistance and high strength-to-weight ratio. Nepomuceno et al. [39] and Zheng et al. [40] emphasized that GFRP bond strength is highly sensitive to surface texture, embedment conditions, and environmental exposure, with durability compromised under aggressive agents like chlorides and alkalis. These findings stress the importance of robust anchorage detailing and adequate embedment lengths. Barrier–deck overhang joints for new construction must satisfy demanding load criteria, especially under MASH TL-5 conditions. The NCHRP Report 1078 [41] and Frosch & Morel [42] highlighted the need for stronger anchorage in thinner decks. Arrington et al. [43] showed that optimized anchorage improves impact resistance even with 5-inch deck thicknesses. Khederzadeh and Sennah [30] validated the performance of sand-coated GFRP bars under static loading, supporting their applicability in such configurations.

Precast or replacement of barrier–deck joints benefits from advanced connectors. Zhao et al. [44] and Nguyen et al. [45] tested grouted sleeves and loop-bar connectors, demonstrating high impact resistance. Yu et al. [46] optimized prefabricated connector geometry, while Alaywan [47] and Azimi [48] demonstrated the feasibility of precast alternatives for accelerated bridge construction. Ecklund and Sritharan [49] further supported this through full-scale testing. Patel et al. [50] proposed an innovative precast barrier system improving constructability and safety. Coleman et al. [51] experimentally studied the effect of beam depth on anchorage strength of hooked and headed bars, while Cho and Choi [52] showed that anchorage of barrier posts or curbs is equally critical. A few studies [53–55] examined failure modes under horizontal loads. Ahmed et al. [56] and Azimi et al. [57] combined steel and GFRP in post-and-curb systems. Ultra-high-performance concrete (UHPC) was examined as a filling material in the barrier–deck overhang joints, and steel dowels for postinstalled barrier construction were explored to enhance prefabricated barrier strength [58,59]. Finally, recent investigations by Rostami et al. [60] and Azimi et al. [61] confirmed that postinstalled headed GFRP bars with epoxy provide reliable anchorage strength, making them viable for deteriorated barrier replacement on existing decks.

The integration of GFRP bars in bridge construction is on the rise, supported by comprehensive research supported by the Ministry of Transportation of Ontario (MTO) to evaluate the practical use of existing products. Following successful crash tests and static load evaluations on a developed MTO TL-5, GFRP-reinforced barriers [26,30], the MTO established Standard Drawing MTO-S110-92, which outlines the dimensions and GFRP specifications for use by designers and contractors. In this design framework, GFRP bars are employed to reinforce the barrier wall both vertically and horizontally, while headed-end bars are utilized to strengthen the junction between the barrier and the deck. The headed section of the bar is anticipated to enhance load-bearing capacity, thereby facilitating a more efficient design approach. The GFRP bars developed, which possess material strength and modulus of elasticity comparable to those used in the crash-tested barriers from other GFRP manufacturers, are recommended for barrier reinforcement per MTO Drawing MTO-S110-92. Given that GFRP bars are proprietary products, they must undergo evaluation for structural performance in the event of vehicle impact [62].

This paper presents an in-depth investigation into the structural performance of newly developed high-modulus (HM) GFRP bars with headed ends used for anchoring concrete bridge barriers to deck slabs. The research is divided into two main phases. The first phase involves experimental testing of six full-scale TL-5 bridge barrier–deck specimens subjected to transverse loading until failure. These specimens represent three anchorage systems: (i) preinstalled headed-end GFRP bars in a thick non-deformable deck slab, (ii) the same anchorage detail applied to a flexible deck overhang, and (iii) postinstalled straight GFRP bars bonded with epoxy to simulate retrofitting deteriorated barriers. Crack development, failure modes, displacements, and strain evolution were thoroughly monitored and analyzed to assess structural behavior. Besides, the experimental capacities compared to the analytical resistances calculated using the equations proposed in the existing literature [63,64]. In the second phase, direct pullout tests were performed on GFRP bars with various embedment conditions to quantify anchorage capacity and compare it to applied forces at the barrier–deck interface due to vehicle impact. The results are used to validate a design equation for diagonal cracking at the barrier–deck junction and to confirm the adequacy of the GFRP anchorage under factored CHBDC loading. This comprehensive approach supports both new construction and rehabilitation of concrete bridge barriers using GFRP reinforcement.

2. Experimental Program

2.1. Test Specimens

A total of six full-scale Test-Level 5 (TL-5) barrier specimens, each with a length of 900 mm, were constructed and subjected to transverse loading that simulated vehicle impact until complete failure. This was conducted to assess their ultimate load-carrying capacities and the modes of failure at the deck barrier joint. In the case of barrier walls installed on top of solid slab bridges and voided slab bridges, the barrier wall is considered connected to a non-deformable thick slab. Specimen B-1 shown in Figure 2a represents this scenario where the barrier wall is fixed to a 500 mm thick slab resting on the laboratory floor to prevent flexural deformation. The dimensions and GFRP arrangement are identical to the TL-5 barrier specified in MTO Standard Drawings [65]. The barrier is reinforced with 15M GFRP vertical bars at the front face at 300 mm spacing and 13M GFRP vertical bars at the back face of the barrier wall at 300 mm spacing. All vertical GFRP bars are embedded into the deck slab with a vertical embedment length of 195 mm. The deck slab was reinforced in the main direction with M20 steel bars at 100 mm spacing. This barrier specimen represents the interior portions between the expansion joints or ends of the barrier wall. However, the exterior portion of the barrier wall is expected to be more critical than the interior portion when impacted by a vehicle. As a result, MTO Standard Drawing specifies vertical bar spacing at the traffic side of the barrier as half the spacing between similar bars at interior locations. Thus, Figure 2b shows a schematic diagram of specimen B-2 that represents the end segment of the TL-5 barrier on which the vertical reinforcement at the front face of the barrier is doubled. In this case, 15M GFRP vertical bars were used at the front face of the barrier at 150 mm spacing, while the vertical reinforcement at the back face of the barrier was kept as 13M at 300 mm spacing.

Specimens B-3 and B-4 represent barrier walls installed over a deck slab overhang in a slab-on-girder bridge. Figure 2c shows a schematic diagram of specimen B-3 that represents the interior segment of the TL-5 barrier. The barrier is reinforced with 15M GFRP vertical bars at the front face at 300 mm spacing and 13M GFRP vertical bars at the back face of the barrier wall at 300 mm spacing. All vertical GFRP bars are embedded into a 250 mm thick deck slab overhang with a vertical embedment length of 195 mm. The deck slab overhang was reinforced in the main direction with M20 steel bars at 100 mm spacing. Figure 2d shows a schematic diagram of specimen B-4 that represents the end segment of the TL-5 barrier on which the vertical reinforcement at the front face of the barrier is doubled. In this case, 15M GFRP vertical bars were used at the front face of the barrier at 150 mm spacing, while the vertical reinforcement at the back face of the barrier was kept as 13M at 300 mm spacing.

Specimens B-5 and B-6 illustrate the scenario of the replacement of the deteriorated bridge barrier in currently existing bridges. Figure 2e shows a schematic diagram of the postinstalled barrier wall in a thick concrete slab at an interior location, on which vertical bars are spaced every 300 mm. The barrier vertical bars are postinstalled into the deck over a vertical embedment length of 195 mm using Hilti epoxy. Figure 2f shows a schematic diagram of the postinstalled barrier wall in a thick concrete slab at an end location, on which vertical bars at the front face are spaced every 150 mm, while the rest of the reinforcement is identical to that in specimen B-5 at the interior location.

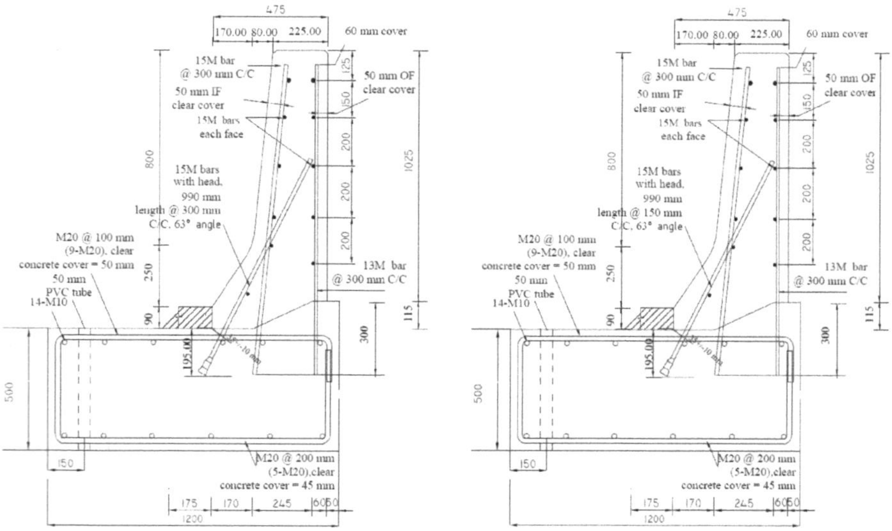

(**a**) Specimen B-1 with front vertical bars at 300 mm spacing (**b**) Specimen B-2 with front vertical bars at 150 mm spacing

Figure 2. *Cont.*

applied sciences

Editorial

Advances in Bridge Design and Construction: Technologies and Applications

Munzer Hassan [1,*] and Khaled Sennah [2]

1 Department of Construction Engineering, École de Technologie Supérieure, Montréal, QC H3C 1K3, Canada
2 Department of Civil Engineering, Toronto Metropolitan University, Toronto, ON M5B 2K3, Canada; ksennah@torontomu.ca
* Correspondence: munzer.hassan@etsmtl.ca

1. Introduction

Bridges are critical elements of transportation networks, which, in turn, play a crucial role in the development of communities, cities, and nations. Throughout human history, the design and construction of bridges have represented the technological development of countries and are points of pride. With the Industrial Revolution, and especially after the Second World War, the construction of bridges experienced a significant boom in all industrial countries. These bridges, built 50 to 75 years ago, have reached their useful lifespan and must be reinforced, rehabilitated, and sometimes monitored in real time while awaiting replacement [1]. Furthermore, the rehabilitation or reconstruction of bridges is subject to climate change and new, more restrictive seismic requirements [2]. To meet the aforementioned needs and properly design and build a new generation of bridges, researchers worldwide are working hard at developing innovative technologies and materials that help realize high-performance and durable bridges with full respect for the environment [3].

In this context, the Special Issue "Advances in Bridge Design and Construction: Technologies and Applications" was launched to showcase state-of-the-art developments, practical applications, and visionary research in bridge engineering. This Special Issue aimed to gather contributions from academic researchers and practicing engineers addressing innovative solutions and breakthroughs across several dimensions of bridge design, construction, maintenance, retrofitting, and monitoring. The theme of this Special Issue was built around addressing emerging challenges, including structural deterioration in aging infrastructure, seismic resilience, the integration of advanced materials like CFRP and GFRP, real-time health monitoring, digital construction techniques such as 3D printing, and environmentally responsible approaches involving recycled materials. These diverse yet interconnected research domains reflect the evolving multidisciplinary nature of modern bridge engineering. A notable observation is the increasing convergence between traditional structural engineering practices and modern technologies such as artificial intelligence, optimization algorithms, and digital fabrication. Furthermore, sustainability is no longer a peripheral concern; it has become central to the bridge design philosophy. Simultaneously, emphasis on resilience, particularly seismic resistance and redundancy in structural systems, highlights a shift toward proactive rather than reactive engineering strategies.

This Special Issue successfully brings together 10 high-quality contributions from various parts of the world, each addressing critical and timely topics. While the articles offer depth in their respective focuses, collectively they bridge critical gaps in understanding and provide actionable insights for practitioners, policymakers, and researchers alike. Beyond summarizing existing knowledge, the published works point to future directions and research needs in areas such as AI-based design, advanced construction methods, performance-based seismic assessment, hybrid material applications, and sustainable lifecycle practices.

As guest editors, we are pleased to observe how the contributions in this issue align with the overarching goals of enhancing the performance, longevity, and environmental compatibility of bridge infrastructure. This Special Issue not only reinforces the significance of multidisciplinary collaboration in addressing engineering challenges but also lays a strong foundation for subsequent research and innovation.

2. An Overview of Published Articles

This section provides a concise summary of the ten published contributions, highlighting the motivation, methods, and key findings of each study. Contribution (1) is titled Ultimate Capacity of a GFRP-Reinforced Concrete Bridge Barrier–Deck Anchorage Subjected to Transverse Loading. This study addresses a practical challenge in bridge safety, which is barrier anchorage design, especially when using non-corrosive materials such as Glass Fiber-Reinforced Polymer (GFRP) bars. The research employed large-scale experimental testing to evaluate the transverse capacity of GFRP-reinforced barrier–deck anchorage systems. The outcomes offer critical insights for engineers seeking to replace traditional steel bars with GFRP bars in bridge safety systems, especially for environments prone to corrosion.

Contribution (2) is titled Influence of Girder Flaring on Load Effect in Girders of Composite Steel Bridges. This analytical study investigates the influence of girder flaring on internal force distribution. Girder flaring is a geometric feature often required due to alignment constraints. Using finite element analysis and parametric modeling, the authors quantified the amplified stresses and redistribution effects caused by varying degrees of flaring. The findings emphasize the need for more refined design practices in composite girder bridges where non-parallel girder configurations may lead to unexpected stress concentrations.

Contribution (3) is titled Neural Network-Based Prediction of Amplification Factors for Nonlinear Soil Behaviour: Insights into Site Proxies. Geotechnical input remains a crucial factor in bridge design, particularly for seismic performance assessment. This study applies machine learning, specifically neural networks, to predict amplification factors related to nonlinear site responses. The authors utilize a database of numerical simulations to train models based on site proxy parameters. The approach enhances accuracy over traditional empirical equations and has implications for bridge site characterization in seismic zones.

Contribution (4) is titled Construction Control Technology and Monitoring Analysis of Walking Incremental Launching Construction of Small-Curvature Steel Box Girder Bridges Across Expressways. Incremental launching is an increasingly popular bridge construction method, but applying it in curved steel box girders presents unique challenges. This contribution reports on a real-world application in China, analyzing construction control methods and monitoring techniques used in the field. Data from sensors and deflection monitoring systems were integrated with theoretical predictions to ensure the geometric precision and safety of the launched structure. The study highlights advancements in real-time construction monitoring and adaptive control strategies.

Contribution (5) is titled Seismic Design and Ductility Evaluation of Thin-Walled Stiffened Steel Square Box Columns. Bridge piers with stiffened steel box sections are promising for seismic applications due to their high ductility and energy dissipation. This

Advances in Bridge Design and Construction: Technologies and Applications

Guest Editors

Munzer Hassan
Khaled Sennah

MDPI

Basel • Beijing • Wuhan • Barcelona • Belgrade • Novi Sad • Cluj • Manchester

Guest Editors

Munzer Hassan
Department of Construction Engineering
École de Technologie Supérieure
Montreal
Canada

Khaled Sennah
Civil Engineering Department
Toronto Metropolitan University
Toronto
Canada

Editorial Office
MDPI AG
Grosspeteranlage 5
4052 Basel, Switzerland

This is a reprint of the Special Issue, published open access by the journal *Applied Sciences* (ISSN 2076-3417), freely accessible at: https://www.mdpi.com/journal/applsci/special_issues/5W5O35I403.

For citation purposes, cite each article independently as indicated on the article page online and as indicated below:

Lastname, A.A.; Lastname, B.B. Article Title. *Journal Name* **Year**, *Volume Number*, Page Range.

ISBN 978-3-7258-6640-3 (Hbk)
ISBN 978-3-7258-6641-0 (PDF)
https://doi.org/10.3390/books978-3-7258-6641-0

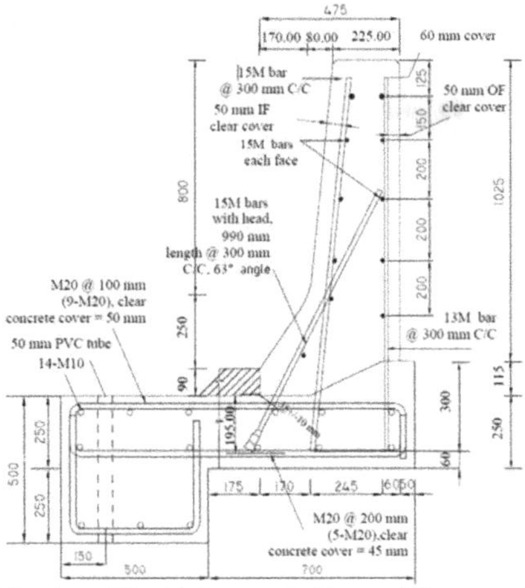

(c) Specimen B-3 with front vertical bars at 300 mm spacing

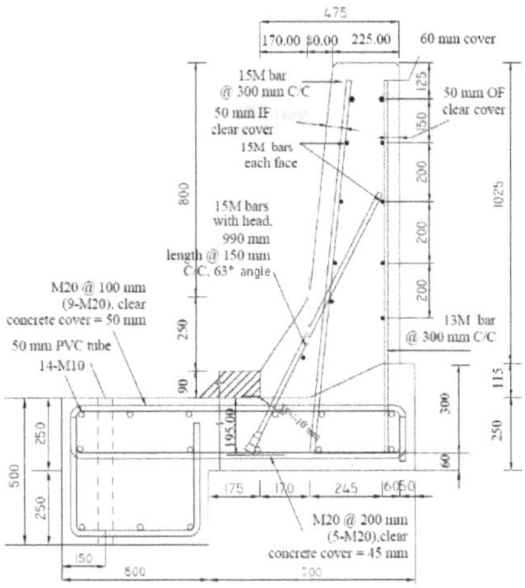

(d) Specimen B-4 with front vertical bars at 150 mm spacing

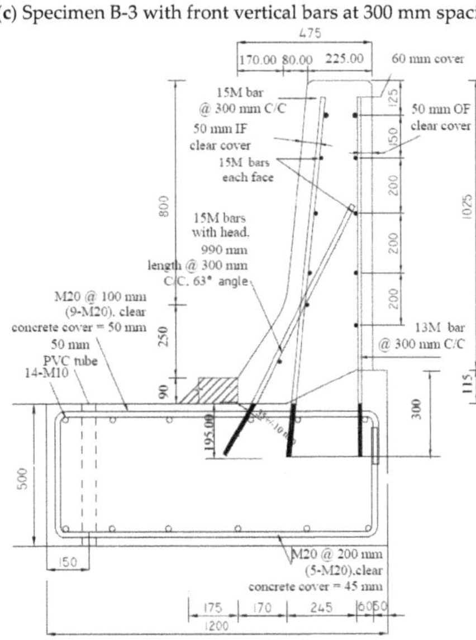

(e) Specimen B-5 with front vertical bars at 300 mm spacing

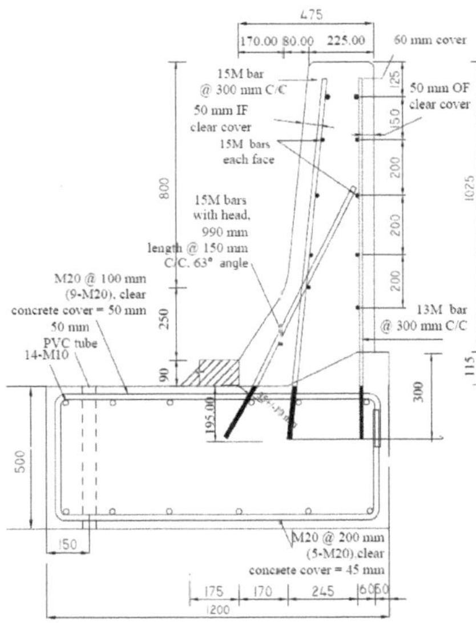

(f) Specimen B-6 with front vertical bars at 150 mm spacing

Figure 2. Concrete dimensions and reinforcement detailing for all specimens.

A preprepared wood formwork was designed and assembled using steel bolts to cast the specimens as shown in Figure 3a. After casting, the barriers were kept moist for the first four days by applying water to the wood formwork, which concealed the entire barrier exterior surface area. The barriers were not covered with any burlap or plastic but were left to cure in normal conditions inside the laboratory at approximately 23 °C and with a relative humidity of about 50% (the concrete cylinders were cured under the same conditions). After a week, the formwork for all barriers was removed. Figure 3b,c show typical specimens after formwork removal. Upon the removal of the forms, no deficiencies were observed, and there were no indications of honeycombing present.

(a) Typical formwork for TL-5 barrier specimens (b) B-1 after casting (c) B-3 after casting

Figure 3. Details for casting of specimens.

Approximately 2 months later, the two postinstalled specimens (B-5 and B-6) were prepared for barrier wall casting. The upper surface of the slab, where the wall was intended to be cast, was intentionally roughened to enhance the adhesion between the two separately poured concrete elements. The holes were first drilled to their specified depths with a concrete hammer drill using a 1.0625-inch bit. Then, the holes were blown out using an air compressor to remove heavy debris and dust; afterward, the hole was scrubbed with a steel brush throughout its entire depth to remove excess remaining debris/dust. The final two steps were repeated a total of five times. Hilti HIT-RE 500 V3 epoxy (Tulsa, OK, USA) [66] was then injected using the HDM 500 manual dispenser. Epoxy was injected at a slow rate to avoid improper consolidation, and the GFRP bars were installed slowly with a rotating motion about their vertical axis while being slowly pushed into the epoxy-filled hole. When drilling for the 63° and 84° angled holes took place, the drill was held at the correct location with the aid of supporting wood and a digital level to maintain a straight anchor hole. The straight GFRP bars at the front face, which were angled at 63°, had their strain gauges installed before anchorage in the slab. Once all bars were installed, the strain gauges were then run through thin plastic tubing leading out of the formwork through a drilled hole, identical to the preinstalled barrier specimens as depicted in Figure 4. Upon installation of all bars, in both specimens, epoxy was left to cure for about a week, before casting the wall section. The concrete was then poured for each wall, where cylinders were also cast at the same time. Concrete cylinders were placed beside the barriers to allow for similar curing conditions. Concrete followed the same curing condition as the previous four barrier samples, and was demolded about a week later.

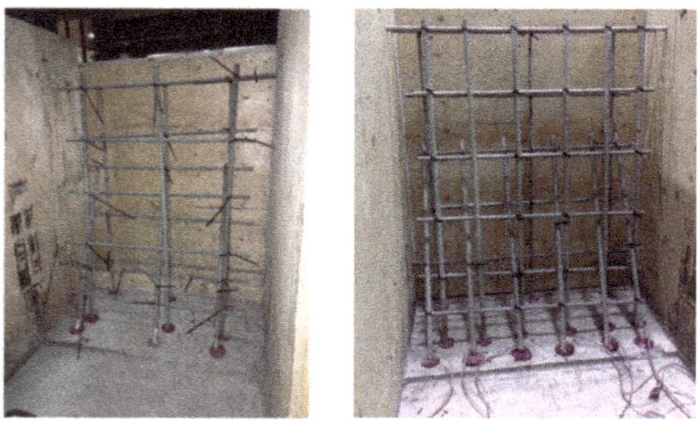

(a) Specimen B-5 (b) Specimen B-6

Figure 4. Postinstalled barrier walls before casting for specimens B-5 and B-6.

2.2. Material Properties

2.2.1. Concrete

All specimens were cast using ready-mix concrete with a target compressive strength of 35 MPa, with a Target slump = 150 mm. The quantities of the Cement Type GU, sand, and 20 mm aggregate were taken as 390, 733, and 1070 kg/m^3, respectively. The water content was 155 L/m^3, the air voids were at 6.5%, the high-range water-reducing admixture, was 400 mL/m^3, while the MasterAir, which is the air-entraining admixture, was 230 mL/m^3.

To determine the concrete's strength, three 100 × 200 mm concrete cylinders were collected from the concrete used in the casting of each barrier. Subsequently, the cylinders were tested on the same day the barrier specimens were loaded to failure. The test specimens were cast with two different patches. The first batch involved casting the four specimens having preinstalled TL-5 barriers (B-1, B-2, B-3, and B-4), as well as the deck slab for specimens B-5 and B-6. Then, the second batch was used for casting the barrier walls for specimens B-5 and B-6. For the postinstalled barrier specimens (B-5 and B-6), the ideal condition would have been to have concrete of the same strength at the time of testing for both the slab and wall. Cylinder testing data showed that the concrete strengths achieved between the two different segments were relatively close, as shown in Table 1. To consider the deviation of the strength value of each cylinder with the average value and the number of tested cylinders, the concrete characteristic strength was calculated using the following equation specified in CHBDC for bridge evaluation [62].

$$f'_c = 0.9\overline{f}_c\left[1 - 1.28\left[\frac{(k_c V)^2}{n} + 0.0015\right]^{0.5}\right] \quad (1)$$

where $\overline{f}_c$ = average cylinder strength, V = coefficient of variation of cylinder strengths, n = number of cylinders tested, and k_c = coefficient of variation modification factor for concrete, taking into account the number of tested cylinders (1.47) for three samples. Table 1 summarizes the values for concrete compressive strength at the time of testing for all specimens.

Table 1. Concrete compressive strength at the time of testing for all specimens.

Specimen	Average Compressive Strength (MPa)		Characteristic Value (MPa)	
	Deck Slab	Barrier Wall	Deck Slab	Barrier Wall
B-1	46.65		37.12	
B-2	52.46		43.27	
B-3	46.39		39.40	
B-4	51.80		43.57	
B-5	47.59	47.91	37.91	40.35
B-6	50.47	46.34	39.05	38.07

2.2.2. GFRP Bars

The GFRP bar for both 13M and 15M is Grade III type with a minimum tensile strength of 1000 MPa, modulus of elasticity of 60 GPa, ultimate tensile strain of 1.67%, and nominal cross-sectional area of 199 mm^2 per CSA-S807:18 Standard [67]. The use of headed-end GFRP bars was proposed in this research to allow for anchorage in concrete at a lower cost than the bend bars. The head of the 15M bar is 100 mm long. It begins with a wide disk of 38 mm diameter that transfers a portion of the load from the bar into the concrete. Beyond this disk, the head tapers in steps to the outer diameter of the bar as shown in Figure 5 to transfer load to the concrete by bearing. This geometry ensures optimal anchorage forces and minimal transverse splitting action in the vicinity of the head [68].

(**a**) Straight bars (**b**) Headed-end bar

Figure 5. Newly developed HM GFRP bar [68].

Compared to traditional anchorage methods such as 180° hooks or bent GFRP bars, headed-end GFRP bars offer notable advantages in constructability and performance. Bent GFRP bars are susceptible to strength reduction at the curved regions due to fiber misalignment, and their fabrication requires off-site forming, increasing logistical complexity. In contrast, headed-end bars provide reliable anchorage through mechanical bearing, maintain fiber continuity, and simplify installation by reducing reinforcement congestion. These benefits make them particularly suitable for barrier–deck junctions and retrofit applications, where space constraints and anchorage reliability are critical. The findings of this study reinforce the structural and practical value of headed-end GFRP bars as a preferred anchorage solution.

2.2.3. Epoxy Adhesive

The epoxy adhesive used for the postinstalled GFRP bars in specimens B-5 and B-6 exhibited a minimum bond strength of 10.8 MPa after 2 days of curing as indicated by the manufacturer [43]. Table 2 summarizes the epoxy specification as provided by the manufacturer. These properties in Table 2 represent the epoxy's key performance indicators for structural anchorage, ensuring adequate bond strength, stiffness, and durability when used with GFRP bars in concrete substrates. The epoxy is loaded with two separate cartridges in one unit, which is mixed at the time of use. The installation of the epoxy into the concrete substrate was accomplished using a specialized Hilti manual dispenser (Hilti Corporation: Tulsa, OK, USA), commonly referred to as a caulking gun. This product, designated as HDM 500, facilitates the simultaneous dispensing of the two components of the epoxy, thereby reducing the likelihood of errors in the mixing ratio.

Table 2. Epoxy specifications [66].

2-Day Cure Bond Strength (MPa)	Compressive Strength (MPa)	Compressive Modulus (MPa)	7 Day Tensile Strength (MPa)	Elongation at Break (%)	Absorption (%)
10.8	82.7	2600	49.3	1.1	0.18

2.3. Test Setup and Instrumentation

The concrete deck of the barrier was anchored to the 1000 mm laboratory rigid floor with two 50 mm diameter threaded rods, which were spaced 600 mm center-to-center from each other. The threaded rods were tightened using specified nuts and half-inch-thick plates acting as washers at the top and bottom of the rods. Above the bolted connection was a steel-intensive setup consisting of W shapes, HSS sections, steel plates, and other custom shapes. All the steel members acted as support for the load cell and hydraulic jack. In front of the hydraulic jack was a W-shaped member reinforced with several stiffeners, and it had a robust axis aligned with the direction of the load. Behind the W-shaped member was a piece of timber stud, cut to a trapezoidal shape, to fill the gaps between the vertical end of the loading system and the tapered surface of the front face of the barrier wall. The mechanism positioned in front of the load jack was employed to create a distributed load pattern. This setup was placed at the correct height, which was 990 mm above the concrete deck. Lastly, behind the barrier was an anchored HSS section with steel plates wedged between the gap of the barrier and the steel HSS. In the case of the barrier resting on the deck slab overhang, the same HSS section was used, but additional solid cylindrical steel sections were placed beneath the barrier. These measures were used to prevent any sliding of the barriers under horizontal loading and to minimize errors in displacement readings, as shown in Figure 6a.

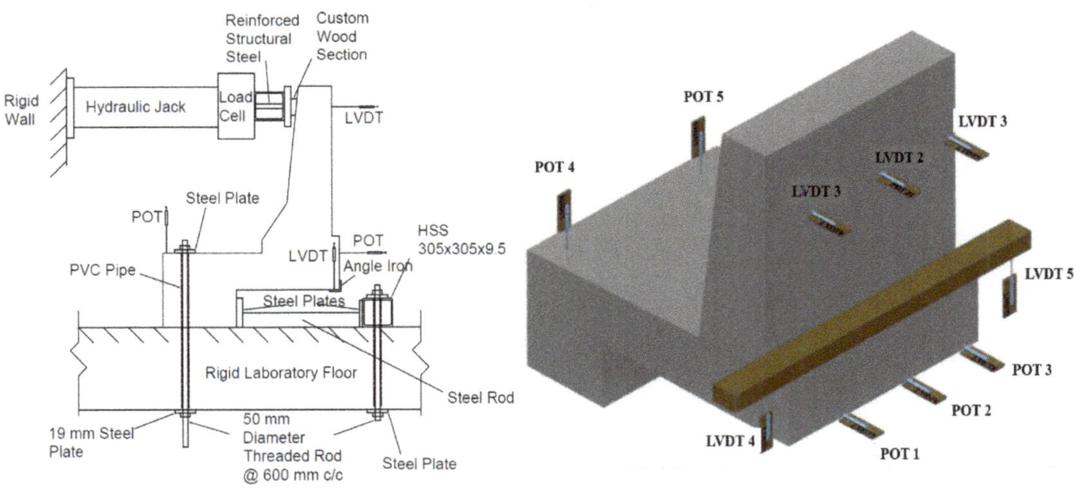

(**a**) Test setup showing the tie-down system (**b**) Locations of LVDTs and potentiometers (POTs)

Figure 6. Test setup.

The data acquisition unit was used to record and store the readings of all sensors during testing. A total of 10 linear variable displacement transducers (LVDTs) and potentiometers (POTs) were used in the case of a cantilevered barrier, and a total of 8 were used in the case of the barrier resting over a thick solid slab. Two POTs were used to measure the deck uplift displacement, while three POTs were used to measure the horizontal displacement of the deck slab at 500 mm from the top of the laboratory floor. Three LVDTs were used to measure displacement at a height of 1490 mm from the laboratory rigid floor level at the rear of the barrier wall, as this was the location directly behind the applied load. This setup of displacement monitoring was present for all barriers; however, barriers with a cantilevered portion had two extra LVDTs. These two extra LVDTs were installed in a position to measure the vertical deflection of the cantilevered portion of the deck slab, as shown in Figure 6b. To keep all LVDTs and POTs in their correct placements throughout testing for each barrier specimen, a wood structure was used for all sensors at the rear of the barrier, and the two POTs measuring uplift of the deck were mounted onto separate 2 × 4 timber studs, which were ultimately mounted onto external W-shaped pedestals. Three concrete strain gauges were installed at the rear of the barrier wall at a height of approximately 625 mm above the rigid floor. Barrier walls were loaded in increments of 10 kN and were held for approximately 2–3 min to allow for crack monitoring and marking. The test stopped when the specimen could not take any increase in applied loading.

3. Test Results and Discussion

Table 3 summarizes the experimental results, including failure load, maximum displacement, developed maximum tensile strain on the GFRP bars, as well as the developed maximum compressive strain on the concrete barrier wall for all specimens.

Table 3. Experimental results.

Criteria	Specimen					
	B-1	B-2	B-3	B-4	B-5	B-6
Experimental failure load (kN)	169	183	130	163	160	187
Experimental failure load per meter (kN/m)	187	203	144	182	178	208
Experimental resisting moment per meter (kN.m/m)	186	201	143	180	176	205
Developed design moment per meter (kN.m/m)	123 *	144 *	103 **	126 **	123 *	144 *
Capacity/demand ratio in design (experimental failure moment/ developed design moment)	1.51	1.40	1.38	1.43	1.43	1.42
Capacity/demand ratio using 0.85 environmental reduction factor	1.28	1.19	1.17	1.22	1.22	1.21
Top front displacement (mm)	26.32	23.74	44.75	66.76	8.92	11.28
Bottom back displacement (mm)	3.81	4.27	8.43	13.42	0.73	3.46
Overhang deflection (mm)	-	-	17.57	27.88	-	-
Front uplift (mm)	0.39	1.62	3.66	5.30	0.97	0.96
GFRP micro-strain	6504	4765	5802	5495	15,848	7226
Concrete micro-strain	−818	−868	−688	−1045	−893	−701
Observed failure mechanism	GFRP-concrete anchorage	Diagonal shear in the wall	GFRP-concrete anchorage	GFRP-concrete anchorage	Concrete breakout	Diagonal shear in the wall

* Omar, 2024 [64], ** Diab et al., 2024 [63].

3.1. Crack Pattern and Failure Modes

Figure 7a,b present various perspectives of the exhibited crack pattern in specimen B-1 after failure. The initial crack was detected on the front side of the barrier wall at the junction with the deck at a jacking load of 60 kN. Subsequently, a flexural crack emerged at the intersection just above the interface of the two tapered sections on the front side of the barrier wall at a jacking load of 85 kN. As loading continued, flexural cracks initiated at the barrier–deck junction when the load reached 100 kN. Although these flexural cracks at the barrier–deck junction extended deeper into the barrier thickness with increased loading, a sudden concrete breakout occurred at the embedded end of the GFRP bar at a load of 140 kN on both sides of the barrier wall. These breakout cracks propagated towards the top surface of the solid slab and the back face of the barrier wall under higher loads until the barrier could no longer withstand a jacking load exceeding 168.63 kN. To assess the damage to the bar head, the concrete surrounding the outer-headed bar was removed using a drill hammer, as illustrated in Figure 7c. Figure 7d shows the bar head with a fractured section at its largest diameter.

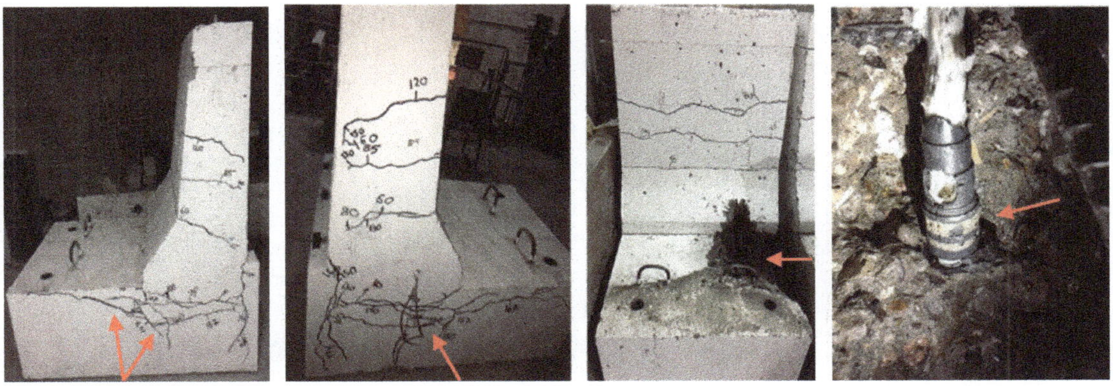

(a) (b) (c) (d)

Figure 7. Views of the crack pattern after failure for specimen B-1: (**a**) Right side concrete breakout; (**b**) Left side concrete breakout; (**c**) Failure of the specimen after jack hammering; (**d**) Close-up view of failed headed GFRP bar after jack hammering concrete at barrier–deck junction.

Specimen B-2 represents a barrier wall connected to a non-deformable concrete slab at the end location. The crack patterns observed in the tested specimen postfailure are illustrated in Figure 8a,b. The initial flexural crack became apparent on the front side of the barrier wall at the junction with the deck at a jacking load of 60 kN. Additional flexural cracks emerged above the junction of the tapered sections on the front side of the barrier wall at a jacking load of 80 kN. As the load increased, significant flexural cracking was noted in the tapered area of the barrier. At a jacking load of 140 kN, flexural cracks at the barrier–deck junction extended deeper into the barrier, indicating signs of flexural failure. However, a sudden diagonal shear failure in the concrete was observed in the upper tapered section of the barrier wall, initiating from the point of load application, at a jacking load of 170 kN on both sides of the barrier wall, as shown in Figure 8a,b. It is important to highlight that the diagonal shear crack extended through the thickness of the barrier to its rear side, resulting in concrete plugging and spalling, as illustrated in Figure 8c. The barrier was able to withstand loads up to 182.63 kN.

(**a**) (**b**) (**c**)

Figure 8. Views of the crack pattern after the test specimen B-2 failed. (**a**) Shear crack on the right side of the barrier wall. (**b**) Shear crack on the left side of the barrier wall. (**c**) Rear side concrete crushing at the end of the shear crack on the back side of the barrier wall.

Figure 9a,b illustrate various perspectives of the crack pattern observed in specimen B-3 after failure. The initial flexural crack was detected at the junction of the tapered sections on the front side of the barrier wall, as well as at the barrier–deck interface and the fixed end of the deck slab at jacking loads of 80 kN and 60 kN, respectively. As the load increased, these flexural cracks extended further, accompanied by additional flexural cracks within the deck slab and barrier wall, as depicted in Figure 9a,b. While flexural cracks were noted in both the barrier wall and the deck slab, a sudden diagonal tension crack emerged in the deck slab at a jacking load of 90 kN. Cracks continued to propagate until the barrier could no longer withstand a jacking load exceeding 129.64 kN. Specimen B-4 represented the end location of a barrier wall supported by a deck slab overhang. The crack patterns of the tested barrier specimen following failure are illustrated in Figure 9c,d. The first flexural crack was observed at the intersection of the tapered portions of the front side of the barrier wall as well as at the barrier–deck junction and at the deck slab fixed end at 50 kN jacking load. Flexural cracks penetrated further at a higher load, along with other flexural cracks in the deck slab as appeared in Figure 9c,d. However, a sudden diagonal tension crack appeared in the deck slab at a jacking load of 90 kN, respectively. Cracks propagated further till the barrier could not sustain a jacking load beyond 163.41 kN.

(**a**) (**b**) (**c**) (**d**)

Figure 9. Views of the diagonal tension crack at the barrier–deck overhang corner after the failure of the tested specimens B-3 and B-4. (**a**) Right side, B-3; (**b**) Left side, B-3; (**c**) Right side, B-4; (**d**) Left side, B-4.

Specimen B-5 represents an interior location of a barrier wall connected to a non-deformable concrete slab using postinstalled, straight-end GFRP bars. Figure 10a,b show different views of the crack pattern of the tested barrier specimen after failure. In this specimen, the first visible crack was observed in the front side of the barrier wall at the barrier–deck junction at 30 kN jacking load, followed by a flexural crack at the intersection of the two tapered portions of the front face of the barrier wall at 70 kN jacking load. These flexural cracks penetrated through the barrier thickness with an increase in the jacking load. However, sudden concrete vertical splitting appeared approximately at the embedded GFRP bar location at a load of 140 kN, followed by signs of concrete breakout close to the end of the embedded bar in the slab at a jacking load of 159 kN. These concrete breakout cracks extended laterally in the slab and towards the back face of the barrier wall at higher loads till the barrier could not sustain a jacking load beyond 159.74 kN. Specimen B-6 represents an end location of the barrier wall connected to a non-deformable concrete slab using postinstalled, straight-end GFRP bars. The amount of vertical reinforcement at the front face was doubled to represent the case of an end segment of the barrier wall as depicted in Figure 4b. Figure 10c,d show different views of the crack pattern of the tested barrier specimen after failure. The first visible crack was observed in the front side of the barrier wall at the barrier–deck junction at 20 kN jacking load, followed by a flexural crack at the intersection of the two tapered portions of the front face of the barrier wall at 60 kN jacking load. These flexural cracks penetrated through the barrier thickness with an increase in the jacking load. Nevertheless, a sudden shear crack appeared at the middle of the top tapered portion of the barrier wall at a jacking load of 180 kN and the specimens could not absorb the jacking load beyond 186.74 kN.

(a) (b) (c) (d)

Figure 10. View of the crack pattern after the test specimens B-5 and B-6 failed. (**a**) Right side concrete breakout, B-5; (**b**) Left side concrete breakout, B-5; (**c**) Diagonal shear failure at top right side of barrier wall, B-6; (**d**) Diagonal shear failure at top left side of barrier wall, B-6.

3.2. Load–Displacement Relationships

Figure 11 illustrates the relationship between the jacking load and displacement for all specimens at the preassigned locations, as shown in Figure 6. For specimen B-1, the average uplift of the deck slab and the horizontal displacement at failure were recorded at 0.39 mm and 3.81 mm, respectively. These measurements are deemed acceptable and do not significantly influence the structural performance of the barrier wall. The maximum lateral deflection of the barrier wall at failure is noted to be 26.32 mm, as referenced in Figure 11a, which is relatively minor considering the presence of a non-deformable deck slab. In the case of specimen B-2, where the sustained load increased due to a greater amount of longitudinal reinforcement at the end location, the average uplift of the deck slab and the horizontal displacement at failure rose to 1.62 mm and 4.27 mm, respectively. It is noteworthy that the maximum lateral deflection of the barrier wall at failure is 23.74 mm, as shown in Figure 11b, which is smaller than that observed for specimen B-1, despite the increase in the ultimate load.

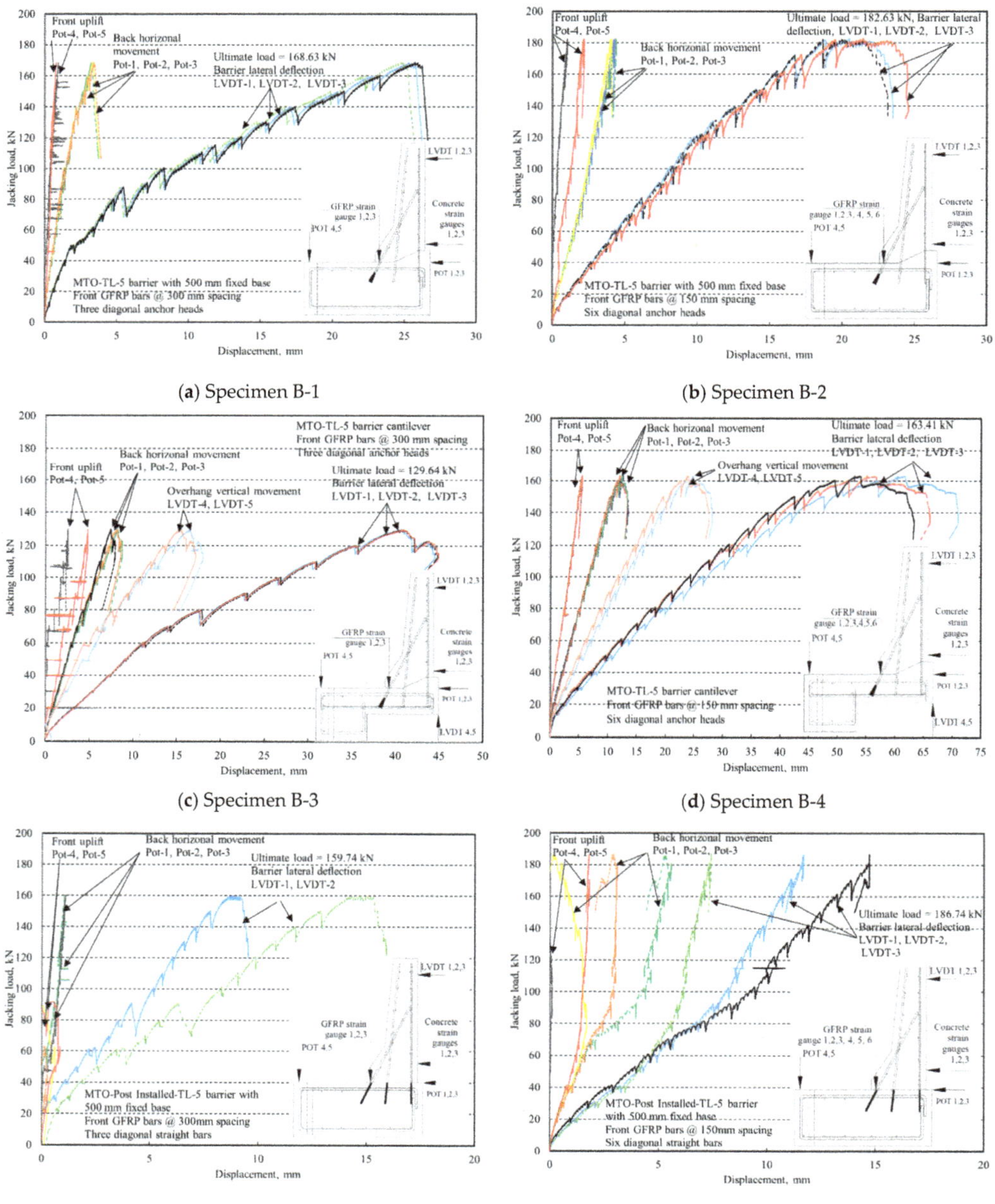

(**a**) Specimen B-1 (**b**) Specimen B-2

(**c**) Specimen B-3 (**d**) Specimen B-4

(**e**) Specimen B-5 (**f**) Specimen B-6

Figure 11. Load–displacement relationships for all specimens.

Regarding the interior location of a barrier–deck slab overhang system, as for specimen B-3, the average uplift of the deck slab and the horizontal displacement at the point of failure were measured at 3.66 mm and 8.43 mm, respectively. These measurements surpass those observed in other specimens (B-1 and B-2) due to the less effective tightening of the threaded rods compared to specimens B-1 and B-2. Nonetheless, these measurements can later be utilized to determine the net lateral deflection of the barrier. It is noteworthy that the maximum lateral deflection of the barrier wall at failure reached 44.75 mm, as illustrated in Figure 11c, while the maximum vertical deflection of the deck slab overhang was recorded at 17.57 mm. In contrast, for the end location represented by specimen B-4, the average uplift of the deck slab and the horizontal movement at failure rose to 5.3 mm and 13.42 mm, respectively. These values exceed those of specimen B-3, attributable to the increased failure load. The maximum lateral deflection of the barrier wall at failure was 66.76 mm, while the vertical deflection of the deck slab overhang was 27.88 mm, as shown in Figure 11d.

A comparison of the responses from barrier walls utilizing postinstalled GFRP bars and those with preinstalled GFRP bars indicates that, in most instances, the responses associated with postinstalled bars were lower than those of the preinstalled bars. This trend was observed for both interior and end locations. In the case of specimen B-5, the average uplift of the deck slab and the horizontal displacement at failure measured 0.97 mm and 0.73 mm, respectively, which are regarded as minimal values. Likewise, the maximum lateral deflection of the barrier wall at failure was recorded at 8.92 mm, as illustrated in Figure 11e. For specimen B-6 at the end location, the average uplift of the deck slab and the horizontal movement at failure were 0.96 mm and 3.46 mm, respectively, with the maximum lateral deflection of the barrier wall at failure reaching 11.28 mm, as shown in Figure 11f. It is important to note that the failure loads for both scenarios were comparable, with a variation of no more than 5%.

Specimen B-5 showed notably lower lateral and vertical displacements compared to deck-overhang specimens. This difference is due to the increased stiffness provided by the thick, non-deformable concrete slab base, which limited barrier rotation and uplift under load. In contrast, the deck-overhang specimens (e.g., B-3, B-4) experienced greater deflections due to cantilever flexibility. The results demonstrate that barrier systems anchored to rigid slabs offer enhanced resistance to displacement, validating the improved stiffness of postinstalled GFRP bars in non-deflectable deck conditions.

3.3. Load–GFRP Bar Tensile Strain Relationships

Strain gauges were fixed at the upper surface of the deck, specifically at the junction between the barrier and the deck, for all specimens to assess the tensile strain developed in the GFRP bars. Figure 12 illustrates the tensile strains observed in the diagonal GFRP bars located at the front face of the barrier under loading conditions. It is important to note that only three strain gauges were utilized for the interior specimens (B-1, B-2, and B-3), while six strain gauges were employed for the specimens at the end locations (B-2, B-4, and B-6). The average tensile strains recorded in the GFRP bars were 6504, 4765, 5802, 5495, 15,848, and 7226 $\mu\varepsilon$ for specimens B-1, B-2, B-3, B-4, B-5, and B-6, respectively. According to the manufacturer's certification data sheet, the ultimate strain for the GFRP bars is 16,700 $\mu\varepsilon$. Consequently, none of the specimens reached the maximum tensile strain at the measurement location, except specimen B-5. Although the experimental ultimate strain observed in specimen B-5 was slightly lower than the specified ultimate strain for the GFRP bars in the manufacturer's datasheet by 5%, tensile rupture did not occur in the GFRP. This is attributed to the fact that the design values provided by the manufacturer are based on statistical data, which indicates that the characteristic tensile strength of the bar is lower than the mean value.

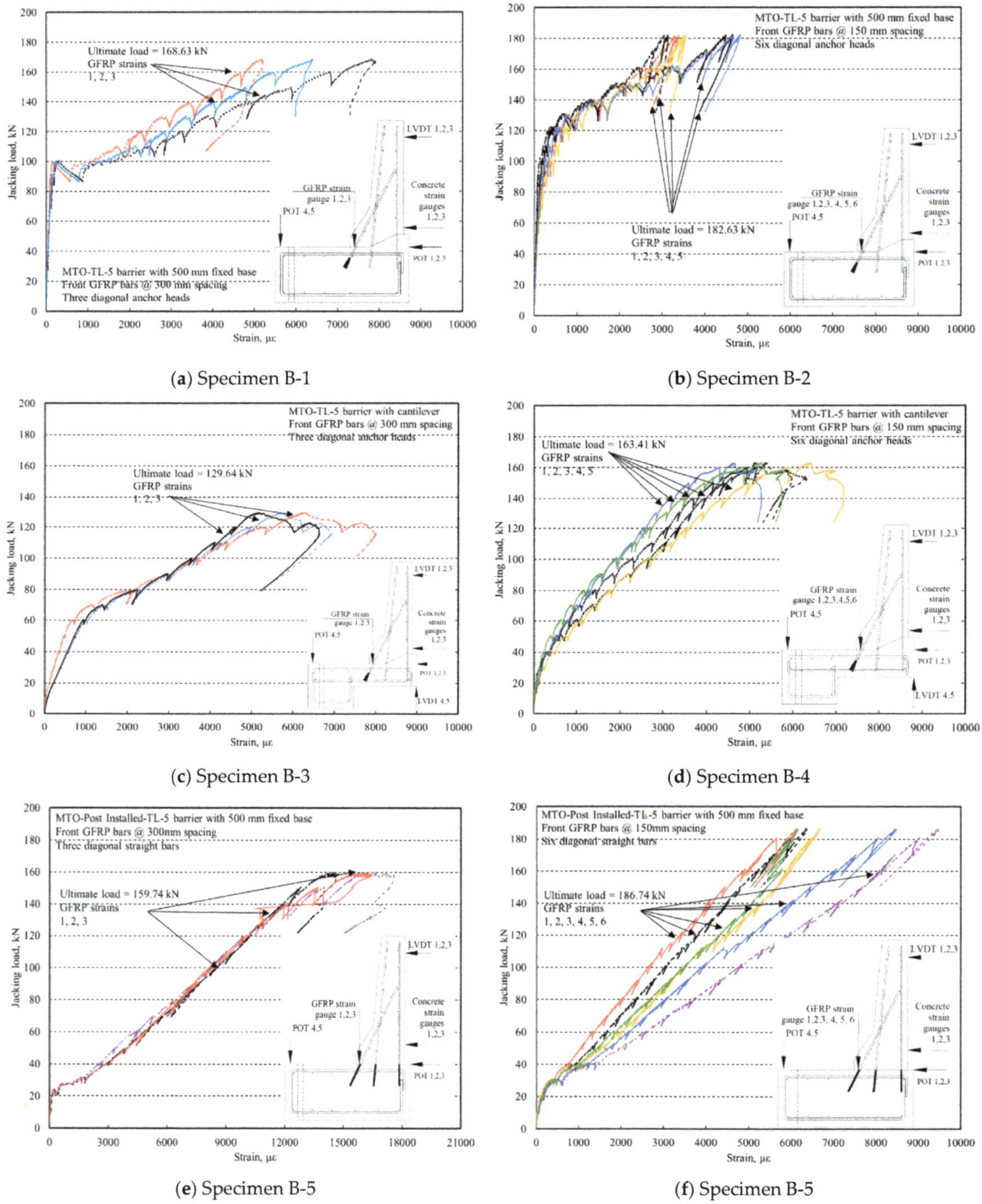

(**a**) Specimen B-1 (**b**) Specimen B-2

(**c**) Specimen B-3 (**d**) Specimen B-4

(**e**) Specimen B-5 (**f**) Specimen B-5

Figure 12. Load–GFRP bar strain relationships for all specimens.

3.4. Load–Concrete Strain Relationships for All Specimens

The developed concrete compressive strains were measured using strain gauges positioned 115 mm above the deck overhang's upper surface and at the barrier wall's front face at three distinct locations, as illustrated in Figure 8c. Figure 13 presents the load–concrete strain relationships for all tested specimens. The average concrete compressive strains at the point of failure were recorded as 818, 868, 688, 1045, 893, and 701 με for specimens B-1, B-2, B-3, B-4, B-5, and B-6, respectively. These values are significantly lower than the ultimate concrete compressive strain at failure, which is 3500 με. The observed low strain values can be attributed to the failure of the barrier–deck junction consistently occurring at locations distant from the measurement points.

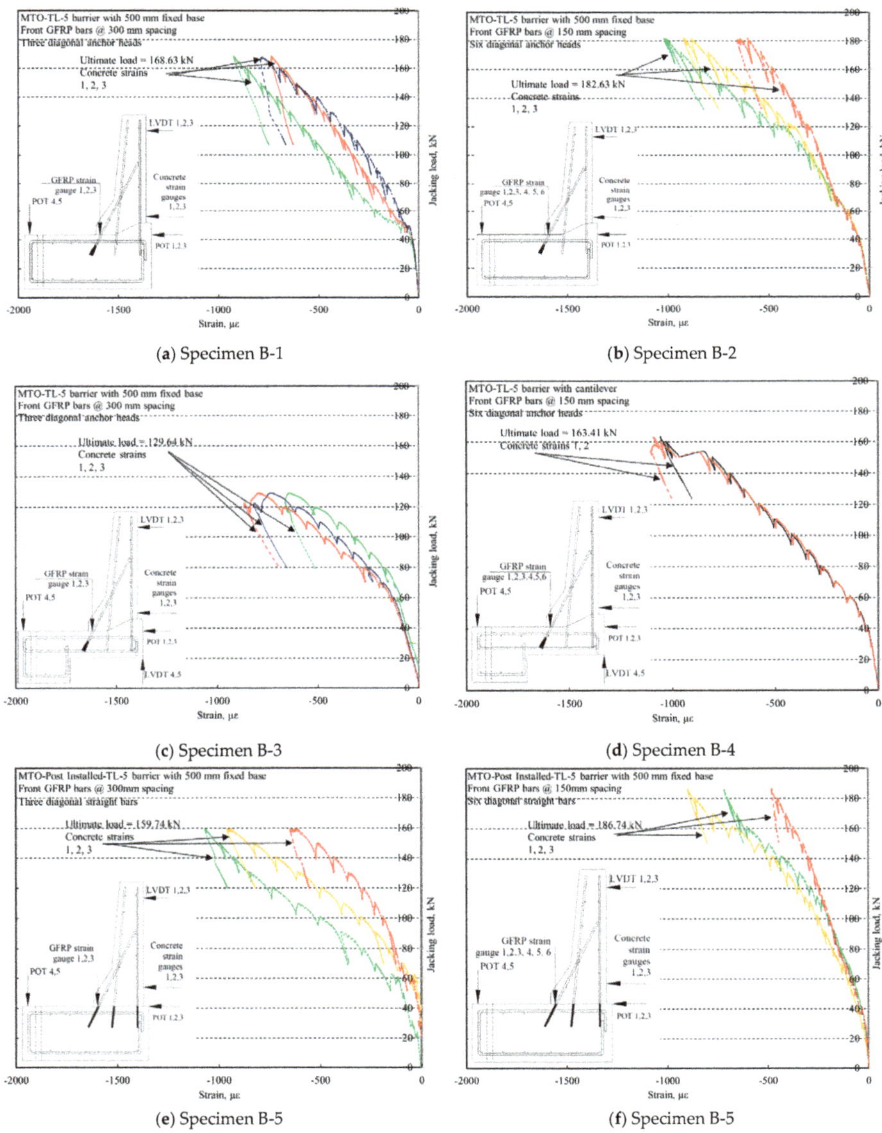

(a) Specimen B-1
(b) Specimen B-2
(c) Specimen B-3
(d) Specimen B-4
(e) Specimen B-5
(f) Specimen B-5

Figure 13. Load–concrete strain relationships for all specimens.

Strain data indicated that neither concrete nor GFRP bars reached their ultimate strain capacities prior to failure. Maximum concrete compressive strains remained well below 3500 $\mu\varepsilon$, and GFRP bar tensile strains were below the 1.67% rupture threshold. These findings confirm that failure occurred due to concrete cracking or GFRP bar pullout, not GFRP bar rupture. In overhang systems, diagonal tension cracks controlled failure, while in thick deck slabs, pullout failure dominated.

3.5. Barrier Maximum Moment Associated with the Diagonal Tension Crack in the Deck Slab Overhang

Upon the cracking of a reinforced concrete joint, the joint behaves as a composite structure composed of both concrete and reinforcement, making its analysis significantly more complex than that of homogeneous materials [69]. The failure at the joint may occur due to diagonal tension cracks, which arise when tensile stresses from external flexural moments are not adequately countered by the reinforcement, and when appropriate reinforcement configurations are lacking. Figure 14a illustrates the diagonal tensile crack in the deck overhang under the barrier wall due to the influence of a transverse load, F_t, while Figure 14b depicts a free-body diagram of the associated forces that develop the diagonal crack. Figure 14c depicts the stress distribution at the corner joint of the barrier wall resulting from the applied moments in both the wall and the deck slab. The bending stress, σ_x, reveals a high tensile stress at the inner corner of the joint, suggesting the potential for crack initiation at this location under minimal loading conditions. Conversely, the stresses perpendicular to the bending stresses, σ_y, generate tensile stresses that lead to diagonal cracking across the corner joint. Such diagonal tension cracks can precipitate abrupt failure unless suitable reinforcement arrangements are implemented in these critical areas [69].

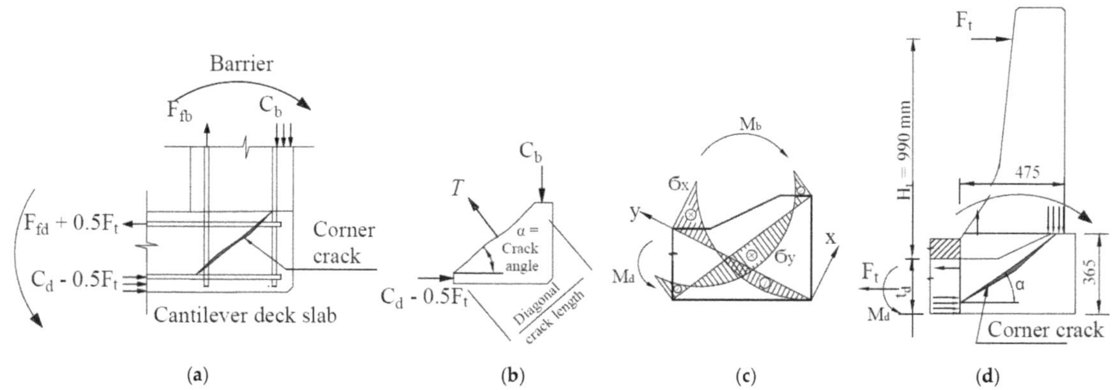

Figure 14. Forces and stresses due to diagonal tension failure at the barrier wall deck slab junction. (**a**) Internal forces; (**b**) Free-body diagram; (**c**) Stress distribution along and normal to the diagonal crack indicating tensile stresses (positive sign) and compressive stresses (negative sign); (**d**) Diagonal crack length measurement.

As shown in Figure 14a, the applied transverse load, F_t, produces a moment at the centerline of the deck overhang (i.e., mid-depth of the deck overhang), M_b, which is equated with the deck overhang internal couple as follows:

$$M_b = F_t\,(H_t + 0.5t_d) = (C_d - 0.5F_t).\ 0.9d_d \tag{2}$$

where H_t is the height of the applied transverse vehicle impact load from the top surface of the deck slab overhang, C_d is the resultant compressive force in the deck section at the inner side of the barrier wall, t_d is the deck overhang thickness, d_d is the effective depth of the deck overhang, and the distance between the resultant compressive and tensile forces is approximately taken as $0.9d_d$.

Analyzing forces parallel to the diagonal crack in Figure 14b leads to the following equation.

$$C_b.\sin\alpha = (C_d - 0.5F_t).\cos\alpha \quad (3)$$

where α is the inclination of the diagonal tension crack and C_b is the resultant compressive force in the barrier wall. To calculate the diagonal crack angle between C_b and $(C_d - 0.5F_t)$ forces, it is conversative to assume their locations are approximately 0.9 times the barrier thickness at the barrier–deck interface and 0.9 times the deck slab thickness at the inner face of the barrier wall, respectively, as depicted in Figure 14d.

Equation (3) is rearranged as follows.

$$C_b = \frac{C_d - 0.5F_t}{\tan\alpha} \quad (4)$$

By analyzing forces normal to the diagonal crack in Figure 14b, the tensile force creating the diagonal crack, T, is calculated as follows:

$$T = C_b.\cos\alpha + (C_d - 0.5F_t).\sin\alpha \quad (5)$$

Substituting Equation (4) in Equation (5) yields the following equation.

$$(C_d - 0.5F_t) = T.\sin\alpha \quad (6)$$

The modulus of rupture of concrete, f_r, is calculated as follows [70].

$$f_r = 0.6\lambda\sqrt{f'_c} \quad (7)$$

where λ = concrete density factor and f'_c = concrete compressive strength.

The tensile capacity for a concrete section of width b at the diagonal crack is calculated as follows, considering a parabolic distribution of the tensile stress along the diagonal crack length, ℓ_{dc}.

$$T = \frac{2}{3} f_r \cdot b \cdot \ell_{dc} \quad (8)$$

By substituting equations 6 and 8 into Equation (2), the maximum moment at the centerline of the deck overhang at the inner side of the barrier wall is as follows.

$$M_b = 0.6\ f_r \cdot b \cdot d_d \cdot \ell_{dc} \cdot \sin\alpha \quad (9)$$

Equation (9) is applied to specimens B-3 and B-4 where the barrier is mounted over a deck overhang. The concrete compressive strengths are taken as the characteristic values in Table 1, with 39.4 MPa and 43.57 MPa for specimens B-3 and B-4, respectively. For a concrete cover to the center of the GFRP bar in the tension side of the deck overhang of 58 mm, the effective depth, d_d, is taken as 192 mm. The width of the section is taken as 1000 mm. Per Figure 14d, the diagonal crack angle was calculated as 37.5°. The maximum moments at the deck overhang centerline when the diagonal tension crack occurs are 142 kN.m/m and 150 kN.m/m, for specimens B-3 and B-4, respectively. The corresponding experimental ultimate moments are taken as the maximum experimental transverse load in Table 3 times (Ht + 0.5td), which is (0.99 + 0.25/2 = 1.115 m). Thus, this leads to corresponding experimental ultimate moments of 161 kN.m/m and 203 kN.m/m for specimens B-3 and B-4, respectively. The ratios between the experimental ultimate moment and the theoretical moment are 1.13 and 1.36 for specimens B-3 and B-4, respectively. When using the average concrete compressive strength rather than the characteristic values in Table 1, the ratios between the experimental ultimate moment and the theoretical moment are 1.04 and 1.24 for specimens B-3 and B-4, respectively. One may observe that Specimen B-3 and B-4 are identical except that the vertical GFRP bars at the front of the barrier are doubled in the latter, and the middle vertical bars pass through the diagonal crack at the deck-barrier corner. This may be the reason the experimental–theoretical moment ratio for specimen B-4 is greater than that of B-3. In any case, the moment ratios of being slightly more than 1 confirm the applicability of the developed equation 8 in the design to preclude the failure due to diagonal tension crack at the barrier–deck corner.

4. Moment Capacity-to-Demand Ratio at the Barrier–Deck Anchorage

Given the width of the barrier of 900 mm and the height of the applied load over the top surface of the deck slab of 990 mm, the experimental resisting moments are calculated as 186, 201, 143, 180, 176, and 205 kN.m/m, respectively, for specimens B-1, B-2, B-3, B-4, B-5, and B-6. These experimental values were then compared to the analytical resistances calculated using the equations proposed in the existing literature. For the TL-5 barrier wall, which is rigidly connected to a non-deformable concrete deck slab, the experimental findings are compared with those calculated by [64] Conversely, for the barrier wall connected to a slab overhang, the results were compared with those presented by [63], considering a 10 m long continuous barrier in the direction of traffic, as detailed in Table 3. Results show that the capacity-to-demand ratios were 1.51, 1.4, 1.38, 1.43, 1.43, and 1.42 for specimens B-1, B-2, B-3, B-4, B-5, and B-6, respectively. These values are much greater than 1, underscoring the adequacy of the GFRP detailing at the barrier–deck joint.

Because the long-term exposure to various environmental conditions may reduce the tensile strength, creep rupture, and fatigue endurance of GFRP bars [71], the 2018 edition of the AASHTO LRFD Bridge Design Guide Specifications for GFRP-Reinforced Concrete [72] specifies environmental factors of 0.8 and 0.7 for concrete not exposed or exposed to earth or weather, respectively. Benmokrane et al. [73] determined that an environmental reduction factor value of 0.85 was appropriate for GFRP bars based on a compiled database of 361 accelerated aging tests of unstressed bars. The 2024 Guide for the Design and Construction of Structural Concrete Reinforced with Fiber-Reinforced Polymer (FRP) Bars, ACI 440.1R-24 [74], adopted this environmental reduction factor for concrete both exposed and not exposed to earth or weather. In this research, an environmental reduction factor of 0.85 was applied to the capacity-to-demand ratios in Table 3. This leads to capacity-to-demand ratios of 1.28, 1.19, 1.17, 1.22, 1.22, and 1.21 for specimens B-1, B-2, B-3, B-4, B-5, and B-6, respectively. These values are still greater than 1, underscoring the adequacy of the GFRP detailing at the barrier–deck joint.

5. Applicability of Static Testing to Dynamic Crash Performance

While static testing does not replicate the strain rate effects and inertia forces present during actual crash events, it remains a widely accepted method for evaluating the ultimate capacity and failure mechanisms of structural components. In fact, static load testing has been used in prior qualification studies of GFRP-reinforced barriers (e.g., [26,28–30,32]) to complement full-scale crash testing and to validate design assumptions under controlled conditions.

The Canadian Highway Bridge Design Code, CHBDC [62], specifies that barrier design is to be based on the use of relevant and existing full-scale crash test data as well as a commonly used method for ensuring that the barrier design satisfies the crash test requirements. CHBDC specifies that changes to the details of a traffic barrier or traffic barrier transition that meet the crash test requirements may be made, provided that any changes affecting the geometry, strength, or behavior of the traffic barrier can be demonstrated not to adversely affect vehicle–barrier interaction. The CHBDC clause for barrier anchorage to the deck states that the suitability of a traffic barrier anchorage shall be based on its performance during crash testing of the traffic barrier. For an anchorage to be considered acceptable, significant damage shall not occur in the anchorage or deck during crash testing. If crash testing results for the anchorage are not available, the anchorage and deck shall be designed to resist the maximum bending, shear, and punching loads that can be transmitted to them by the traffic barrier, except that these loads need not be taken as greater than those resulting from the CHBDC traffic loads (i.e., equivalent transverse vehicle impact loading). This can be achieved by conducting static tests to collapse on the anchorage and/or performing analysis using an approved numerical method such as finite-element computer modeling.

The commentaries of CHBDC suggest the following examples where changes may potentially be acceptable: (a) changes to an anchorage system, where failure of the anchorage is not exhibited during crash testing and the strength of the replacement anchorage is determined to be equivalent to that of the original, and (b) substituting materials with properties identical or superior to the original material as far as behavior during crash testing is concerned.

In conclusion, although dynamic effects such as inertia and strain-rate sensitivity were not directly simulated, the static testing methodology employed in this research is consistent with accepted engineering practices for validating ultimate capacity and failure mechanisms. The GFRP-reinforced systems demonstrated performance exceeding factored design loads, and their structural adequacy was further corroborated by previous full-scale crash tests cited in the study. Therefore, the findings remain applicable to real-world crash scenarios, especially in conjunction with the design provisions and interpretive clauses of the CHBDC that allow for such validation methods.

6. Pullout Strength of GFRP Bars

To quantify the pullout capacity of the GFRP bars used in the experimental program, two cases were considered: preinstalled and postinstalled. In the case of preinstalled bars, a reinforced concrete slab of 1300 mm width, 7500 mm length, and 300 mm thickness, having an average concrete strength of 40 MPa, was prepared and cast. Steel bars of 15M in size and spacing of 300 mm center-to-center were used to reinforce the slab in both directions at the top and bottom. The cast concrete slab had three separate regions to allow for testing different bar configurations, namely: (i) preinstalled GFRP bars with straight ends, (ii) preinstalled GFRP bars with headed ends, and (iii) preinstalled GFRP bars with headed-end and unbonded straight portion. These regions are marked as Groups 1 through 3 in Figure 15a. Group 1 of bar pullout testing consisted of five straight GFRP bars, which are fully bonded to concrete with an embedment length of 195 mm; Group 2 incorporates the same type and number of headed fully bonded bars for the entire 247 mm embedment length. Group 3 consisted of five headed-end GFRP bars, which were not fully bonded to the concrete. The total embedment length of the bar was taken as 247 mm from the top of the slab; however, only the bottom 100 mm was fully bonded to the concrete. To achieve the unbonded condition, pipe insulation foam was used to wrap around the top 147 mm of the embedded length of the bar as depicted in Figure 15b. The reasoning behind using these conditions is to investigate the true pull-out strength contribution of the added anchor head.

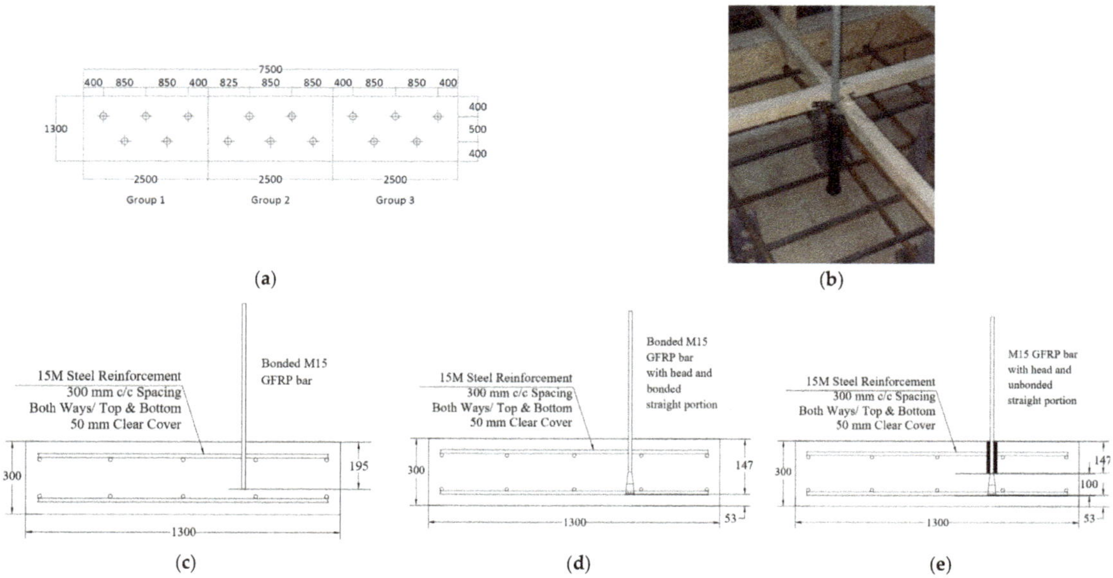

Figure 15. Layout and slab formwork with steel and GFRP reinforcement before casting for preinstalled GFRP bars: (**a**) Plan of the concrete slab showing preinstalled GFRP bar locations; (**b**) Headed bar with unbonded straight portion; (**c**) Group 1: GFRP bars with straight ends; (**d**) Group 2: Headed-end GFRP bars; (**e**) Group 3: Headed-end GFRP bars with unbonded straight portion.

As for postinstalled bar pullout capacity testing, the same slab used for preinstalled bar testing was utilized. After testing the preinstalled GFRP bars, two bar groups were installed in the deck slab in locations marked with solid circles in Figure 16. Every group utilizes the same #5 bar, though with varying embedment lengths. Specifically, Groups 1 and 2 employed embedment lengths of 150 mm and 195 mm, respectively. Each group comprised 5 bars, resulting in a cumulative total of 15 bars, which includes an additional 5 bar specimens.

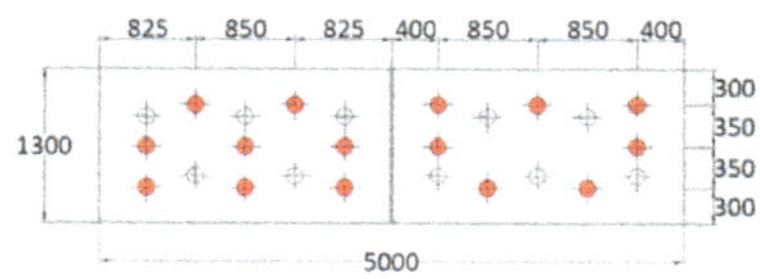

Figure 16. Plan of concrete slab showing GFRP postinstalled bar locations.

In the process of installing the postinstalled GFRP bars, a hole was initially created using a concrete impact drill. A drill bit with a diameter of 27 mm was selected to accommodate the #5 bar size and ensure a proper bond between the epoxy, bar, and concrete components. A mark was placed on the drill bit to denote the necessary depth for each group of bars. Following the drilling, the hole was thoroughly cleaned; this involved using compressed air to remove all concrete debris, followed by scrubbing the entire hole with a brush and applying compressed air again. This cleaning procedure was conducted five times before the epoxy injection. After injecting epoxy to approximately three-quarters of the hole's depth, the GFRP bar was carefully positioned by applying pressure while simultaneously rotating the bar in a back-and-forth motion. Figure 17 illustrates the steps of installing the GFRP bars within the slab.

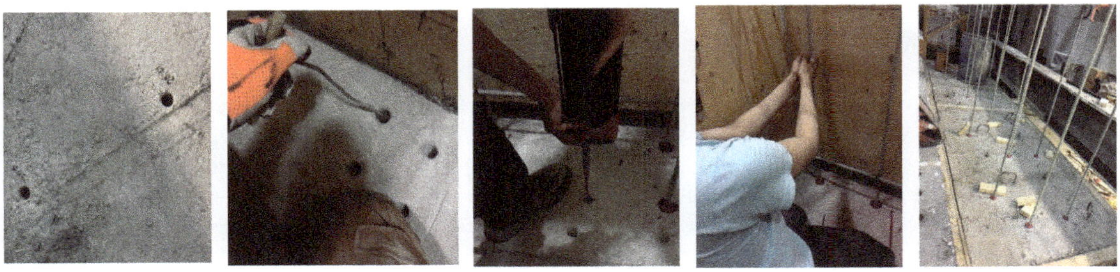

(a) (b) (c) (d) (e)

Figure 17. Installation procedure for postinstalled GFRP bars: (**a**) Postinstalled holes after drilling; (**b**) Cleaning of postinstalled barrier anchor holes; (**c**) Injecting epoxy into the drilled hole; (**d**) Installing the GFRP bar while slowly rotating left and right; (**e**) Final view of the preinstalled bars in the slab.

The pullout test follows the ASTM E488/E488M-15 test method [75], which describes all appropriate equipment and setup needed to experiment. The test setup consisted of a series of HSS sections resting on metal bearing plates that were spaced in a manner so that there was a minimum clearance radius of about two times the bar embedment length from the bar being tested to each support. Three HSS sections were stacked on top of one another on one side of the bar and in a parallel fashion on the opposite side of the bar, all of which were resting on four square metal plates. On top of these six HSS sections were two longer HSS sections, which were perpendicular to the supporting HSS members, while the GFRP bar of interest was centered between them. On top of these two sections is a series of plates, a rubber pad, a load cell, a hydraulic jack, and pull-out grips. This setup is demonstrated in Figure 18. Once the rigorous steel setup is complete, a potentiometer (POT) was placed adjacent and connected to the bar, while the tip of the needle was on the top of the concrete, as shown in Figure 18.

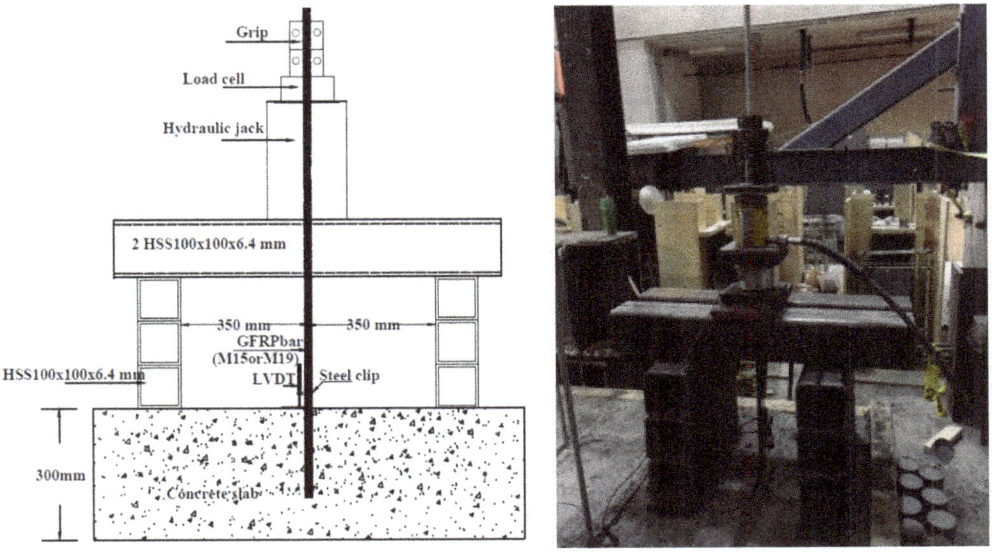

(**a**) Schematic diagram of the test setup (**b**) Front view of the test setup

Figure 18. Pullout test setup.

Test results showed that the straight bars failed primarily due to bar crushing at the grip location to the extent that the part split laterally along the free-standing length between the grips and the top of the concrete slab, as depicted in Figure 19a. Although the bars did not manage to achieve their nominal ultimate tensile capacity, and the bar did not visually appear to slip out of the concrete, the test results showed an average pullout force of 146.76 kN, as listed in Table 4, which is about 73% of the tension capacity of the bar. After pullout tests, concrete core samples were taken at the bar location, as shown in Figure 19b. Subsequently, the core samples were sliced in half using a concrete saw, as shown in Figure 19c. The intent of slicing the samples directly in half (cutting the GFRP bar in half as well) was to examine whether any slippage or other forms of failure within the concrete matrix occurred during the test. Figure 19d shows that the bar had no visual slippage at its end in the concrete slab. The bar appeared to be fully bonded to the concrete surrounding after the test.

(a) (b) (c) (d)

Figure 19. Failure modes for Group 1 for preinstalled GFRP bars. (**a**) Bar rupture at grip area; (**b**) Coring of the slab; (**c**) Slicing the core; (**d**) Sliced core without bar slip.

Table 4. Summary of bar pullout strength and associated failure modes.

Type of Construction	Group	Embedment Length (mm)	No.	Failure Load (kN)	Average Failure Load (kN)	Type of Failure
Preinstalled bars	(1): GFRP bars with straight ends	195	A	150.88	146.76 STDEV = 2.80 COV = 0.0193	Bar crushing at the grip
			B	146.02		Bar crushing at the grip
			C	144.67		Bar crushing at the grip
			D	145.48		Bar crushing at the grip
	(2): Headed-end GFRP bar	247	A	157.89	153.5 STDEV = 8.46 COV = 0.0551	Bar crushing at the grip
			B	161.94		Bar crushing at the grip
			C	151.69		Bar crushing at the grip
			D	142.51		Bar crushing at the grip

Table 4. *Cont.*

Type of Construction	Group	Embedment Length (mm)	No.	Failure Load (kN)	Average Failure Load (kN)	Type of Failure
Preinstalled bars	(3): Headed-end GFRP bars with unbonded straight portion	247	A	137.38	131.04 STDEV = 7.65 COV = 0.0583	Bar slip from head
			B	122.00		Bar slip from head
			C	137.38		Bar slip from head
			D	127.40		Bar slip from head
Postinstalled bars	(1): GFRP bars with straight ends	150	A	158.70	141.9 STDEV = 26.48 COV = 0.1866	Bar crushing at the grip
			B	146.83		Bar crushing at the grip
			C	158.97		Bar slip
			D	103.10		Bar slip
	(2): GFRP bars with straight ends	195	A	163.56	151.89 STDEV = 12.40 COV = 0.0817	Bar slip, concrete cone breakout
			B	153.58		Bar crushing at the grip
			C	156.01		Bar crushing at the grip
			D	134.41		Bar slip

Test results showed that the headed-end bars with a fully bonded straight portion failed primarily due to bar crushing at the grip location to the extent that the part split laterally along the free-standing length between the grips and the top of the concrete slab. Despite the bars failing to reach their nominal ultimate tensile capacity and the absence of any visible slippage from the concrete, the test results indicated an average pullout force of 153.5 kN, representing approximately 76% of the bar's tension capacity. Table 4 summarizes the pullout capacity of each bar in Group 2, along with the corresponding mode of failure. As shown in Figure 20a, the bar appeared to be fully bonded to the concrete surrounding after the test. Also, no fracture in the anchor head was observed. Moreover, no shear failure was visually observed at the bar-head interface. As for the headed-end bars with the unbonded straight portion, failure was primarily due to bar slippage from the end head as shown in Figure 20b at an average pullout force of 131.04 kN (about 66%). Comparing the results of Groups 2 and 3, one may observe that the bonded straight portion of the bar increased the pullout capacity by 10% beyond the pullout force of the headed end alone.

As for the postinstalled GFRP bars, test results indicated that the straight bars, installed with an embedment length of 150 mm, predominantly failed due to slippage at the interface between the bar and adhesive. Additionally, the bars ruptured at the anchorage point outside the concrete specimen, as illustrated in Figure 21a,b. Visual evidence of bar slippage at the bar-adhesive interface was noted, while the adhesive appeared to be fully bonded to the surrounding concrete following the test. Furthermore, no slippage was observed between the adhesive and the concrete. An increase in embedment length to 195 mm led to a rise in tensile force; however, the mode of failure remained largely consistent, as demonstrated in Figure 21a and summarized in Table 4.

(a) Group 2 (b) Group 3

Figure 20. Sliced core sample for Groups 2 and 3 for preinstalled GFRP bars: (a) Complete bond between GFRP bar and its head; (b) View of the sliced core samples for Group 3 showing bar slip from the head.

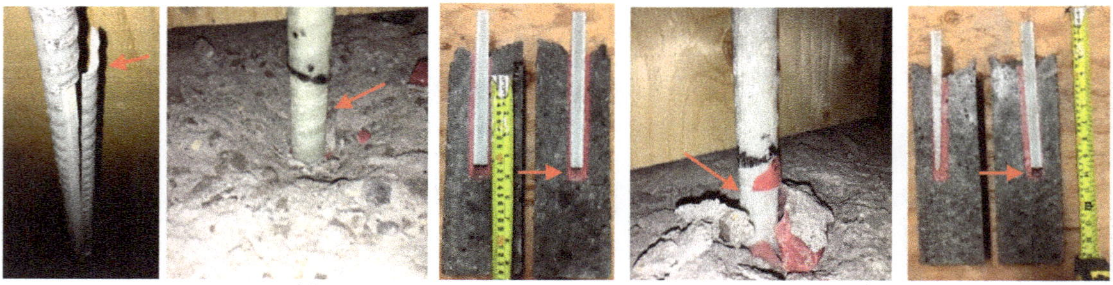

(a) (b)

Figure 21. Failure modes for Groups 1 and 2 for postinstalled GFRP bars. (a) Bar rupture, bar slippage from concrete, and sliced specimen in Group 1; (b) Bar slippage and sliced bar in Group 2.

It can be inferred that the pullout strengths of postinstalled GFRP bars are comparable to those of GFRP-headed bars. A significant finding was that all test results indicated a pullout strength of no less than 100 kN, thereby ensuring the suitability of headed GFRP bars for use in bridge construction, per the guidelines set forth by the Ministry of Transportation of Ontario.

The reported experimental pullout strength investigation complements the structural performance results of the full-scale barrier–deck tests. This enables direct comparison between measured pullout capacities and the tensile demand from transverse vehicle impact, as predicted by applied moment equations presented elsewhere [63,64]. Presenting global performance first, followed by anchorage-specific validation, provides a logical flow from system behavior to component-level reliability. This structure enhances the practical relevance of the results for design verification and future research applications.

While previous studies have explored the bond behavior and pullout strength of headed-end GFRP bars, this study advances the field by integrating high-modulus (HM) headed-end GFRP bars into full-scale TL-5 barrier–deck systems and validating their performance under realistic transverse loading conditions. The research uniquely quantifies the contribution of the bonded straight portion to pullout strength, demonstrating about a 10% increase compared to unbonded configurations. Additionally, it benchmarks the performance of postinstalled bars against preinstalled alternatives, confirming their viability for retrofit applications. These findings are supported by statistically consistent pullout test results and capacity-to-demand ratios exceeding CHBDC requirements, even after applying environmental reduction factors. This comprehensive approach bridges the gap between component-level anchorage behavior and system-level structural performance, offering a validated and constructible solution for both new and rehabilitated bridge barriers.

7. Conclusions

This study comprehensively evaluated the structural performance and anchorage behavior of high-modulus (HM) GFRP-reinforced TL-5 concrete bridge barriers under transverse loading. Both preinstalled and postinstalled GFRP bar configurations were examined through full-scale testing and direct pullout experiments. The findings validate the use of GFRP reinforcement for new construction and retrofit applications, offering a corrosion-resistant and structurally reliable alternative to conventional steel reinforcement.

The following are the key findings from this research.

- **Structural Adequacy**: All six full-scale TL-5 barrier–deck specimens exceeded the factored design moments per CHBDC, with capacity-to-demand ratios ranging from 1.38 to 1.51. Even after applying an environmental reduction factor of 0.85, all ratios remained above 1.17, confirming the robustness of the GFRP anchorage systems.
- **Embedment Length**: A 195 mm embedment length was sufficient for both preinstalled and postinstalled GFRP bars, meeting or exceeding the required anchorage strength in all configurations.
- **Headed-End Performance**: Headed-end GFRP bars demonstrated superior pullout strength compared to straight-end bars. When combined with a bonded straight portion, the pullout capacity increased by approximately 10%, consistently surpassing the 100 kN threshold required by MTO.
- **Failure Mechanisms**: Observed failure modes on the tested actual-size barrier–deck specimens included concrete breakout at the deck-barrier corner or diagonal shear in the barrier wall. No GFRP bar rupture was recorded, and strain measurements confirmed that neither concrete nor GFRP bars reached their ultimate strain limits.
- **Postinstalled Bar Viability**: Postinstalled straight GFRP bars bonded with epoxy exhibited comparable performance to preinstalled headed bars, validating their use in retrofitting deteriorated barriers.
- **Deck Overhang Behavior**: Specimens with deck overhangs experienced diagonal tension cracking at the barrier–deck corner as the governing failure mode. The developed design equation for the resisting moment due to diagonal cracking (Equation (9)) showed strong agreement with experimental results, with moment ratios of 1.13 and 1.36, confirming its conservatism and applicability in design.
- **Reinforcement Detailing**: Doubling the vertical GFRP bars at the traffic face significantly enhanced load resistance, crack control, and deflection performance, particularly at barrier ends.
- **Design Implications for Code and Standard Updates:** The results support updating MTO-S110-92 Standard Drawing and CHBDC to recognize 195 mm embedment lengths for GFRP bars and to permit headed-end and postinstalled GFRP anchorage systems as structurally adequate and constructible alternatives to traditional bent bars.

Further research is required to investigate the long-term durability and vehicle impact resistance of postinstalled GFRP anchorage systems to simulate real-world service conditions. The long-term performance of epoxy-bonded GFRP anchors under thermal cycling, freeze–thaw exposure, and elevated temperature conditions requires further research. While this study focused on experimental validation, future research should incorporate finite element modeling to simulate stress distributions, crack development, and anchorage behavior in GFRP-reinforced barrier–deck systems. Such models would enable parametric analyses, support the refinement of the proposed design equation for diagonal cracking, and extend the applicability of the findings to a broader range of geometries and loading scenarios.

Author Contributions: G.D.: Methodology, Validation, Formal analysis, Investigation, Data curation, Writing—original draft, Visualization, Project administration. K.S.: Conceptualization, Investigation, Data curation, Writing—review and editing, Project administration. H.M.A.: Formal analysis, Investigation, Writing—review and editing. A.D.: Formal analysis, Writing—review and editing. All authors have read and agreed to the published version of the manuscript.

Funding: This research received no external funding.

Data Availability Statement: Data will be made available on request.

Acknowledgments: The financial support from the NSERC Engage program of the Natural Science and Engineering Council of Canada and the in-kind contribution of B&B FRP Manufacturing Inc. are greatly appreciated.

Conflicts of Interest: Author Gledis Dervishhasani was employed by the company Skampis Ltd. The remaining authors declare that the research was conducted in the absence of any commercial or financial relationships that could be construed as a potential conflict of interest.

References

1. Manning, D. Corrosion Performance of Epoxy-Coated Reinforcing Steel: North American Experience. *J. Constr. Build. Mater.* **1996**, *10*, 349–365. [CrossRef]
2. Smith, J.; Virmani, Y. Performance of Epoxy-Coated Rebars in Bridge Decks. *Public. Roads* **1996**, *60*, 2750012.
3. Lai, D.; Raven, R. Performance of Epoxy Coated Reinforcement in Bridge Barriers Subjected to Direct Salt Splashing. In Proceedings of the 8th International Conference on Short and Medium Span Bridges, Niagara Falls, ON, Canada, 3–6 August 2010; Paper # 76. pp. 424–434.
4. Islam, S.; Afefy, H.M.; Sennah, K.; Azimi, H. Bond characteristics of straight- and headed-end ribbed-surface GFRP bars embedded in high-strength concrete. *Constr. Build. Mater.* **2015**, *83*, 283–298. [CrossRef]
5. Lu, J.; Afefy, H.M.; Azimi, H.; Sennah, K.; Sayed-Ahmed, M. Bond performance of sand-coated and ribbed-surface glass fiber reinforced polymer bars in high-performance concrete. *Structures* **2021**, *34*, 10–19. [CrossRef]
6. Khederzadeh, H.; Sennah, K. Pullout Strength of Pre-installed GFRP Bars in Concrete. In Proceedings of the CSCE Annual Conference, Canadian Society for Civil Engineering, Montreal, QC, Canada, 29 May–1 June 2013; pp. 1–10.
7. Ahmed, E.A.; El-Sayed, A.K.; El-Salakawy, E.; Benmokrane, B. Bend strength of FRP stirrups: Comparison and evaluation of testing methods. *J. Compos. Constr.* **2010**, *14*, 3–10. [CrossRef]
8. Achillides, Z.; Pilakoutas, K. Bond Behavior of Fiber Reinforced Polymer Bars under Direct Pullout Conditions. *J. Compos. Constr.* **2004**, *8*, 173–181. [CrossRef]
9. Baena, M.; Torres, L.; Turon, A.; Barris, C. Experimental Study of Bond Behavior between Concrete and FRP Bars using a Pull-out Test. *Compos. Part B Eng.* **2009**, *40*, 784–797. [CrossRef]
10. Won, J.; Park, C.; Kim, H.; Lee, S.; Jang, C. Effect of Fibers on the Bonds between FRP Reinforcing Bars and High-Strength Concrete. *Compos. Part B Eng.* **2008**, *39*, 747–755. [CrossRef]
11. Mazaheripour, H.; Barros, J.; Cruz, S.; Soltanzadeh, F. Analytical Bond Model for GFRP Bars to Steel Fiber Reinforced Self-Compacting Concrete. *J. Compos. Constr.* **2013**, *17*, 1–16. [CrossRef]
12. Maranan, G.; Manalo, A.; Karunasena, W.; Benmokrane, B. Bond Stress-Slip Behavior: Case of GFRP Bars in Geopolymer Concrete. *J. Mater. Civ. Eng.* **2015**, *27*, 1–9. [CrossRef]
13. Maranan, G.; Manalo, A.; Karunasena, W.; Benmokrane, B. Pullout Behavior of GFRP bars with Anchor in Geopolymer Concrete. *J. Compos. Struct.* **2015**, *132*, 1113–1121. [CrossRef]
14. Ametrano, D. Bond Characteristics of Glass Fibre Reinforced Polymer Bars Embedded in High Strength and Ultra-High Strength Concrete. Master's Thesis, Department of Civil Engineering, Ryerson University, Toronto, ON, Canada, 2011.
15. Mak, C. Experimental and Theoretical Investigation of Glass Fibre Reinforced Polymer Tension Lap Splices in Ultra High Performance Concrete. Master's Thesis, Department of Civil Engineering, Ryerson University, Toronto, ON, Canada, 2011.

16. Yoo, D.-Y.; Kwon, K.-Y.; Park, J.-J.; Yoon, Y.-S. Local Bond-Slip Response of GFRP Rebar in Ultra-High-Performance Fiber-Reinforced Concrete. *Compos. Struct.* **2015**, *120*, 53–64. [CrossRef]
17. Ahmed, M.S.; Sennah, K.; Afefy, H.M. Structural Behavior of Full-Depth Deck Panels Having Developed Closure Strips Reinforced with GFRP Bars and Filled with UHPFRC. *J. Comp. Sci.* **2024**, *8*, 468.
18. Ahmed, M.S.; Sennah, K.; Afefy, H.M. Fatigue and Ultimate Strength Evaluation of GFRP-Reinforced, Laterally Restrained, Full-Depth Precast Deck Panels with Developed UHPFRC Filled Transverse Closure Strips. *Appl. Sci.* **2024**, *14*, 19.
19. Firas, S.A.; Gilles, F.; Robert, L.R. Bond Between Carbon Fibre-reinforced Polymer (CFRP) Bars and Ultra-High Performance Fibre Reinforced Concrete (UHPFRC): Experimental study. *Constr. Build. Mater.* **2011**, *25*, 479–485.
20. Benmokrane, B.; Mohamed, M.H.; Manalo, A.; Cousin, P. Evaluation of Physical and Durability Characteristics of New Headed Glass-Fiber-Reinforced-Polymer (GFRP) Bars. *ASCE J. Compos. Constr.* **2017**, *21*, 1–12. [CrossRef]
21. Vint, L.; Sheikh, S. Investigation of Bond Properties of Alternate Anchorage Schemes for Glass Fiber-Reinforced Polymer Bars. *ACI Struct. J.* **2015**, *112*, 59–68.
22. Vint, L. Investigation of Bond Properties of Glass Fibre Reinforced Polymer (GFRP) Bars in Concrete under Direct Tension. Master's Thesis, Department of Civil Engineering, University of Toronto, Toronto, ON, Canada, 2012.
23. Mohamed, H.; Zaki, M.; Benmokrane, B. Newly Developed FRP-Headed Rebars for Infrastructure Applications. In Proceedings of the 3rd International Structural Specialty Conference, Canadian Society for Civil Engineering, Edmonton, AB, Canada, 6–9 June 2012; pp. 1–9.
24. Lu, J.; Afefy, H.M.; Azimi, H.; Sennah, K.; Sayed-Ahmed, M. Bond characteristics of glass fiber reinforced polymer bars in high-strength concrete. *Struct. Build.* **2022**, *175*, 748–764. [CrossRef]
25. Johnson, D.; Sheikh, S. Experimental Investigation of Glass Fiber-Reinforced Polymer-Reinforced Normal-Strength Concrete Beams. *ACI Struct. J.* **2013**, *113*, 1165–1174. [CrossRef]
26. Sennah, K.; Khederzadeh, H. Development of Cost-Effective PL-3 Concrete Bridge Barrier Reinforced with Sand-Coated GFRP Bars: Vehicle Crash Test. *Can. J. Civ. Eng.* **2014**, *41*, 357–367. [CrossRef]
27. Sennah, K.; Hedjazi, S. Structural Qualification of a Developed GFRP-Reinforced PL-3 Concrete Bridge Barrier Using Vehicle Crash Testing. *Int. J. Crashworthiness* **2018**, *24*, 1–18.
28. Sennah, K.; Tropynina, E.; Ibrahim, Z.; Hedjazi, S. Structural Qualification of a Developed GFRP-Reinforced PL-3 Concrete Bridge Barrier Using Ultimate Load Testing. *Int. J. Concr. Struct. Mater.* **2018**, *12*, 1–22. [CrossRef]
29. Rostami, M.; Sennah, K.; Afefy, H.M. Ultimate capacity of barrier-deck anchorage in MTQ TL-5 barrier reinforced with headed-end, high-modulus, sand-coated GFRP bars. *Can. J. Civ. Eng.* **2018**, *45*, 263–275. [CrossRef]
30. Khederzadeh, H.; Sennah, K. Development of Cost-Effective PL-3 Concrete Bridge Barrier Reinforced with Sand-Coated GFRP Bars: Static Load Tests. *Can. J. Civ. Eng.* **2014**, *41*, 368–379. [CrossRef]
31. Sennah, K.; Mostafa, A. Performance of a Developed TL-5 Concrete Bridge Barrier Reinforced with GFRP Hooked Bars: Vehicle Crash Testing. *ASCE J. Bridge Eng.* **2018**, *23*, 1–20. [CrossRef]
32. Rostami, M.; Sennah, M.; Mostafa, A. Experimental Study on the Transverse Load Carrying Capacity of TL-5 Bridge Barrier Reinforced with Special Profile GFRP Bars. *ASCE J. Compos. Constr.* **2019**, *23*, 1–16. [CrossRef]
33. Hasaballa, M.; El-Salakawy, E. Seismic performance of beam-column joints reinforced with GFRP headed bars. In Proceedings of the 3rd Asia-Pacific Conference on FRP Is Structures (APFIS 2012), Sapporo, Japan, 2–4 February 2012.
34. El-Badry, M.; Moravvej, M.; Joulani, P. Performance of a Hybrid FRP-Reinforced Bridge Truss Girder System – Experimental Assessment. In Proceedings of the 4th International Conference on Smart Monitoring, Assessment, and Rehabilitation of Civil Structures, SMAR 2017, Zurich, Switzerland, 13–15 September 2017; pp. 1–9.
35. Lu, C.; Qi, Z.; Zheng, Y.; Xuan, G.; Yan, Y. Long-term tensile performance of GFRP bars in loaded concrete and aggressive solutions. *J. Build. Eng.* **2023**, *64*, 105587. [CrossRef]
36. Zhu, P.; Li, Z.; Zhu, Y.; Wu, Y.; Qu, W. Prediction of the Long-Term Tensile Strength of GFRP Bars in Concrete. *Buildings* **2023**, *13*, 1035. [CrossRef]
37. Hao, Z.H.; Zeng, J.J.; Chen, G.M.; Dai, J.G.; Chen, J.F. Durability of FRP-to-concrete bonded joints subjected to 110 months of accelerated laboratory and field exposure. *Eng. Struct.* **2024**, *305*, 117681. [CrossRef]
38. Sun, J.-H.; Su, N.-J.; He, Z.-Q.; Jia, R.-X.; Liu, Y.; Lin, F.-K.; Ka, T.A. Durability of concrete-encapsulated GFRP bars subjected to seawater immersion. *Case Stud. Constr. Mater.* **2024**, *20*, e03249. [CrossRef]
39. Nepomuceno, E.; Sena-Cruz, J.; Correia, L.; D'Antino, T. Review on the bond behavior and durability of FRP bars to concrete. *Constr. Build. Mater.* **2021**, *287*, 123042. [CrossRef]
40. Zheng, Y.; Ni, M.; Lu, C.; Chu, T.; Wu, X. Bond Behavior of GFRP-Concrete Under Long-term Exposure to Aggressive Environments. *J. Adv. Concr. Technol.* **2020**, *18*, 730–742. [CrossRef]
41. National Academies of Sciences, Engineering, and Medicine. *MASH Railing Load Requirements for Bridge Deck Overhang. NCHRP Research Report 1078*; The National Academies Press: Washington, DC, USA, 2023. [CrossRef]

42. Frosch, R.J.; Morel, A.J. *Guardrails for Use on Historic Bridges: Volume 2—Bridge Deck Overhang Design*; Report No. FHWA/IN/JTRP-2016/34; Joint Transportation Research Program, Purdue University: West Lafayette, IN, USA, 2016; 109p.
43. Arrington, D.R.; Holt, J.M.; Bligh, R.P. Concrete Roadside Barrier Anchorage Details for 5-in., Cast-in-Place Decks on Prestressed Beams. *Transp. Res. Rec.* **2012**, *2313*, 33–41. [CrossRef]
44. Zhao, H.; Xie, Z.; Yang, B.; Li, L.; Wang, R.; Chen, W. Impact resistance performance of precast reinforced concrete barriers with grouted sleeve and steel angle-to-plate connections. *Eng. Struct.* **2024**, *316*, 118533. [CrossRef]
45. Nguyen, Q.T.; Maki, T.; Mutsuyoshi, H.; Ishihara, Y. Experimental Study on Connection Between Precast Concrete Barrier and Existing Bridge Deck by Loop Steel Bars and Post-installed Anchor. In *International Symposium of the International Federation for Structural Concrete*; Lecture Notes in Civil Engineering; Springer: Cham, Switzerland, 2023; Volume 350. [CrossRef]
46. Yu, X.; Huang, Y.; Yang, Y.; Chen, Y.; Luo, Y.; Jia, B. Optimal design for the connector of bridge prefabricated concrete barriers. *Sci. Prog.* **2021**, *104*, 1–19. [CrossRef]
47. Alaywan, W.R. *Performance and Analysis of Concrete Bridge Railing Using Conventional and Composite Reinforcement Materials*; Report No. FHWA/LA.16/547; Louisiana Transportation Research Center: Rouge, LA, USA, 2018; 161p.
48. Azimi, H.; Grégoire, M.; Liu, Z.; Bahous, A. A Precast Barrier Alternative Connected to Moment Slab for Accelerated Construction. In Proceedings of the 10th International Conference on Short and Medium Span Bridges, Quebec City, QC, Canada, 31 July–3 August 2018; pp. 1–9.
49. Ecklund, A.; Sritharan, S. *Precast Concrete Bridge Barriers for Accelerated Bridge Construction*; Technical Report; Iowa State University: Ames, IA, USA, 2018; 103p.
50. Patel, G.; Sennah, K.; Azimi, H.; Lam, C.; Kianoush, R. Development of Precast Concrete Barrier Wall System for Bridge Decks. *J. Prestress. Concr. Inst. (PCI)* **2014**, *59*, 83–102. [CrossRef]
51. Coleman, Z.; Jacques, E.; Roberts-Wollmann, C. Effect of Beam Depth on Anchorage Strength of Hooked and Headed Bars. *ACI Struct. J.* **2023**, *120*, 197–206. [CrossRef]
52. Cho, D.-Y.; Jeong, H.-S.; Choi, J.-W. A Study of the Bolt Connection System for a Concrete Barrier of a Modular Bridge. *Int. J. Eng. Technol. Innov.* **2018**, *8*, 107–117.
53. Matta, F.; Nanni, A. Connection of Concrete Railing Post and Bridge Deck with Internal FRP Reinforcement. *J. Bridge Eng.* **2009**, *14*, 66–76. [CrossRef]
54. Meda, A.; Rinaldi, Z.; Spagnuolo, S.; Grecco, R. Full-scale tests on bridge kerbs subjected to horizontal actions. *Struct. Concr.* **2021**, *22*, 813–826. [CrossRef]
55. Cattaneo, S.; Crespi, P. Response of Connections between Concrete Corbels and Safety Barriers. *Materials* **2019**, *12*, 4103. [CrossRef]
56. Ahmed, E.; Matta, F.; Benmokrane, B. Steel Post-and-Beam Barrier with GFRP-Reinforced Concrete Curb and Bridge Deck Connection. *J. Bridge Eng.* **2013**, *18*, 1189–1197. [CrossRef]
57. Azimi, H.; Sennah, K.; Sayed-Ahmed, M.; Nikravan, N.; Louie, J.; Hassaan, A.; Al-Bayati, N. Bridge Deck-Guardrail Anchorage Detailing for Sustainable Construction. *ASCE J. Bridge Eng.* **2014**, *19*, 04014039. [CrossRef]
58. Farghal, H.; Arvan, H.; Azizinamini, A. *Use of All Lightweight Concrete in Conjunction with UHPC Connection for Prefabricated Barrier System*; Technical Report No. ABC-UTC-2016-C4-FIU02; Florida International University: Miami, FL, USA, 2025; 39p.
59. Khodayari AMantawy, I.M.; Azizinamini, A. Experimental and Numerical Investigation of Prefabricated Concrete Barrier Systems Using Ultra-High-Performance Concrete. *Transp. Res. Rec.* **2023**, *2677*, 624–634. [CrossRef]
60. Rostami, M.; Sennah, K.; Azimi, H.; Afefy, H. Behavior and design of GFRP bar adhesive anchors under direct tension for deteriorated concrete bridge barrier replacement. *Adv. Struct. Eng.* **2025**. [CrossRef]
61. Azimi, H.; Sennah, K.; Tropynina, E.; Goremykin, S.; Lucic, S.; Lam, M. Anchorage Capacity of Concrete Bridge Barriers Reinforced with GFRP Bars with Headed Ends. *ASCE J. Bridge Eng.* **2014**, *19*, 1–15. [CrossRef]
62. *CAN/CSA-S6:19*; Canadian Highway Bridge Design Code. Canadian Standards Association: Toronto, ON, Canada, 2019.
63. Diab, A.; Sennah, K.; Rostami, M.; Dervishhasani, G.; Sayed Ahmed, M. Development of live load moment and tension force in bridge deck slab overhang due to transverse vehicle impact for CHBDC 2025. In Proceedings of the CSCE Annual Conference, Niagara Falls, ON, Canada, 5–7 June 2024.
64. Omar, O. *Investigation of Moment and Tension Force in Bridge Deck Slab Overhang in Slab-on-Girder Bridge Due to Traffic Loading*; M.Eng. Project Report; Civil Engineering Department, Toronto Metropolitan University: Toronto, ON, Canada, 2024.
65. MTO. *Structural Manual*; Ministry of Transportation Ontario, Transportation Infrastructure Management Division, Standards and Contracts Branch Structures Office: Ontario, ON, Canada, 2018.
66. Hilti Canada Corporation. HIT-RE 500 V3 Epoxy Adhesive Anchoring System. In *North American Product Technical Guide—Volume 2: Anchor Fastening Technical Guide*, 17th ed.; Hilti Corporation: Tulsa, OK, USA, 2016.
67. *CSA-S807:18*; Specification for Fibre Reinforced Polymers. Canadian Standard Association: Toronto, ON, Canada, 2018.
68. B&B FRP Manufacturing Inc. MST-BAR® GIII Material Specification. B&B Manufacturing Inc. 2016. Available online: https://www.bandbfrp.com/ (accessed on 23 November 2016).

69. Nilsson, I.H.E.; Losberg, A. Reinforced concrete corners and joints subjected to bending moment. *J. Struct. Div.* **1976**, *102*, 1229–1254. [CrossRef]
70. *CAN/CSA-A23.3:24*; Design of Concrete Structures. Canadian Standards Association: Toronto, ON, Canada, 2024.
71. D'Antino, T.; Pisani, M.A.; Poggi, C. Effect of the Environment on the Performance of GFRP Reinforcing Bars. *Compos. Part B Eng.* **2018**, *141*, 123–136. [CrossRef]
72. AASHTO. *AASHTO LRFD Bridge Design Guide Specifications for GFRP-Reinforced Concrete*; American Association of State Highway and Building Officials (AASHTO): Washington, DC, USA, 2018.
73. Benmokrane, B.; Brown, V.L.; Ali, A.; Mohamed, K.; Shield, C. A Reconsideration of the Environmental Reduction Factor CE for GFRP Reinforcing Bars. *ASCE J. Compos. Constr.* **2020**, *24*, 06020001. [CrossRef]
74. *ACI 440.1R-24*; Guide for the Design and Construction of Structural Concrete Reinforced with Fiber-Reinforced Polymer (FRP) Bars. American Concrete Institute: Chicago, CA, USA, 2024.
75. *ASTM E488/E488M-15*; Standard Test Methods for Strength of Anchors in Concrete Elements. American Society for Testing and Materials: Pennsylvania, PA, USA, 2015.

applied sciences

MDPI

Article

Influence of Girder Flaring on Load Effect in Girders of Composite Steel Bridges

Faress Hraib [1] and Sami W. Tabsh [2,*]

1 ABNA, 4140 Lindell Boulevard, St. Louis, MO 63108, USA; fhraib@abnacorp.com
2 Department of Civil Engineering, College of Engineering, American University of Sharjah, Sharjah 26666, United Arab Emirates
* Correspondence: stabsh@aus.edu

Abstract: A flared or splayed girder bridge is a structure made up of a concrete slab on girders with linearly varying spacing along the length. For such an irregular bridge, the girder distribution factors in the AASHTO LRFD Bridge Design Specifications are not applicable. In lieu of using a refined method of analysis, the study at hand proposes a simple approach for computing the dead and live load effect in the girders. To do so, fifteen composite steel girder bridges are analyzed by the finite element method to determine the influence of the girder flaring angle, girder spacing, number of girders, deck slab thickness, span length, girder stiffness, and presence of cross-bracing on the load distribution within the bridge. This study showed that the tributary width concept is a reliable approach for determining the dead load effect on the splayed girders, especially for the case of shored construction. The girder distribution factors for flexure in the AASHTO specifications can be reasonably utilized for such irregular bridges if the girder spacing at the location of each truck axle is considered, leading to a maximum of 14% difference on the conservative side between the AASHTO approach and the finite element analysis. On the other hand, the lever rule can provide a good estimate of the live load distribution among the splayed girders when subjected to shear, as the maximum safe deviation from the finite element outcome in this situation is less than 10%.

Keywords: finite element method; flared girders; gravity loads; highway bridges; composite steel girders; girder distribution factor; splayed girders; structural analysis

1. Introduction

The complexity of transportation networks has been increasing with time to accommodate growing volumes of traffic due to population growth and societal needs. Often, this issue requires the use of irregular bridge layouts to account for different traffic levels, highway separations, grades, and curvatures. In some cases, special geometries of bridges are needed to provide for the gradual change in the number of traffic lanes. Such a situation may demand widening or narrowing of the bridge along the span length, which often results in a bridge with linearly varying superstructure width. A bridge at a highway partial separation (on-off ramp) is one example of such bridges. To satisfy the variation in a bridge's width, flared girders are often used in slab-on-girder bridges, which significantly impact the load distribution within the primary members. In general, splayed, flared, or non-parallel girders are essentially beams with varying spacing between them along the span of the bridge. In this study, the minimum girder spacing at the narrow end and maximum girder spacing at the wide end of the bridge are denoted by S_1 and S_2, respectively, as shown in Figure 1 for a typical simply supported splayed girder bridge.

A comprehensive literature search on the subject revealed that this type of bridges is not adequately covered in previously published research studies addressing the structural analysis of highway bridges.

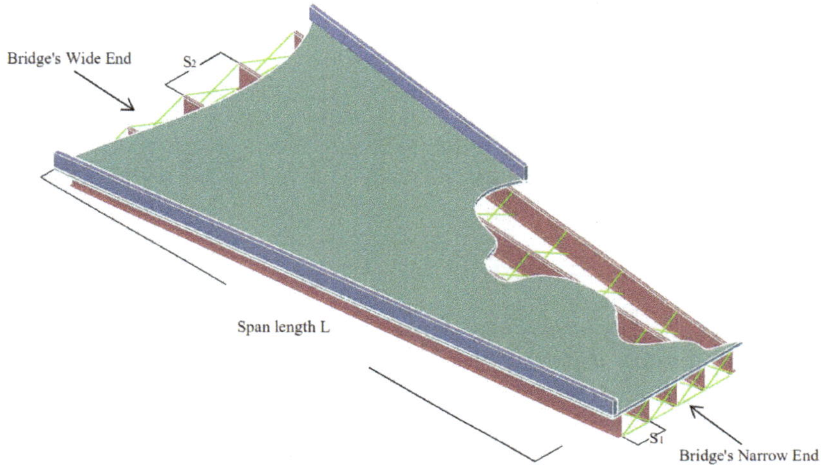

Figure 1. Typical simply supported composite steel bridge containing flared girders.

Bridge design codes and specifications do not allow the load effect in the structural members of irregular bridges to be determined by the simplified deck slab analysis approaches and formulas for the girder distribution factors. The current methodology for analyzing a flared girder bridge requires the development of a three-dimensional (3D) model using a refined method, such as the finite element procedure. Fully detailed 3D modeling of such an irregular bridge is somewhat complex, error-prone, and time-consuming, yields results that are hard to interpret, and is not always the first preferred choice for bridge engineers. Hence, there is a need to develop an approach that does not require the use of a refined method for analyzing a flared girder bridge, while allowing for the determination of the effect of girder splayedness on the distribution of gravity loads within the structure with reasonable accuracy. This study aims to quantify the effect of a gradual linear change in the girder spacing along the span length on the live and dead load distribution in exterior and interior girders of composite steel girder bridges.

The objectives of this study are as follows: (1) investigate the effect of splayed girders on dead and live load distribution represented in interior and exterior girder distribution factors for composite steel girder bridges; (2) study the influence of different bridge parameters on the girder distribution factors for both flexure and shear in splayed girder bridges; and (3) recommend a simple approximate approach to determine the girder distribution factor for flexure and shear in splayed girder bridges.

The concrete slab-on-steel girder bridges are very popular around the world due to the versatility of their application, simplicity of their structural system, uncomplicated design requirements, economy, and ease of construction. Therefore, only this type of bridge is addressed in this study. Simply supported short- and medium-length bridges up to 50 m are considered, and all the considered bridges are flared symmetrically on either side of the bridge's centerline. In all cases, linearly variable deck widths with the corresponding girder splayedness ratio are addressed. Multiple parameters are considered in this study, including the span length (ranging from 30 m to 50 m), girder spacing (ranging from 1.5 m to 5.25 m), deck slab thickness (ranging from 150 mm to 300 mm), girder web depth (ranging from 1.4 m to 2.0 m), number of girders (ranging from 3 to 7), and cross-bracing spacing (ranging from 5 m to 40 m). Different loading scenarios are accounted for in order to maximize the effect of live load in both flexure and shear, using one, two, and three transversely positioned trucks in separate lanes to account for the live load effect. All bridges are modeled within the linearly elastic range using a finite element analysis software to carry out the parametric study.

2. Background

Many highway bridge design codes and standards, such as the AASHTO LRFD Highway Bridge Design Specifications [1], do not account for variable girder spacing in their simplified girder distribution factor (GDF) formulas. This leads to more conservative designs with high-cost implications for splayed girder bridges if the larger girder spacing is used to analyze the entire length of the bridge structure. The variable girder spacing along the bridge length can also have a significant effect on the deck slab structural design requirements along the bridge.

The spacing of the girders has the most significant effect on the structural behavior of both the steel girders and the concrete deck in slab-on-girder bridges. Buckler et al. [2] tested different bridges with different girder spacing to examine the structural behavior of the deck slab and found out that there is a significant increase in deflections, compressive stresses, and tensile stresses in the deck slab due to the increase in girder spacing. A study by Barr and Amin [3] showed that the shear load effect in the interior girder is more sensitive than that in the exterior girder due to changes in the girder spacing. Since the AASHTO girder distribution factors cannot be used for irregular bridges with parallel girders that are unequally spaced, Tabsh and Sahajwani [4] suggested an approximate method to calculate the GDF. The method proposed by the authors was based on analyzing transverse strips of the deck slab directly under the wheel loads as continuous beam on elastic spring supports that simulate the flexibility of the girders. The GDF values resulting from the approximate approach were close and slightly more conservative when compared with the corresponding finite element analysis for the cases of flexure and shear.

Using an approach that utilizes a grillage analysis, Song et al. [5] examined a two-span continuous box-girder bridge with a maximum flare of 6.25% at one of the two bridge ends relative to the other. The results were compared with AASHTO GDF expressions when applied to a bridge having a constant girder spacing equal to the maximum girder spacing of the examined bridge at the wide end. The result showed high conservatism in the load effect in the girders, especially for the exterior ones. The NCHRP report 592 by Puckett et al. [6] suggest considering the largest girder spacing in a splayed girder bridge if it is analyzed using the lever rule. Utah's department of transportation bridge management manual recommends using a girder spacing at 2/3 of the span length on the wider bridge end to approximate a splayed girder bridge by a structure having uniform spacing [7]. Other state departments of transportation, such as Kansas [8], propose using AASHTOWare [9] to model a constant girder spacing bridge to represent the splayed girder bridge, but with the maximum girder spacing in order to analyze the interior girders. The Washington state department of transportation has a computer software (PGSuper) that can model a prestressed concrete bridge with different girder spacings at each end of the bridge [10]. Most of the recent research concerning the structural analysis of irregular highway bridges has focused on dynamic behavior under seismic loading [11–13].

The live load in highway bridges is primarily the weight of vehicular traffic moving on the bridge. It is critical because of its heavy weight, concentration at few points, and high uncertainty compared to dead loads. In the AASHTO LRFD Bridge Design Specifications [1], the design live load is specified as HL-93, which consists of either a design truck or a design tandem—whichever gives a larger effect—together with a design lane unform load. In the AASHTO specifications, the fraction of the live load that is carried by an individual girder is often represented by formulas referred to as the girder distribution factor. This factor is different for flexure and shear, single-lane and multi-lane bridges, and interior and exterior girders. The equations of the GDF include parameters that are related to the stiffness and geometry of the bridge, such as the girder spacing, bridge length, deck slab thickness, and the girder's moment of inertia.

The girder distribution factor for flexure in the interior girder, $(GDF_M)^{int}$, of slab-on-girder bridges can be computed as follows:

For one live loaded lane

$$(GDF_M)^{int}_{1-lane} = 0.06 + \left(\frac{S}{4300}\right)^{0.4}\left(\frac{S}{L}\right)^{0.3}\left(\frac{K_g}{L\,t_s^3}\right)^{0.1} \quad (1)$$

For two or more live loaded lanes

$$(GDF_M)^{int}_{2-lanes} = 0.075 + \left(\frac{S}{2900}\right)^{0.6}\left(\frac{S}{L}\right)^{0.2}\left(\frac{K_g}{L\,t_s^3}\right)^{0.1} \quad (2)$$

where S = girder spacing (mm), L = bridge span length (mm), t_s = deck slab thickness (mm), and K_g = girder longitudinal stiffness factor (mm^4), obtained as follows:

$$K_g = n\left(I + Ae^2\right) \quad (3)$$

in which I = moment of inertia of the non-composite girder (mm^4), A = bare girder cross-sectional area (mm^2), e = eccentricity between centroid of the girder and the deck slab (mm), and n = modular ratio between the girder and the deck slab materials.

The girder distribution factor for shear in the interior girder, $(GDF_V)^{int}$, of slab-on-girder bridges is represented as follows:

For one live loaded lane

$$(GDF_V)^{int}_{1-lane} = 0.2 + \frac{S}{7600} \quad (4)$$

For two or more live loaded lanes

$$(GDF_V)^{int}_{2-lanes} = 0.2 + \frac{S}{3600} - \left(\frac{S}{10700}\right)^2 \quad (5)$$

With regard to the load effect in the exterior girder, the GDF for such girder is usually computed in terms of the GDF of the interior girder as follows:

$$(GDF)^{ext} = e * (GDF)^{int}$$

where for the case of flexure

$$e = 0.77 + (d_e/2800) \quad (6)$$

and for the case of shear

$$e = 0.60 + (d_e/3000) \quad (7)$$

in which d_e = distance from the face of the edge parapet to the center of the exterior girder (mm). Other requirements on the subject are included in the AASHTO LRFD Bridge Design Specifications [1].

Many researchers working on live load distribution in girder bridges have found the GDF formulas presented by the AASHTO LRFD specifications to be reasonably accurate for bridges without cross-bracing or diaphragms. However, it appears that the GDF expressions can be more conservative for bridges with large girder spacings and long spans, while they can be somewhat unconservative for bridges with small girder spacings and short spans [14]. Limitations and restrictions on using these equations are applied, and in some cases, other methods such as the lever rule or rigid body rotation should be used to determine the girder distribution factor. The lever rule is a simple method that is based on assuming an internal hinge develops in the deck slab over each interior girder and then using statics to solve for the reaction of the considered girder as a fraction of the truck load. The rigid body rotation approach is a way of determining a lower bound on the girder distribution factor of exterior girders in bridges containing diaphragms or cross-bracing by assuming an infinitely rigid deck slab on elastic supports represented by the girders. Complex analytical methods of analysis for slab-on-girder bridges, such as the finite element method, finite difference procedure, or grillage analogy, are the most accurate methods for establishing the distribution of live load on bridges. Three-dimensional finite element models can provide very accurate results if the model is reasonable, detailed and representative of the actual structural behavior. However, it can be time-consuming for the designer and prone to error in modeling and interpretation of the output because of the sensitivity of the results to the type of elements in use, nature of connections, and meshing intensity.

3. Approach

The approach that is followed to investigate the dead and live load effects in the flared girders requires modeling a number of irregular bridges by finite elements and then carrying out a parametric study to determine the resulting load distribution due to changes in girder spacings, span lengths, deck slab thicknesses, cross-bracing spacings, girder depths, and number of girders. For the bridges considered, the dead load in the girders and live load girder distribution factors are calculated for the interior and exterior girders when subjected to critical flexural and shear effects at various locations within the bridge. The obtained results are then compared with corresponding results based on simplified approaches found in codes, specifications, or the available literature on the structural design of slab-on-girder bridges. The approach that gave the closest results to those of the finite element method is adopted.

3.1. Finite Element Modeling

All the considered composite steel girder bridges in this investigation are modeled and analyzed within the elastic range by the ANSYS Version 14 software [15]. Over the past few decades, researchers working in the field of structural analysis of slab-on-girder bridges have used various approaches in their modeling that employ different types of finite elements and connectivity [4,16–26]. In this study, the girders are modeled with four-node shell elements for both flanges and the web (SHELL181). The material for all girders is steel which has a 200 GPa modulus of elasticity and 0.3 Poisson ratio. The uncracked concrete deck slab is modeled using multiple layers of eight-node solid elements (SOLID185) to account for the strain variation within these elements through the thickness. The deck slab material is concrete with a 25 GPa modulus of elasticity, which corresponds to approximately 30 MPa compressive strength, and 0.2 Poisson ratio. Diaphragms in the form of cross-bracing are modeled using two-node beam elements (BEAM188). The diaphragm material is steel with the same modulus of elasticity and Poisson ratio as the girders.

All the girders in the considered bridges are simply supported; they are simulated by a pin at one end and a roller at the opposite end. This means that the longitudinal translation of each girder is restrained at the pin location and unrestrained at the roller location. Live load is represented by the truck component of the AASHTO HL-93 live load, which is a common practice since neglecting the lane load component leads to conservative live load distribution results. Note that the truck is composed of three axles with 1.8 m gauge width, of which the front axle weighs 35 kN while the middle and rear axles each weigh 145 kN, with the distance between the front and middle axles being 4.3 m and between the middle and rear axles ranging between 4.3 and 9 m. The finite element model in this paper was verified by Hraib [27] using the results of the laboratory test on a full-scale bridge conducted by Fang at the University of Texas [28] and the field testing of the Creek Relief bridge in Texas carried out by Schonwetter [29]. Figure 2 provides a summary of the type of finite elements used to model the composite steel bridge girder.

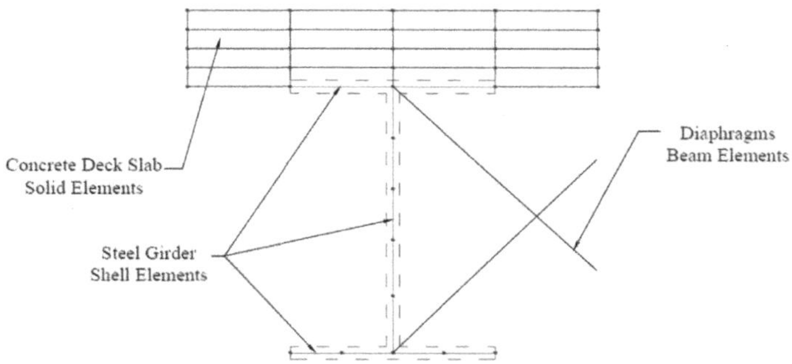

Figure 2. Finite element model used in the study.

3.2. Information on Parametric Study

A simply supported, splayed girder bridge with selected dimensions and material properties is chosen as a reference bridge. The considered parameters are the girder spacings (small spacing S_1 and large spacing S_2), and span length (L), slab thickness (t_s), cross-bracing spacing (D), girders' web depth (d), and number of girders (n). The small and large girder spacing at the two extreme ends of the bridge were within the range of 1500–3000 mm and 3750–5250 mm, respectively. The span length varied between 30,000 and 50,000 mm, while the concrete deck slab thickness varied between 150 and 300 mm. The spacing of the cross-bracing was as small as 5 m and as large as the bridge length, i.e., no intermediate cross-bracing was present. The number of girders within the superstructure was 3, 5 and 7. While the dimensions of top and bottom flanges of the steel girders were kept constant, the depth of the web varied between 1400 and 2000 mm. The deck slab overhang ranged between 500–1250 mm at the supported narrow width of the bridge and 1000–1750 mm at supported wide width of the bridge. The range of variables considered in the study represent practical values that are encountered in real-life applications.

In the parametric study, one parameter is increased and decreased beyond the reference value, while all other parameters are kept unchanged. As the girders in this study are splayed by changing the spacing at the beginning and end of the bridge, a girder spacing parameter is changed four times using two different aspect ratios. The first aspect ratio is $(S_2 - S_1)/L$, which represents the degree of splayedness of the bridge, and by varying this parameter twice—once up and once down—the girder spacing is changed at one end and kept constant at the other. The second aspect ratio is the S_1/S_2 ratio, where the degree of splayedness ($(S_2 - S_1)/L$) is kept constant by changing the girder spacing of the end that was kept constant through the first aspect, which also generates two different bridges. The overhang cantilever width from the edge of the parapet to the centerline of the exterior girder (OH) is taken one-third of the interior girder spacing in all bridges, which is in line with common practice. The symbol OH_1 represents the overhang width at the narrower girder spacing end and OH_2 represents the overhang at the wider girder spacing end. A total of fifteen bridges are modeled by the computer program ANSYS to carry out the parametric study. The steel girders used in all the bridges considered consist of a 30 mm × 300 mm top flange, a 60 mm × 600 mm bottom flange, and a 15 mm thick web with varying girder depth (d), as shown in Figure 3a. The chosen cross-bracing consists of two or three equal-angle sections of 150 mm × 15 mm, as shown in Figure 3b. The concrete parapet cross-section is 300 mm wide and 1000 mm in height.

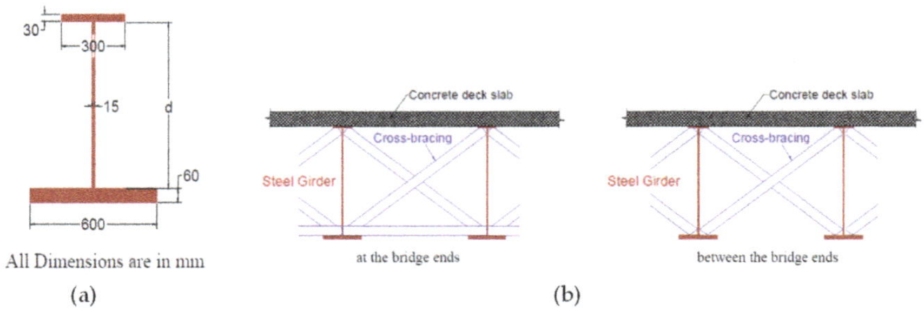

Figure 3. Properties of girders and geometry of cross-bracings: (**a**) girder dimensions; (**b**) cross-bracing configuration at the end supports and within the span.

The fifteen bridges considered, shown in Table 1, are modeled using the finite element software ANSYS to determine the flexural and shear girder distribution factors for the critical interior and exterior girders. In all 15 cases, the line of support at the abutments is taken perpendicular to the central girder, which is the shortest in length among all the girders within the superstructure. In addition, two more bridges that are similar to the reference bridge B1 are considered, with the exception that the line of support at the abutments is either perpendicular to one of the exterior girders (B1-a) or one of the first interior girders (B1-b). The truck position is changed in each model to maximize the live load effect in the transverse and longitudinal directions for bending moment and shear force. The truck rotational angle within the bridge plan is also examined to maximize this effect, as elaborated in detail later in the paper.

Table 1. Characteristics of bridges considered in parametric study.

Bridge No.	S_1 (mm)	S_2 (mm)	L (mm)	t_s (mm)	D (m)	n	d (mm)	OH_1 (mm)	OH_2 (mm)
B1	2250	4500	40,000	220	40	5	1700	750	1500
B2	3000	4500	40,000	220	40	5	1700	1000	1500
B3	1500	4500	40,000	220	40	5	1700	500	1500
B4	1500	3750	40,000	220	40	5	1700	500	1250
B5	3000	5250	40,000	220	40	5	1700	1000	1750
B6	2250	4500	40,000	150	40	5	1700	750	1500
B7	2250	4500	40,000	300	40	5	1700	750	1500
B8	2250	4500	40,000	220	5	5	1700	750	1500
B9	2250	4500	40,000	220	10	5	1700	750	1500
B10	2250	4500	40,000	220	40	3	1700	750	1500
B11	2250	4500	40,000	220	40	7	1700	750	1500
B12	2250	4500	40,000	220	40	5	1400	750	1500
B13	2250	4500	40,000	220	40	5	2000	750	1500
B14	2250	4500	30,000	220	30	5	1700	750	1500
B15	2250	4500	50,000	220	50	5	1700	750	1500

3.3. Location of Maximum Dead Load Effect

Determination of the dead load effect in a flared girder within a bridge is a straightforward task. It starts by isolating an individual girder bounded by the tributary width of the slab supported by the girder. The major component of dead load includes the self-weight of the girder and part of the deck slab above it. As per the specifications, the weight of the parapets and wearing surface can be assumed to be shared equally by all the girders if they are placed after the concrete in the deck slab has cured and hardened. In addition to the distributed load along the length, the individual girder is also subjected to concentrated loads from the weight of the cross-bracings at their locations. If the latter load is taken as a small percentage of the girder's self-weight, as is normally done in practice, the resulting dead load on the girder becomes uniformly distributed in the shape of a trapezoid along the girder's length. The shear force and bending moment in the girder can then be computed by determining the support reactions, isolating a free body diagram of part of the girder at some distance away from one of the supports, and using statics if the structure is statically determinate. For a simply supported bridge, the maximum shear in the girder is equal to the support reaction at the bridge's end, where the girder spacing is largest. The critical bending moment does not occur at the midspan due to the varying width of the bridge deck slab, which causes the critical flexural effect in the girder to be shifted toward the region with larger girder spacing, where the shear is zero. Figure 4 shows a single interior girder isolated from a simply supported splayed girder bridge and loaded by its self-weight, exclusive of the superimposed dead load, together with the corresponding shear and bending moment diagrams. The shift in the location of the maximum moment from midspan represents the location of zero shear. Conducting a three-dimensional finite element analysis on various splayed girder bridge systems yielded virtually the same result as the line element approach. Note that for the reference bridge B1, the self-weight of the steel girder is 5.43 kN/m, while the weight of the concrete deck slab over the tributary width is 11.65 kN/m at the smallest girder spacing and 23.3 kN/m at the largest girder spacing. For the interior girder in the reference bridge B1, this loading causes a maximum shear of 480.3 kN at the left support and a maximum bending moment of 4438.8 kN/m at 19,165 mm from the left support.

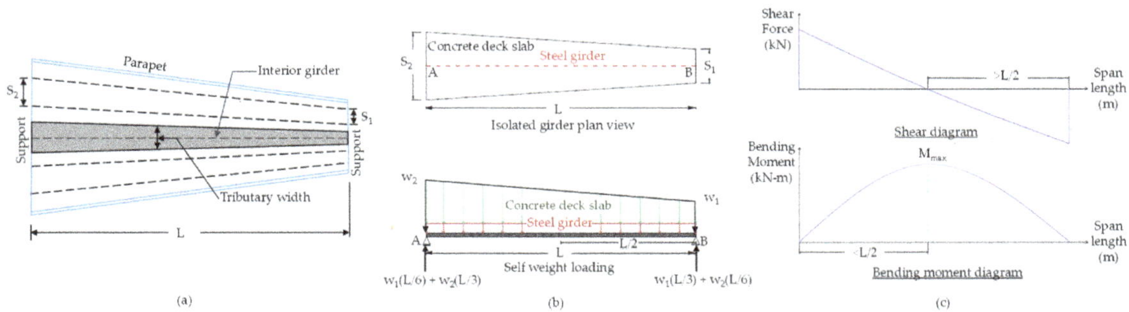

Figure 4. Dead load effect in the interior girder of reference bridge B1: (**a**) entire bridge plan; (**b**) single interior girder and loading; (**c**) shear and moment bending diagrams of interior girder.

3.4. Truck Positioning and GDF

In this study, the girder distribution factor is calculated for flexure and shear in the exterior and interior girders due to the truck component of AASHTO HL-93 live load. The truck(s)' position in the longitudinal and transverse direction is very critical to maximize the effect of live load in the bridge primary elements. Single truck positioning in the longitudinal and transverse directions, which maximizes the GDF for flexure and shear in the girders, is explained. Multiple truck positioning in the transverse direction follows a similar pattern, provided that each of the side-by-side trucks stays a minimum of 600 mm from the edge of the 3600 mm lane, as per the AASHTO requirements.

3.4.1. Longitudinal Truck Position

Based on the AASHTO LRFD specifications [1], no more than one truck is allowed to be placed within each lane on the bridge, but multiple side-by-side trucks in the transverse direction are permitted provided that appropriate multiple presence factors are considered. Since maximum shear force in the girders occurs at the support of the bridge with the larger girder spacing, critical truck positioning in this case requires placing the rear axle just off the support with the intermediate and front axles positioned at 4300 mm and 8600 mm away from the support, respectively. To maximize the bending moment in a simply supported bridge with parallel girders, the truck is generally placed at the midspan of the bridge, and by using an HL-93 truck, the middle axle is positioned at the midspan of the bridge. This moment is not necessarily the largest moment, but since the dead load moment occurs at the midspan of a regular bridge with a constant width, it is common practice to consider the combined load effect at the midspan. However, this is not the case in a splayed girder bridge, because the varying width of the bridge causes the maximum bending moment in the girders to be shifted away from the midspan toward the region with the larger girder spacing. In lieu of using an influence surface approach to determine the critical longitudinal truck location for maximum flexural effect in a splayed girder bridge, one truck is positioned symmetrically within the width of the bridge and the tensile stress at the bottom of the central girder is determined as a function of the longitudinal truck position. For reference bridge B1, the location of maximum flexure in the girder, determined by 3D finite element analysis of the entire bridge system under the critical longitudinal truck position, was found to be at 18,925 mm from the abutment that supports the wide part of the bridge, as shown in Figure 5. The obtained value is less than 1.3% of the 19,165 mm value, which represents the location of maximum flexural dead load effect. Based on the above finding, it seems practical to assume that the critical middle axle location of the truck(s) along the longitudinal axis for maximum flexure in a splayed bridge coincides with the location of maximum dead load flexural effect. Such an assumption is reasonable since the critical load effect in the girder is a combination of the influence of both dead and live loads. Furthermore, the longitudinal location of the truck(s) on the bridge is generally insensitive to the magnitude of the dimensionless girder distribution factor as long as the longitudinal truck location does not vary by much from the correct location.

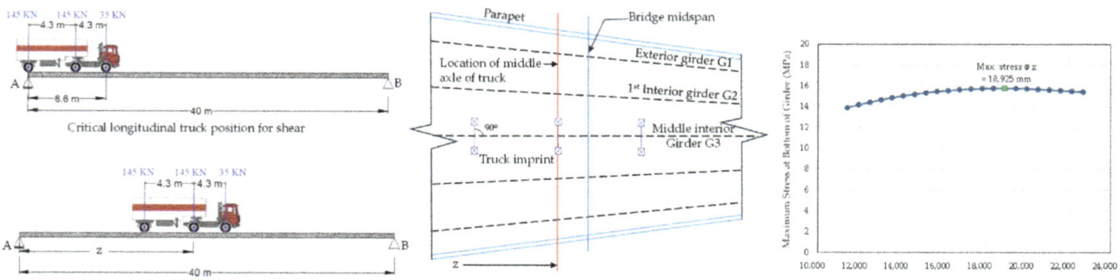

Figure 5. Critical longitudinal truck position in the reference bridge B1: (**a**) elevation; (**b**) plan; (**c**) finite element results.

3.4.2. Transverse Truck Position

The bridge is loaded by single or multiple side-by-side trucks placed within the transverse lanes to maximize the live load flexural and shear effects in the exterior or critical interior girder. The AASHTO LRFD specifications [1] account for the lower probability of simultaneous presence of multiple trucks over a bridge by employing a multiple presence factor. To find the maximum live load effect in the exterior or interior girders, different locations should be checked in the transverse direction for maximum effect in the member under consideration. The truck is first placed at a minimum clear distance from the edge of the interior face of the parapet and is then moved transversely in small increments toward the parapet on the opposite side of the bridge until the considered girder reaches the greatest live load effect.

In this study, variable spacing of the girders has an effect on truck positioning in the transverse direction, where both the position of the truck and the truck plan orientation angle affect the structural analysis results. By giving the truck the correct angle in the proper position, the live load effect can be maximized. Finite element analysis of the bridge systems showed that the truck orientation on the bridge influences both the interior and exterior girders but has a higher effect on the exterior girder, especially when the overhang width is large. By loading a splayed girder bridge with a truck oriented perpendicular to the line of supports, with a minimum wheel distance of 600 mm from the parapet, as required by AASHTO, only the wheel at the front or rear axle of the truck will maintain this minimum distance, while the other two axles will have a greater distance from the parapet, thus resulting in a lower flexural effect on the exterior girder. If the truck is oriented parallel to the parapet with a constant wheel distance of 600 mm from the parapet, the flexural effect of the truck on the exterior girder will be maximized because all of the truck wheels will contribute more toward the load effect in the girder. With regard to an interior girder, maximum flexural load effect is attained if the truck axle is oriented perpendicular to the centerline of the girder under consideration, as shown in Figure 6.

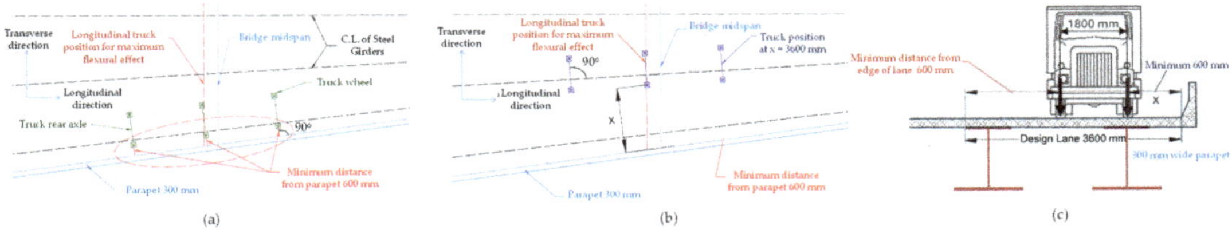

Figure 6. Critical transverse position for a single truck: (**a**) plan for exterior girder; (**b**) plan for first interior girder; (**c**) cross-section.

To maximize the shear effect due to live load, the truck is placed just off the support at the end of the bridge in the longitudinal direction. However, like in flexure, the truck should be moved in the transverse direction starting from 600 mm away from the face of the parapet until the maximum effect is reached in the girder under consideration. Finite element analysis showed that rotating the truck to maximize the shear effect was not critical in the interior girders, where keeping the truck perpendicular to the end support produced the maximum shear effect for such girders. Alternatively, orienting the truck in a position parallel to the parapet resulted in the maximum shear effect in the exterior girder.

3.4.3. Girder Distribution Factor

At each step of transverse truck positioning for the case of flexure at midspan, longitudinal normal stresses at the bottom flange of all girders due to flexure are recorded along a straight line through the girders, representing the truck's chosen longitudinal position. The

GDF for flexure is then calculated for each girder as a ratio of its stress at the extreme fibers to the summation of stresses in all girders at the critical location, since in elastic analysis, the normal stress due to flexure is proportional to bending moment. The GDF for shear is computed using the ratio of the support reaction of the girder under consideration to the summation of the reactions of all girders at the loaded end of the bridge. Equations (8) and (9), presented below, are used in this study to calculate the GDF for flexure and shear, respectively, in both interior and exterior girders, using the finite element results with consideration of the AASHTO multiple truck presence factor [1].

Flexure in interior or exterior girders:

$$GDF_j = \frac{Nm\sigma_j}{\sum_{i=1}^{n} \sigma_i} \quad (8)$$

Shear in interior or exterior girders:

$$GDF_j = \frac{NmR_j}{\sum_{i=1}^{n} R_i} \quad (9)$$

where N = number of loaded lanes, m = multiple presence factor, σ_j = normal stress due to flexure at the bottom flange of girder j (MPa), σ_i = normal stress due to flexure at the bottom flange of girder i (MPa), R_j = support reaction at girder j (N), R_i = support reaction at girder i (N), and n = number of girders. For the case of flexure, all stresses obtained from the 3D finite element analysis at the bottom flange of the girders, where there are five nodes, are weighted average stresses. For the case of shear, the vertical reactions are taken exactly at the supported nodes underneath the individual girders.

4. Detailed Analysis of Reference Bridge B1

In this section, the method outlined earlier for determining the girder distribution factor is illustrated in detail for the reference splayed girder bridge B1. The 40 m simply supported bridge is composed of five girders that are spaced at 4.5 m at the wide end and 2.25 m at the narrow end, with the overhang width being one-third of the girder spacing. The bridge has a 220 mm thick concrete slab and contains diaphragms in the shape of cross-bracing only at the supports. The structural layout, cross-section dimensions, and finite element model are shown in Figure 7.

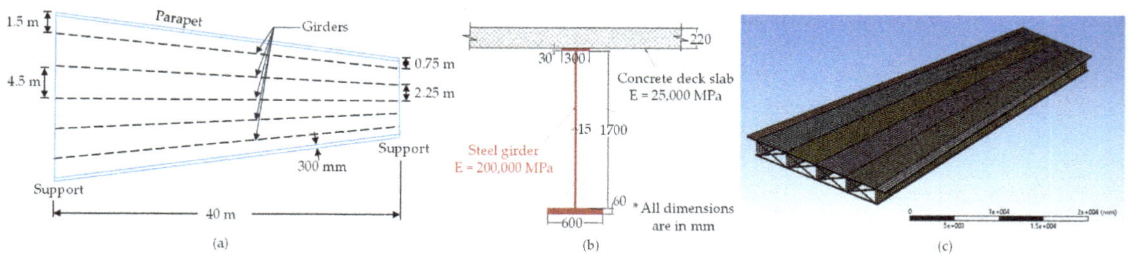

Figure 7. Details of reference bridge B1: (**a**) plan; (**b**) cross-section; (**c**) finite element mesh.

Each splayed highway bridge modeled in this study was loaded for maximum flexure and shear effect by moving one, two, and three trucks within their lanes in the transverse direction to maximize the live load effect in the girders. Moving the side-by-side trucks incrementally from one parapet to the opposite side leads to changes in the flexural and shear GDF values among the bridge girders. Figure 8 shows the load effect in the reference bridge due to two side-by-side trucks longitudinally positioned near the midspan, considering the longitudinal normal stresses in the bottom flanges and vertical shear stresses in the webs of the girders.

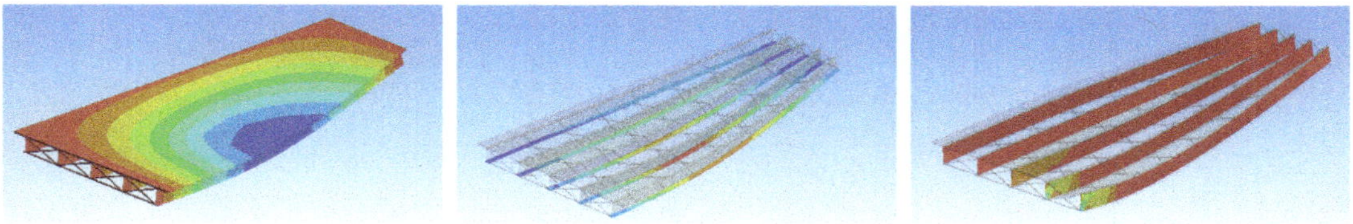

(a) (b) (c)

Figure 8. Load effect in reference bridge B1 by finite element analysis: (**a**) entire bridge; (**b**) normal stresses in bottom flanges; (**c**) shear stresses in webs.

For the reference bridge under consideration, Figure 9 shows the flexural GDF values plotted against the distance of the truck from the parapet in cases of one, two, and three trucks with consideration of the appropriate multiple presence factors. The results indicate that critical flexure in the first interior girder G2 led to the highest GDF (equal to 0.808) when compared with the central interior girder G3. Furthermore, the exterior girder G1 had the highest GDF among all the girders—equal to 0.871. For flexure, the overall GDF value was governed by two-lane loading for the exterior girder, and by three-lane loading for the interior girders. Note that the girder designation (G1, G2, and G3) has been presented earlier in Figure 5b.

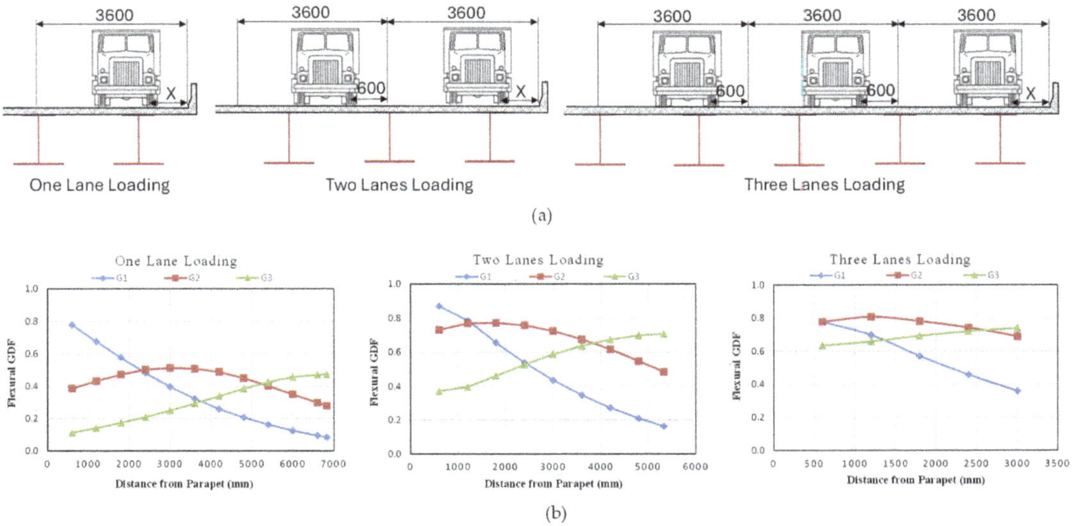

Figure 9. Flexural analysis of reference bridge B1: (**a**) transverse truck positioning; (**b**) GDF due to one, two and three loaded lanes.

For the case of shear load effect in the girders, two cases of in-plan truck orientation were checked because maximization of the GDF for the interior girders required a different truck placement to that for the exterior girders. One of the two truck placements consisted of the trucks' lines of wheels being parallel to the parapet (Case 1), while the second consisted of the trucks' axles being perpendicular to girder under consideration (Case 2). The results, shown in Figure 10 as a function of the lateral distance from the parapet, show that the maximum shear GDF for the exterior girder is governed by one-lane loading, while for the interior girder, it is governed by two-lane loading. This finding is not in agreement with the flexural effect, where three-lane loading governs the GDF for the interior girders. Similar to the flexure case, shear GDF in first interior girder governs the central interior girder. Compared to the flexural GDF, the shear GDF is larger in magnitude, which is an expected result since there is little differential vertical deflection among the girders near the support, leading to little load sharing. The critical shear GDF is 1.036 for the exterior girder and 1.313 for the interior girder, resulting in a 26.7% difference between the two values.

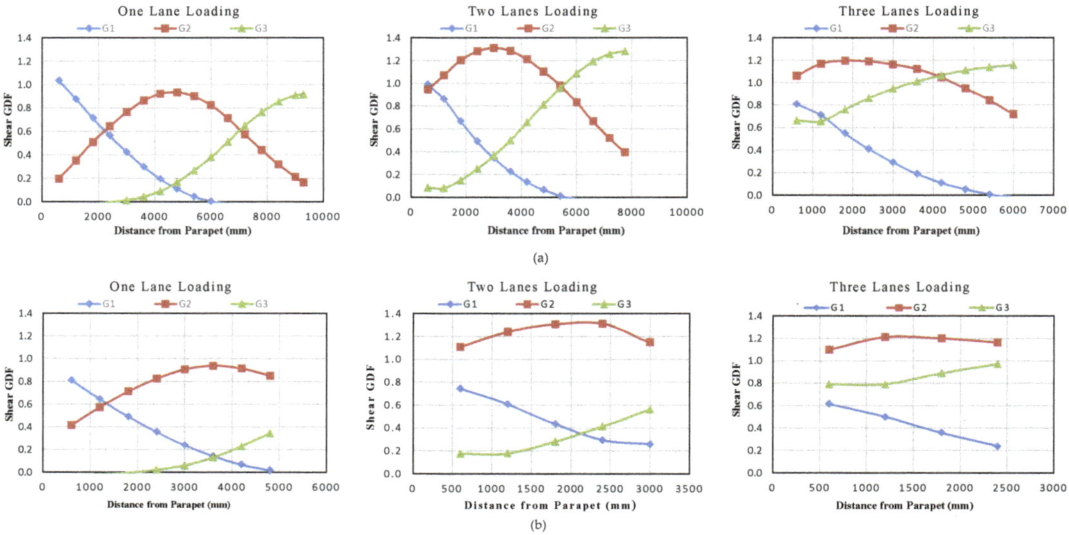

Figure 10. Shear GDF of reference bridge B1 due to one, two and three loaded lanes: (**a**) Case 1, truck lines of wheels placed parallel to parapet; (**b**) Case 2, truck axles placed perpendicular to girder under consideration.

All bridges considered in this study and shown in Table 1 are symmetrical in plan around the centerline of the bridge, which results in the central girder(s) being the shortest and the exterior girders being the longest in length. Therefore, to ensure that the results of this study are applicable to different in-plan splayedness orientations, two bridges similar to the reference bridge are modeled, but with a different splayedness orientation, as shown in Figure 11. For the first bridge (B1−a), the girder that is perpendicular to the line of supports is the exterior girder. In the second bridge (B1−b), the first interior girder is the one perpendicular to the line of supports. Finite element models are developed for each of the two alternative splayed girder bridges, B1−a and B1−b, and the same loading procedure followed in the reference bridge for maximizing flexural effect is applied for these two bridges, with consideration of the same longitudinal truck(s)' position. Results from the 3D finite element analysis showed the most critical GDF values, whether in the exterior or in the first interior girders, negligibly changed (by less than 1%) in the two alternative bridges B1−a and B1−b compared to the reference bridge B1. Based on the above, it can be concluded that lack of symmetry of the girder splayedness is not an important factor in the analysis of splayed girder bridges.

After presenting the finite element results for the reference bridge, an approximate procedure is proposed and demonstrated for computing the GDF for both flexure and shear in interior and exterior girders. This method avoids the need for a refined method of analysis as it utilizes a one-dimensional beam analysis. The procedure is based on the AASHTO LRFD formulas used to calculate the GDF for flexure and shear in slab-on-parallel-girder bridges. The objective of this procedure is to find out if such a simple method can be reasonably used to determine the load effect in the girders of such irregular bridges without the use of three-dimensional finite element analysis.

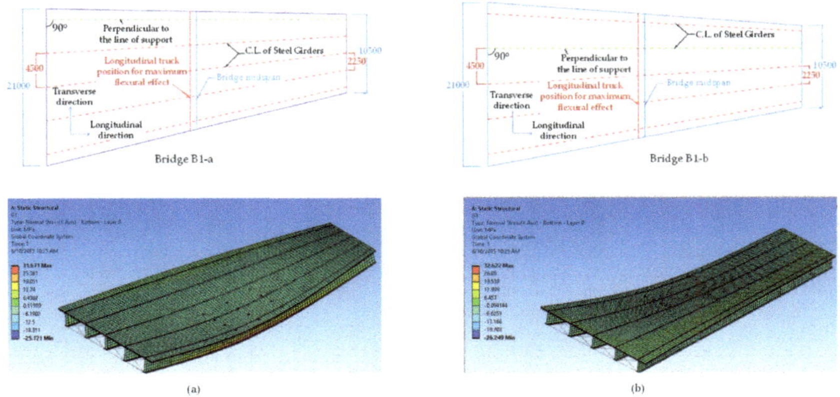

Figure 11. Alternatives to reference bridge B1: (**a**) bridge B1−a, exterior girder perpendicular to line of support; (**b**) bridge B1−b, first interior girder perpendicular to line of support.

The general concept of the approximate approach is to compute the GDF for interior or exterior girders considering the actual girder spacing at each truck axle position along the bridge using the AASHTO LRFD formulas [1]. Each of the three GDF values at the front, middle, and rear axles is then multiplied by the corresponding axle load, and structural analysis of a single composite girder is carried out to determine the live load flexural effect on the interior girders from Equations (1) and (2) or the shear effect on the interior girders from Equations (4) and (5) at the critical location. For the live load effect on the exterior girders, Equations (7) and (8) can be used to determine the GDF for cases of flexure and shear, respectively. The same single composite girder is loaded again with the full AASHTO truck load, without multiplying it by the GDF, and structural analysis is used again to obtain the critical flexural or shear effect in the girder at the critical location. Finally, the equivalent GDF is obtained as the ratio of the maximum live load effect due to multiplying the load by the GDF corresponding to the live load effect under the full truck load without any factors. The outlined approach consolidates the three hand-calculated GDF values at the three different axle locations into one quantity at the critical location for flexure or shear. Such an approach makes it convenient to compare the AASHTO results with the finite element findings obtained with the help of Equations (8) or (9).

To test the proposed approximate method, the equivalent GDF values for the reference bridge B1 are calculated and compared with the finite element outcome. Figure 12a shows the flexural results of the equivalent GDF by the AASHTO formulas (referred to as "Equ AASHTO") with the finite element outcome (referred to as "FEM"). On the other hand, Figure 12b presents the shear results of the equivalent GDF by the AASHTO formulas and the finite element result, together with the equivalent GDF by the lever rule (Referred to as "Equ LR"). To apply the lever rule, a section of the superstructure composed of the slab supported on the girders is taken, a hinge is assumed to develop in the slab over each interior girder, side-by-side axles of trucks are applied with their appropriate multiple presence factors, and the statically determinate problem is solved for the reaction of the considered girder as a fraction of the axle load using statics.

For flexure in the reference bridge, the results show that the equivalent GDF values for the critical interior and exterior girders are about 4% and 3% higher than the corresponding finite element results. For shear, the finite element results are 5% and 18% higher than the equivalent GDF by AASHTO formulas for the interior and exterior girders, respectively. Use of the lever rule for determining the shear GDF led to higher values than those obtained with the AASHTO formulas, thus leading to values that were closer to the finite element results. For the shear GDF in interior girders, the lever rule matched the finite element outcome and yielded about 6% higher values for the exterior girders.

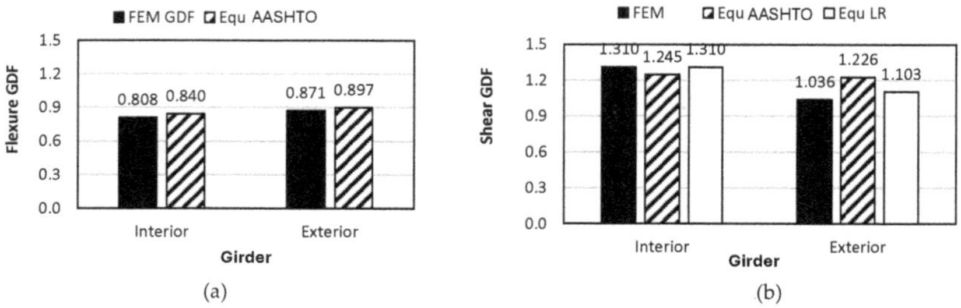

Figure 12. Comparison of finite element results with hand calculation for reference bridge B1: (**a**) flexural effect; (**b**) shear effect.

Based on the above, the suggested approach of determining the live load effect in interior and exterior girders subjected to flexure or shear seems reasonable and safe. However, before fully accepting the approximate approach, validation of the suggested procedure should be extended to bridges with different geometries, which is discussed in the next section.

5. Parametric Study

To verify the general applicability of the outlined equivalent GDF approach for flexure and shear in interior and exterior girders, a parametric study is followed. This study considers different aspects that strongly affect the behavior of splayed girder bridges. These parameters are represented in the 15 bridges (B1–B15) listed in Table 1, which include variations in girder spacing S, slab thickness t_s, cross-bracing spacing D, number of girders n, girder web depth d, and span length L. Table 2 shows the ranges of these parameters, each varied from the specified values of the reference bridge B1. For each parameter, two additional configurations are created—one with a lower value and one with a higher value—while keeping the rest constant. This results in a total of 15 splayed girder bridges with different parameters to be considered in this study. It can be said that each parameter is studied using three bridges, including the reference bridge.

Table 2. Range of variables considered in the parametric study.

Parameter	Girder Spacing Parameter: $(S_2 - S_1)/L$ *		Girder Spacing Parameter: S_1/S_2		Slab Thickness t_s (mm)	Cross-Bracing Spacing D (m)	No. of Girder n	Girder Depth d (mm)	Span Length L (mm)
	S_1 (mm)	S_2 (mm)	S_1 (mm)	S_2 (mm)					
Reference (B1)	2250	4500	2250	4500	220	40	5	1700	40,000
Lower value	1500	4500	1500	3750	150	5	3	1400	30,000
Upper value	3000	4500	3000	5250	300	10	7	2000	50,000

* Bridge length = 40,000 mm.

Since the girder spacing is the most critical parameter that greatly impacts the live load distribution, two splayedness measures are considered in the parametric study. The first one is the splayedness ratio $(S_2 - S_1)/L$, which represents the degree of splayedness of the bridge, and the second one is the girder spacing ratio S_1/S_2, which considers the girder spacing without the span length. In the latter case, the splayedness ratio $(S_2 - S_1)/L$

is kept constant while varying the S_1/S_2 ratio. Hence, four bridges in total are considered for the effect of girder spacing—two for each aspect.

5.1. Flexure in Girders Due to Live Load

In this section, the flexural effect in the splayed girders is considered and discussed with consideration of various parameters, such as the girder spacing, deck slab thickness, spacing of cross-bracing, and other parameters.

5.1.1. Effect of Girder Spacing

As mentioned before, the influence of the girder spacing on the structural behavior was examined using two distinct approaches—first, by examining the splayedness ratio $(S_2 - S_1)/L$ effect, and second, by exploring the girder spacing ratio S_1/S_2 effect—in order to gain a better insight into the effect of the change in the girder spacing along the bridge span.

Bridges B2 and B3 in Table 1 were developed based on the reference bridge B1 to observe the effect of the splayedness ratio $(S_2 - S_1)/L$ on the structural behavior. Figure 13a provides a plan view showing bridge B1 in black, B2 in blue, and B3 in green. Starting with the reference bridge B1, the girder spacing at the wide end was kept constant, while the spacing at the narrow end was adjusted to vary the splayedness ratio—increasing it by 33% to form bridge B2 and decreasing it by 33% form bridge B3. It should be noted that increasing the splayedness ratio leads to a reduction in girder spacing along the span, while decreasing the ratio leads to increased girder spacing. The finite element results in Figure 13b show that increasing the splayedness ratio by 33% leads to a 7.254% reduction in the flexure GDF for interior girders, while decreasing it by 33% leads to an almost equal increase in the GDF. For exterior girders, increasing the splayedness ratio by 33% results in an 8.12% reduction in the flexure GDF value, while decreasing it by 33% causes the flexure GDF to increase by 6.93%, as shown in Figure 13c. The bridges were also analyzed using the AASHTO-based equivalent GDF approach outlined earlier, allowing us to find the flexure GDF values with a simple approach and compare them to the finite element outcome. Figure 13b,c show that for the interior and exterior girders, the equivalent AASHTO flexure GDF values for the three bridges are very close to the finite element values, with reasonable conservatism, amounting to a maximum percentage difference of less than 5% for all the considered bridges.

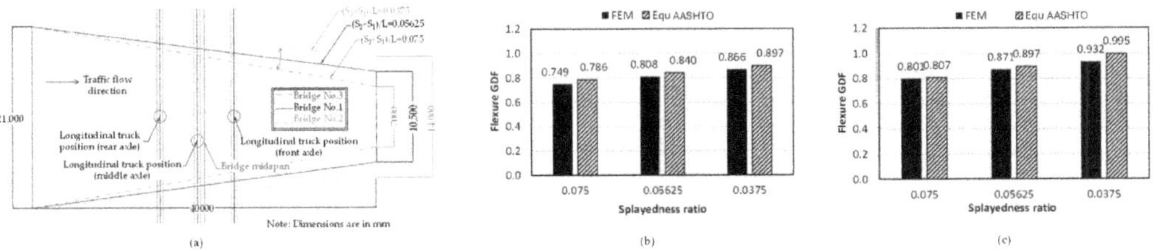

Figure 13. Effect of girder splayedness ratio on flexural GDF: (**a**) plan of bridges B1, B2, and B3; (**b**) onterior girder; (**c**) exterior girder.

Bridges B4 and B5 were developed by modifying reference bridge B1 with the aim of assessing the effect of the girder spacing ratio S_1/S_2 on the structural response. To effectively examine this ratio without interference from the splayedness ratio, the $(S_2 - S_1)/L$ ratio for all three bridges was set to be constant—equal to 0.05625. Figure 14a shows a plan view for bridges B1, B4, and B5. By decreasing the girder spacing of the reference bridge at any section along the span by 750 mm and increasing it by 750 mm, the S_1/S_2 ratios for bridges B1, B4, and B5 become 0.5, 0.4, and 0.571, respectively. The finite element results for the interior girders, shown in Figure 14b,c, indicate that increasing the S_1/S_2 ratio by 14.3% leads to a 15.05% increase in the GDF, while decreasing the S_1/S_2 ratio by 20% leads to a 16.39% reduction in the GDF. For the exterior girders, increasing the S_1/S_2 ratio by 14.3% leads to a 15.17% growth in the GDF, while decreasing the $S1/S_2$ ratio by 20% leads to a 17.41% reduction in the GDF. Note that although the S_1/S_2 ratio increased by 14.3% and then decreased by 20%, these changes resulted in almost equal absolute percentage changes in the GDF for interior and exterior girders, as they caused the same changes—increasing or decreasing—in girder spacing value at any particular section along the bridge length. However, the small difference between the percentage growth and reduction in the GDF—whether in the interior or exterior girder—is due to minor changes in the truck(s)' longitudinal location between bridges B1, B4, and B5 in order to maximize the GDF for flexure. Figure 14b,c also include the equivalent AASHTO GDF for the interior and exterior girders for all three bridges. It is clear that the GDF values obtained by finite element analysis and the AASHTO formulas are very close in the interior and exterior girders, with an adequate level of conservatism. The only case in which there is a somewhat large deviation (about 11%) between the two analysis methods is observed for exterior girders in the bridge in which the S_1/S_2 = 0.5714; nevertheless, this deviation is on the safe side.

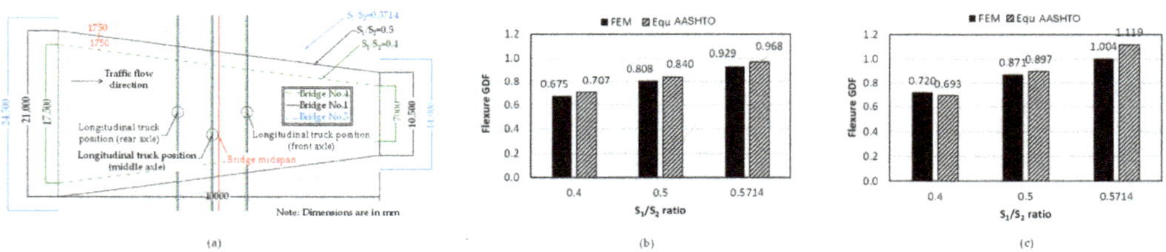

Figure 14. Effect of girder spacing ratio on flexural GDF: (**a**) plan of bridges B1, B4, and B5; (**b**) interior girder; (**c**) exterior girder.

Before the discussion about the influence of girder spacing on the live load distribution characteristics of splayed girder bridges under flexure is completed, we consider one more parameter on the subject. Figure 15 shows the change in the flexural GDF as a function of the average girder spacing, determined by $S_{avg} = (S_1 + S_2)/2$, for bridges B1, B2, B3, B4, and B5. As expected, the average increase in the girder spacing always leads to higher GDF values. For the bridges considered, in which the overhang width is equal to one-third of the girder spacing, the exterior girder GDF was consistently larger than the corresponding interior girder GDF. The equivalent AASHTO GDF expressions closely match the finite element results, particularly for bridges with short and medium average girder spacings. Note that other splayed girder bridges with a smaller deck slab overhang length than the considered value in the study may lead to a more critical load effect in the interior girders than in the exterior ones.

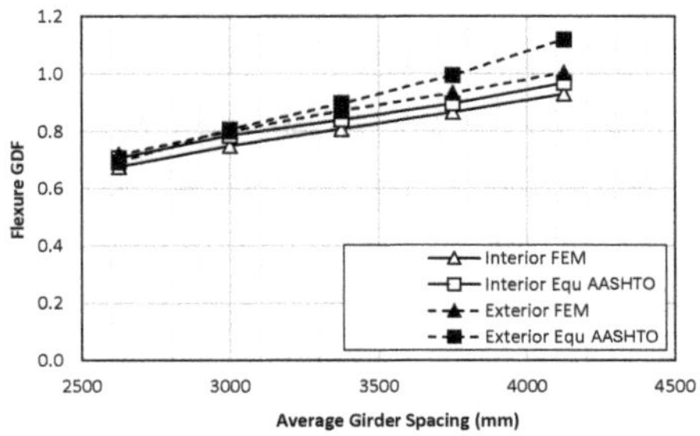

Figure 15. Effect of the average girder spacing on flexural GDF.

5.1.2. Effect of Deck Slab Thickness

The influence of slab thickness on girder splayedness is examined in this study by using a 33.3% lower and 33.3% higher value of the concrete deck slab thickness compared with reference bridge B1. Bridges B6 and B7 are identical to bridge B1, except for their slab thickness being 150 mm and 300 mm, respectively. The finite element models for the considered bridges showed that for the interior girders, increasing the 225 mm slab thickness by 33.3% leads to a 5.2% decrease in the flexure GDF value, while decreasing it by the same percentage results in a 6.7% increase in the GDF value. This is because increasing the slab thickness results in a stiffer member, which transfers the truck(s)' load more uniformly across the supporting interior girders and leads to less differential deflection between the girders. This results in more load sharing among the girders, decreasing GDF in bridge B7. In contrast, the thinner slab in bridge B6 results in less load sharing, which leads to a higher GDF. However, the exterior girders did not act in a similar way to the interior girders, where increasing or decreasing the slab thickness by 33.3% compared to the reference bridge resulted a change in GDF of only about 1%. The reason the GDF of the exterior girders was not significantly affected by the slab thickness can be explained by the fact that the load on these elements is mainly due to a single truck placed near the overhang or in the vicinity of the girder region. This mechanism is equivalent to a statically determinate element (in this case, a cantilevered overhang) where slab thickness does not contribute to the load effect. Figure 16 shows the finite element and equivalent AASHTO GDF values versus the slab thickness for critical interior and exterior girders in the considered bridges. The equivalent AASHTO GDF approach showed the same trend for the interior girders as the finite element results but exhibited a different tendency for the exterior girders. This is because the GDF is based on the AASHTO LRFD formulas, where the exterior GDF is a product of the interior girder GDF and a factor related only to the overhang distance, resulting in a constant value for all the considered bridges. Comparing the outcome of the refined method with that of the hand calculations showed that the equivalent AASHTO GDF for both the interior and exterior girders yielded slightly conservative results compared to the finite element values in most cases. As the slab thickness increased, the difference between the finite element and equivalent AASHTO GDF factors diminished.

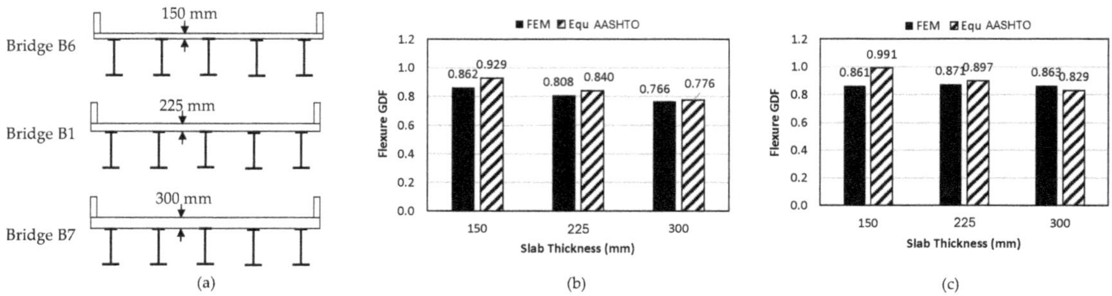

Figure 16. Effect of deck slab thickness on flexural GDF: (**a**) cross-section of bridges B1, B6, and B7; (**b**) interior girder; (**c**) exterior girder.

5.1.3. Effect of Cross-Bracing

The AASHTO formulas used to calculate the GDF were developed based on bridge models not containing diaphragms or cross-bracings [30]. Therefore, the reference bridge in this study is assumed to have cross-bracing only at the ends of the bridge (40 m). However, two other bridges with different cross-bracing spacings are developed to address the cross-bracing effect on the structural behavior of splayed girder bridges. Bridge B8 has a cross-bracing spacing equal to 5 m, while bridge B9 has a cross-bracing spacing equal to 10 m. Figure 17 shows the change in GDF due to the movement of two trucks in adjacent lanes in the transverse direction for two bridges—one in which the bracings are located at the supported ends only (reference bridge B1), and the other in which the bracings are located at 5 m intervals along the bridge (bridge B8). It is clear from the results that the fraction of flexural live load in the interior girder for the bridge without bracing is higher than in the corresponding bridge with bracing. In contrast, the presence of cross-bracing in bridge B8 resulted in higher GDF values in the exterior girders than those resulting from the same bridge without cross-bracing (bridge B1). Adding cross-bracing to a bridge results in more uniform deflection across all girders, thus increasing load distribution among the interior girders and leading to a lower GDF. However, increasing the rigidity of the cross-section with the addition of cross-bracing increases the effect of rigid body rotation on the exterior girder as a result of the eccentricity of the truck load in the transverse direction about the bridge centerline, which results in a higher GDF in the exterior girders.

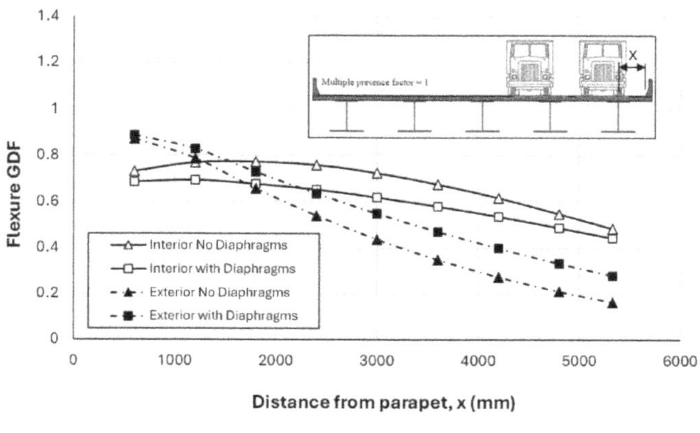

Figure 17. Finite element-based flexural GDF as a function of the distance from the parapet for bridges B1 and B8 under two side-by-side trucks.

Finite element analysis showed that the presence of intermediate cross-bracing affects the GDF values, although the spacing of the cross-bracing is not a significant parameter. Adding cross-bracing to the reference bridge slightly increased the GDF of the exterior girder by up to 2% while reducing GDF of the interior girder by about 8%. Moreover, the analysis showed that rigid body rotation of splayed girder bridges is less pronounced compared to that in parallel girder bridges. This is because transverse eccentric truck positioning has a reduced effect due to the resistance of flared girders to superstructure twisting. Note that the AASHTO equivalent GDF expressions do not account for the presence or spacing of cross-bracing. Therefore, all three bridges—B1, B8, and B9—with different cross-bracing spacings have the same equivalent AASHTO GDF, as shown in Figure 18. Comparing this value for interior or exterior girders with the finite element outcome showed that the equivalent AASHTO GDF is a good predictor of the flexural live load distribution. Note that the rigid body rotation method for finding the equivalent GDF for the exterior girders resulted in a high value equal to 1.023, far exceeding the findings of the finite element analysis and the AASHTO approximate approach. Hence, such an approach is not recommended for splayed girder bridges since it results in an unreasonably higher load effect than the refined method.

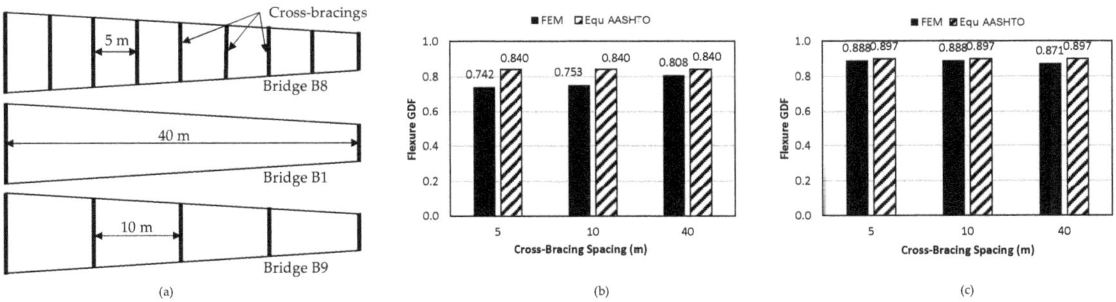

Figure 18. Effect of cross-bracing spacing on flexural GDF: (**a**) plan of bridges B1, B8, and B9; (**b**) interior girder; (**c**) exterior girder.

5.1.4. Effect of Other Parameters

In this section, the number of girders, girder depth, and the span length of a splayed girder are considered to check the validity of using the AASHTO flexural GDF at the respective axle locations to determine the bending moment in the girders.

By reducing the number of girders in reference bridge B1 from five to three and increasing them from five to seven, while keeping all parameters—including the girder spacing—constant, bridges B10 and B11 are generated, respectively. It is clear from the finite element results of Figure 19 that the number of girders in a splayed girder bridge affects the GDF of the exterior girders more than the interior girders (8.5% versus 2.5% maximum difference between the extreme values). Although the AASHTO GDF formulas do not consider the number of girders explicitly in a bridge, the specifications suggest using the lever rule if the number of girders is three, as is the case in bridge B10. However, for consistency, the equivalent GDF is used for all bridges in the study, which results in an equal equivalent GDF in all three bridges with different numbers of girders. The results show that the AASHTO GDF provides a good predictor of the flexural effect in the girders, especially when the number of girders is not excessively large. It should be noted that use of the lever rule to compute an equivalent GDF resulted in much higher values than the finite element outcome, where in the interior girders, the GDF value was equal to 1.14, and in exterior girders, it was equal to 1.045.

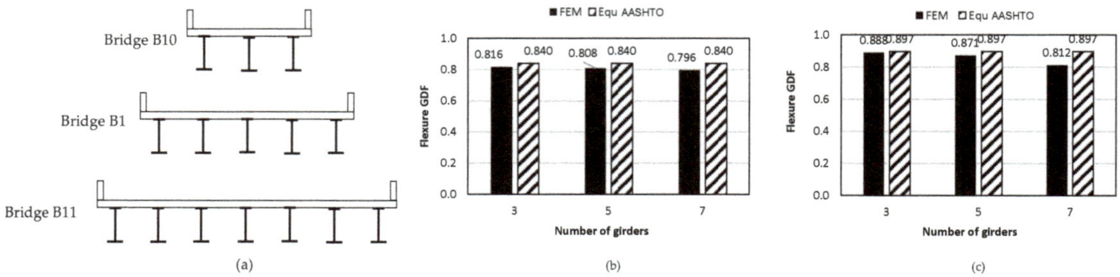

Figure 19. Effect of number of girders on flexural GDF: (**a**) cross-section of bridges B1, B10, and B11; (**b**) interior girder; (**c**) exterior girder.

In this study, the stiffness of the girder was adjusted by changing the depth of the steel girder web in the considered reference bridge. Bridges B12 and B13 were created by reducing the web depth of the 1700 mm girder in the reference bridge by 300 mm and increasing it by the same amount, respectively. Note that the girder level of rigidity in AASHTO is accounted for by the stiffness factor K_g, presented in this paper as Equation (3). The finite element findings shown in Figure 20 confirm the increase in the GDF with increased girder depth, especially for the interior girders, though the increase in GDF is not very significant. The AASHTO equivalent GDF follows the same trend and provides a safe upper limit to the results of the refined analysis.

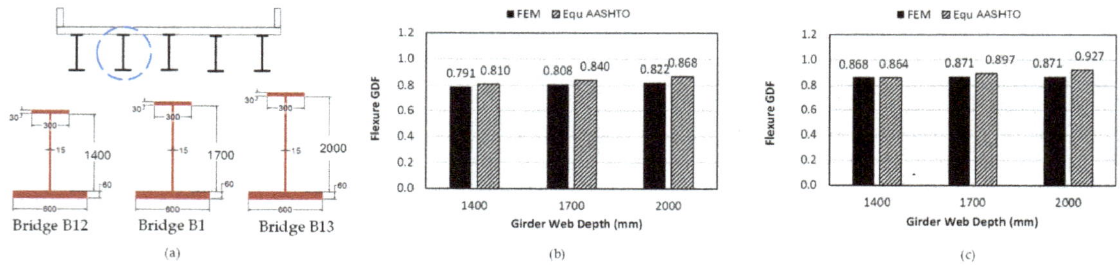

Figure 20. Effect of girder web depth on flexural GDF: (**a**) girder dimensions of bridges B1, B12, and B13; (**b**) interior girder; (**c**) exterior girder.

B14 and B15 were generated by decreasing and increasing the 40 m simple span length of reference bridge B1 by 25%, respectively. Note that changing the bridges' span length affects the splayedness angle of the bridge and consequently the girder spacing at the truck axle critical locations for maximum flexural effect, albeit not by much. Figure 21 shows that an increase in the span length has a favorable influence on the interior girders' GDF, but not on that of the exterior girders. The AASHTO GDF expressions at the location of the axles are also capable of matching the finite element trend for the interior girders, but not the exterior girders, since the AASHTO exterior girder GDF is a result of a factor that is a function of the overhang width multiplied by the GDF of the interior girder. Nevertheless, the difference between the finite element and AASHTO results is insignificant.

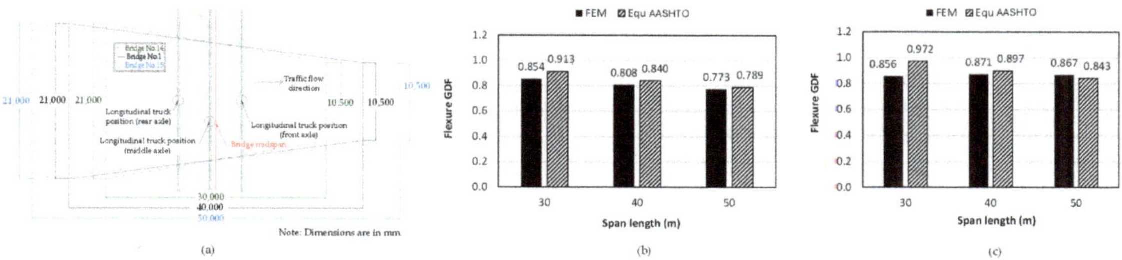

Figure 21. Effect of span length on flexural GDF: (**a**) plan of bridges B1, B14, and B15; (**b**) interior girder; (**c**) exterior girder.

5.1.5. Summary of Flexural Results

In summary, the presented results showed that for all of the fifteen splayed girder bridges analyzed in flexure, the equivalent GDF based on the AASHTO formulas for flexure were, for the most part, slightly larger than the GDF values obtained by finite element analysis. Figure 22 compares both methods with regard to the flexural GDF of critical interior and exterior girders. While there are very few cases in which the equivalent AASHTO GDF slightly underestimates the GDF values obtained by finite element analysis, these cases are nevertheless very close to each other. Hence, it can be concluded that for all practical purposes, the use of the flexural GDF AASHTO formulas with girder spacings corresponding to the locations of the truck axles is a good approach for analyzing splayed girders in flexure. This approach, which depends on a one-dimensional beam analysis, is simple and straightforward and eliminates the need for refined analysis, which is often costly and time-consuming, and yields outputs that are difficult to interpret.

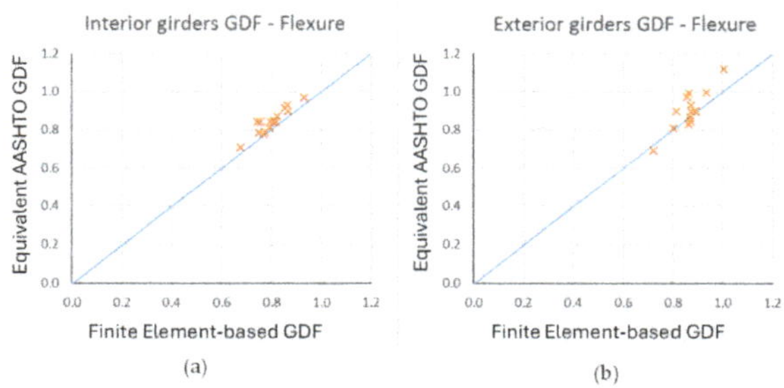

Figure 22. Comparison between the finite element and AASHTO flexural GDF: (**a**) interior girder; (**b**) exterior girder.

5.2. Shear in Girders Due to Live Load

In this section, the internal shear effect in the splayed girders is considered and discussed with consideration of the previously considered parameters in the flexural study, such as girder spacing, deck slab thickness, spacing of cross-bracing, number of girders, girders depth, and span length.

5.2.1. Effect of Girder Spacing

The shear GDF expressions in the AASHTO Bridge Design Specifications [1] only include girder spacing in the determination of the shear load effect in girders. In this study, the same bridges used to examine the flexure effect are analyzed to determine the shear effect by positioning the truck(s) longitudinally on the bridge near the support at the wider girder spacing end in order to maximize the vertical reaction under the critical girder.

Investigation of the effect of splayedness in the reference bridge B1 showed that an equivalent shear GDF based on the lever rule can be more effective in predicting the shear effect in splayed girders than using the AASHTO LRFD expressions. Therefore, the equivalent shear GDF was computed once based on the AASHTO LRFD expressions and once again based on the lever rule. Both values were then compared with the finite element model results for all fifteen bridges listed in Table 1.

The effect of the splayedness ratio $(S_2 - S_1)/L$ is examined for the shear GDF by studying bridges B1, B2, and B3. However, since the changes in this ratio at the wide end of the bridge (the longitudinal critical position for shear) are minimal, there is almost no change in the girder spacing between the three considered bridges. This leads to insignificant changes in the finite element results of the shear GDF. Likewise, the equivalent shear GDF based on the lever rule showed negligible change between the considered bridges, as the loaded region near the end support was relatively small in comparison with the entire surface area of the bridge.

When comparing the interior girder shear GDF values computed by the three different methods, it became obvious that the AASHTO LRFD expressions yielded0 slightly lower values compared with the finite element results, while the lever rule produced almost the same outcome. For the exterior girders, the AASHTO LRFD shear GDF values were higher than the finite element results by as much as 19%, while the lever rule values were much closer to the finite element results, with a maximum difference on the conservative side equal to 7.8%.

The girder spacing ratio S_1/S_2 was also examined for the shear GDF by studying bridges B4, B1, and B5. The finite element results demonstrated that for interior girders, increasing the S_1/S_2 ratio by 14.3% resulted in a 10.54% increase in GDF value, while decreasing the S_1/S_2 ratio by 20% led to a 12.88% reduction in GDF value. The same trend was observed for the exterior girders, where increasing the S_1/S_2 ratio by 14.3% resulted in an increase in GDF value of 7.8%, while reducing the S_1/S_2 ratio by 20% resulted in an 11.86% decrease in the GDF value. It can be observed that all splayed bridges acted like regular bridges with parallel girders, in which increasing girder spacing caused a linear increase in the values of the shear GDF.

Figure 23a,b show the influence of the splayedness and girder spacing ratios on the shear GDF, respectively, for the interior and exterior girders. For bridges B1, B4, and B5, the GDF is computed based on finite element analysis, the AASHTO LRFD expressions, and the lever rule. Compared to the finite element results, the AASHTO expressions yielded slightly lower GDF values for the interior girders and up to 32.8% higher values for the exterior girders. On the other hand, using the lever rule led to reasonably close and conservative values in comparison with the refined analysis for both interior and exterior girders. Based on the outcome of the study concerning the variation in the girder spacing, the lever rule is recommended for determining the shear load effect in interior and exterior girders of splayed girder bridges in lieu of using a complex refined method.

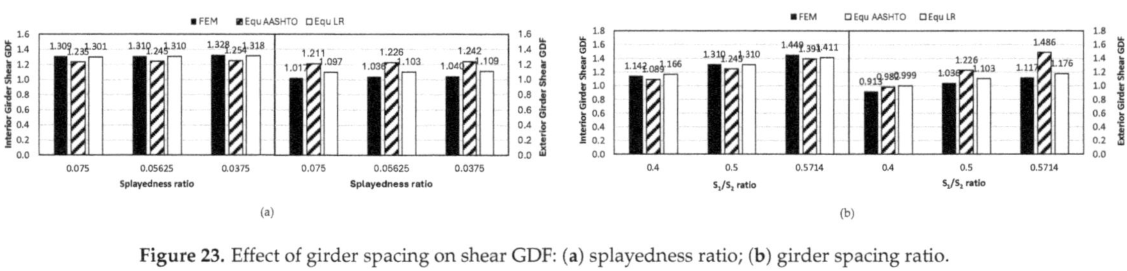

Figure 23. Effect of girder spacing on shear GDF: (**a**) splayedness ratio; (**b**) girder spacing ratio.

5.2.2. Effect of Other Parameters

All the other five parameters (slab thickness, cross-bracing spacing, girder stiffness, number of girders, and span length) were investigated through the analysis of the other ten bridges, B6–B15, as shown in Table 1. As shown in Figure 24, the refined analysis demonstrated that these parameters have very minimal influence on the shear GDF values, which is an expected finding. For the shear load effect in the interior girders, the difference between the lever rule and finite element outcome ranges between −4.9% and 9.5%. The corresponding difference between the lever rule and finite element analysis for the exterior girders ranges between −0.8% and 9.4%. Compared to the finite element findings, the overall results demonstrate that the lever rule gives much more sensible results than the AASHTO LRFD shear GDF expressions for the majority of the considered cases.

5.2.3. Summary of Shear Results

The results presented in the previous sections showed that for the 15 bridges analyzed for shear, use of the lever rule yields more reasonable and slightly conservative shear GDF values than the AASHTO LRFD GDF expressions. Figure 25 compares the finite element shear GDF values with the corresponding shear GDF values based on the lever rule for both interior and exterior girders. For almost all the considered cases, the shear GDF values obtained using the lever rule were reasonably very close to the finite element analysis findings. Thus, for all practical purposes, it can be concluded that the GDF based on the lever rule is a good indicator of the live load shear effect in the main elements of splayed girder bridges.

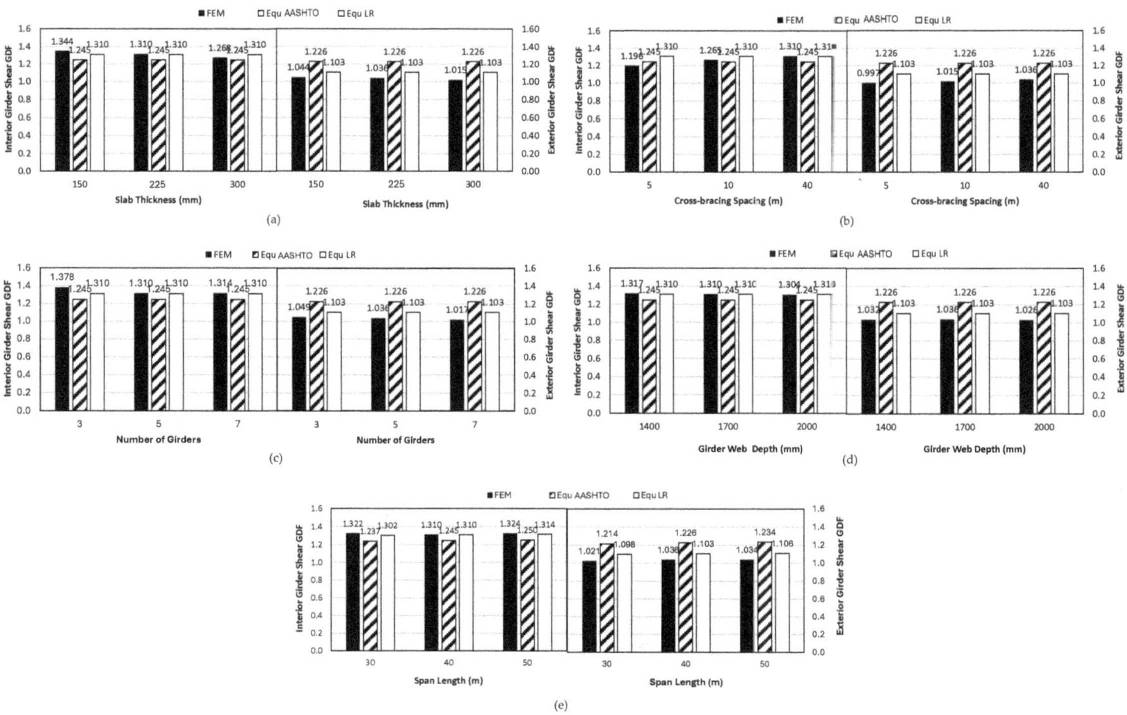

Figure 24. Effect of girder spacing on shear GDF: (**a**) slab thickness; (**b**) cross-bracing spacing; (**c**) number of girders; (**d**) girder web depth; (**e**) span length.

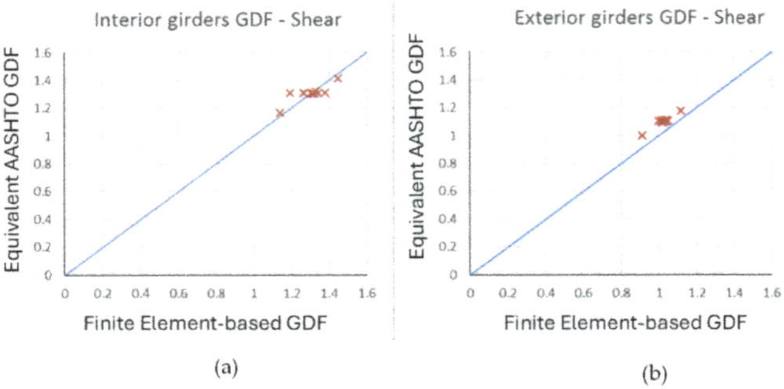

Figure 25. Comparison between the finite element and AASHTO shear GDF: (**a**) interior girder; (**b**) exterior girder.

5.3. Dead Load Effect in Girders

In this study, both shored and unshored construction methods are studied and applied to the reference bridge to understand the behavior of splayed girder bridges under the influence of dead load. Finite element models are developed for both cases and the results are compared with hand calculations for the composite girder of the bridge loaded with the same dead loads based on the tributary area concept.

In shored construction, the composite girder resists all dead load components, including the girder and deck slab self-weight, parapet weight, and future wearing surface (FWS). In the analytical approach, a single composite girder of the bridge is loaded with the concrete deck slab self-weight, where the deck slab width is taken as half of the girder spacing from each side. This approach results in a linearly variable slab width along the span length equal to the girder spacing, resulting in a trapezoidal load on the composite girder. The steel girder self-weight is also represented by a uniform load along the composite girder's length assuming the girder was supported during erection. Regarding the superimposed load from parapet weight and future wearing surface, the AASHTO LRFD Bridge Design Specifications [1] suggest distributing it equally among the girders if sufficient curing of the concrete slab has taken place. Therefore, the parapet weight is represented by a uniform load along the span and its magnitude is taken equal to the summation of parapet loads at both edges of the bridge divided by the number of girders. However, due to the bridge splayedness, the FWS load varies along the girder length, and it is represented by a trapezoidal load with a magnitude at any section equal to the total FWS load per unit width at that section divided by the number of girders. Cross-bracing, if present, is represented by point loads along the span of the composite girder; these loads vary in magnitude from one point to another based on the cross-bracing weight, where it changes due to the variable girder spacing along the span. The composite girder dead load for the reference bridge B1 is shown in Figure 26. The composite girder is then analyzed as a line element using statics, and the moment at the midspan is computed to find the elastic stress at the bottom flange of the girder in the composite section at the midspan. The ensuing stress due to analytically calculated dead load is compared with the one obtained from the finite element (FE) results at the midspan, as shown in Table 3.

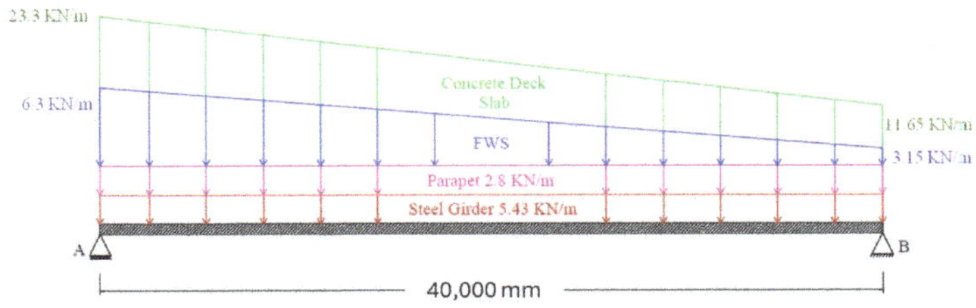

Figure 26. Dead load on a single composite girder of bridge B1 for the case of shored construction.

Table 3. Maximum tensile stress in shored composite girders of reference bridge B1 at midspan due to future wearing surface and parapet weight.

Girder	Analytical Calculation	FE Exterior Girder (G1)	FE First Interior Girder (G2)	FE Intermediate Girder (G3)
Stress (MPa)	77.77	78.57	74.51	73.44

It is clear from the results included in Table 3 that the stresses are very close to each other. The magnitude of the stress resulting from the analytical approach using the transformed composite girder geometric properties is greater than the corresponding stress from the finite element model by 4.2% for the first interior girder and by 5.5% for the intermediate girder, but lower by 1% for the exterior girder. This finding indicates that the analytical approach provides a reasonable approximation of the dead load effect in individual girders for the shored construction method in splayed girder bridges.

In the unshored girder construction case, the composite girder resists loads resulting from FWS and parapet weight, while the non-composite girder resists all other dead loads. Therefore, in the analytical approach, the composite girder is only loaded by the trapezoidal load representing the FWS and the uniform load representing the parapet weight, as shown in Figure 27 for reference bridge B1. Both load components are assumed to be distributed equally among the bridges' girders, as suggested by the AASHTO LRFD specifications, and computed in the same manner as presented previously in the shored construction case. The remaining dead load due to self-weight and the weight of the concrete deck slab within the tributary width is assumed to be carried by the non-composite steel girders. Structural analysis of the bare steel girder subjected to loads on the non-composite section is not considered here since there is no interaction from other girders; hence, the analytical solution in this case would match the finite element results.

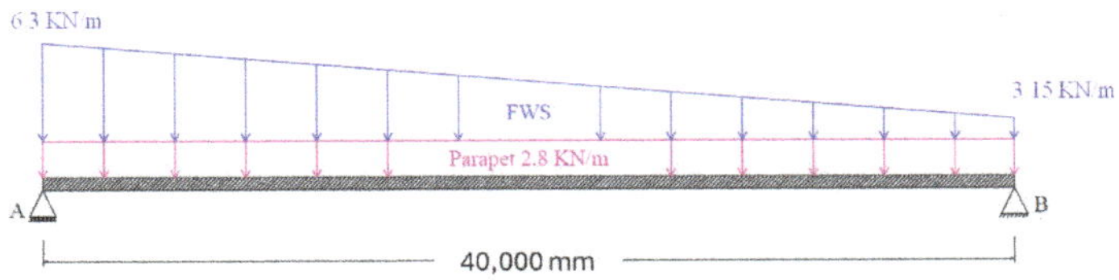

Figure 27. Dead load on a single composite girder of bridge B1 for the case of unshored construction.

The composite girder is then analyzed, and the midspan moment and corresponding stress are computed again in the same manner as in the shored construction method. The analytical results, based on the analysis of a line element and mechanics of elastic material principles, are compared with the finite element findings at the midspan, as shown in Table 4.

Table 4. Maximum tensile stress in unshored composite girders of reference bridge B1 at midspan due to future wearing surface and parapet weight.

Girder	Analytical Calculations	FE Exterior Girder (G1)	FE First Interior Girder (G2)	FE Intermediate Girder (G3)
Stress (MPa)	19.23	21.00	20.25	19.20

The results of the maximum tensile stress in the girders, presented in Table 4, for the case of unshored construction are close to each other. While the stress value obtained with the analytical approach is almost the same as that obtained with the finite element model for the intermediate girder, it is 5.3% lower for the first interior girder and 9.2% lower for the exterior girder. This outcome shows that distributing the dead load equally among all girders, as suggested by the AASHTO LRFD specifications, in the unshored construction method might not be that accurate for splayed girder bridges. This finding could be due to the load from the parapet weight having greater influence on the girders within its vicinity, leading to larger dead load imposed on the exterior and first interior girders than on the central girder. In any case, this dead load component on the composite girder is a relatively smaller fraction of the total dead load, which, when combined with the remaining load on the non-composite girders, will result in closer agreement between the analytical and the finite element results.

6. Conclusions and Recommendations

The results of this study on splayed girder bridges lead to the following conclusions and recommendations:

6.1. Conclusions About Flexural Effect

- The tributary width concept that is often used in practice for regular bridges is reliable for determining the flexural dead load effect on the splayed interior and exterior girders. However, the AASHTO recommendation of equally sharing the superimposed dead load among the girders is not always very accurate, as this study showed that exterior girders receive higher loads from the parapet weight than interior girders.
- Increasing the splayedness ratio leads to reductions in GDF for flexure in both interior and exterior girders. The relationship between the GDF and splayedness ratio is almost linear.
- The splayedness ratio and girder spacing ratio cannot be used as standalone indices in a splayed girder bridge without supporting them with the actual girder spacing. Therefore, to judge the magnitude of the flexural GDF of a splayed girder bridge, a splayedness parameter plus one girder spacing at a specific location within the bridge must be provided.
- The effect of the slab thickness on the flexural GDF for splayed girder bridges is moderate on the interior girders and negligible on the exterior girders. In addition, girder stiffness has little effect on the GDF for both interior and exterior girders. However, an increase in the number of girders beyond five can lead to some reduction in the GDF of the exterior girder, due to the possibility of rigid body rotation over the increased width of the bridge and large eccentric truck loading.
- There is a significant effect on the interior girders due to the addition of cross-bracing to splayed girder bridges, which is similar to the presence of the cross-bracing effect in regular bridges. Where the flexure GDF for interior girders dropped by more than 8% due to adding cross-bracings at 5 m spacing, it increased the exterior girder GDF by about 2%. The location of cross-bracing spacing in reference to the longitudinal truck

axle positions is important, as axles directly above a cross-bracing are more equally distributed among the girders compared to axles located between cross-bracings.

- The AASHTO GDF for flexure yielded reasonable results compared to the FE models, where for the interior girders, the difference was between 2–13% on the conservative side. The high values of the flexure GDF compared with the finite element results, computed following the AASHTO specifications, were due to the presence of cross-bracing, the effect of which is not considered in the AASHTO formulas. The GDF results for exterior girders obtained using the AASHTO formulas were also very reliable, where the percentage difference was between −4% and 14% when compared with the finite element analysis outcomes. The results representing slight conservatism were recorded in bridges with extreme parameter values that exceeded the AASHTO limits.

6.2. Conclusions About Shear Load Effect

- The tributary width concept is a good approach for finding the shear dead load effect in splayed interior and exterior girders. As in the case of flexure, superimposed dead load distribution among the girders is not shared equally, as permitted by AASHTO, since exterior girders collect higher loads from the parapet weight than interior girders.
- Increasing the splayedness ratio causes a negligible effect on the GDF for shear in both interior and exterior girders. On the other hand, the increase in the smaller-to-larger girder spacing ratio, while keeping the splayedness ratio constant, results in a significant increase in the GDF for shear. As in the case of flexure, both the splayedness angle and the girder spacing ratios cannot be used as standalone indices in a splayed girder bridge without specifying the magnitude of the (smaller or larger) girder spacing.
- The effect of slab thickness on the shear GDF for splayed girder bridges is moderate in the interior girders and negligible in the exterior girders. Girder stiffness and the number of girders (if above five) have little influence on the shear GDF for both interior and exterior girders.
- Adding cross-bracing to splayed girder bridges reduces the GDF for shear in the interior girders, particularly if the cross-bracing spacing is small. The effect of using cross-bracing in splayed girder bridges on the exterior girders shear GDF is not significant since the bracing impacts these girders from one side only.
- The use of the lever rule to predict shear in interior or exterior splayed girders yields reliable results compared to the finite element outcome. Girder distribution values from the lever rule yielded results that deviated from the finite element findings by −4.9% to 9.5% for the critical interior girder and by −0.8% to 9.4% for the exterior girder. On the other hand, the AASHTO GDF for shear could not reasonably predict the shear live load effect in the girders, as the outcome from this approach yielded overly conservative values.

6.3. Recommendations

Based on the results of the finite element analysis, the AASHTO Bridge Design Specifications GDF expressions, and the lever rule, the following recommendations with regard to flexure and shear in girders that are part of splayed girder bridges can be drawn:

- The tributary width concept is a reliable approach for determining the dead load effect on the splayed interior and exterior girders. The parapet weight, when placed on the hardened concrete deck, tends to load the exterior girders more than the interior girders; hence, assuming such a load to be equally distributed among the girders is not a valid assumption.

- The girder distribution factors for flexure in the AASHTO LRFD specifications can be reasonably used for splayed girder bridges if the specific girder spacing at the location of each axle of the truck in the longitudinal direction of the bridge is considered.
- The lever rule can provide a good estimate of the live load distribution among splayed girders when subjected to shear.

Author Contributions: Conceptualization and methodology, S.W.T.; software, validation, and analysis, F.H.; investigation and discussion, F.H. and S.W.T.; resources, S.W.T.; writing—original draft preparation, F.H.; writing—review and editing, S.W.T.; supervision and project administration, S.W.T. All authors have read and agreed to the published version of the manuscript.

Funding: Publication of this study was funded by the Open Access Program and the College of Engineering at the American University of Sharjah, UAE. The opinions included in the study are those of the authors and do not reflect the views of the funding agencies.

Institutional Review Board Statement: Not applicable.

Informed Consent Statement: Not applicable.

Data Availability Statement: The original contributions presented in this study are included in the article. Further inquiries can be directed to the corresponding author.

Acknowledgments: The first author gratefully acknowledges the Civil Engineering Department at the American University of Sharjah, UAE, for providing a graduate teaching assistantship during his master's degree, conducted under the supervision of the second author. This research was carried out as part of that graduate program.

Conflicts of Interest: Author Faress Hraib was employed by the company ABNA. The remaining author declares that the research was conducted in the absence of any commercial or financial relationships that could be construed as a potential conflict of interest.

References

1. AASHTO. *LRFD Bridge Design Specifications*, 9th ed.; American Association of State Highway and Transportation Officials: Washington, DC, USA, 2020.
2. Buckler, J.G.; Barton, F.W.; Gomez, J.P.; Massarelli, P.J.; McKeel, W.T., Jr. *Effect of Girder Spacing on Bridge Deck Response*; Virginia Transportation Research Council: Charlottesville, VA, USA, 2000; Final Rep., VTRC 01-R6.
3. Barr, P.J.; Amin, M.N. Shear live-load distribution factors for I-girder bridges. *J. Bridge Eng.* **2006**, *11*, 197–204. [CrossRef]
4. Tabsh, S.W.; Sahajwani, K. Approximate analysis of irregular slab-on-girder bridges. *J. Bridge Eng.* **1997**, *2*, 11–17. [CrossRef]
5. Song, S.-T.; Chai, Y.; Hida, S.E. Live-load distribution factors for concrete box-girder bridges. *J. Bridge Eng.* **2003**, *8*, 273–280. [CrossRef]
6. Puckett, J.A.; Mertz, D.; Huo, X.S.; Jablin, M.C.; Peavy, M.D.; Patrick, M.D. *Simplified Live Load Distribution Factor Equations*; Transportation Research Board: Washington, DC, USA, 2007; NCHRP Rep.
7. Utah Department of Transportation. *Bridge Management Manual*; Utah Department of Transportation: Salt Lake City, UT, USA, 2014.
8. Kansas Department of Transportation. *LRFD Bridge Design*; Kansas Department of Transportation: Topeka, KS, USA, 2009.
9. *A. DARWin-ME*, Version 1.0; American Association of Highway and Transportation Officials: Washington, DC, USA, 2011; Volume 444.
10. WSDOT. *PGSuper Software 2.1 Release*; BridgeSight Inc.: South Lake Tahoe, CA, USA, 2009.
11. Akbari, R.; Maalek, S. A review on the seismic behaviour of irregular bridges. *Proc. Inst. Civ. Eng. Struct. Build.* **2017**, *171*, 552–580. [CrossRef]
12. Soberón, M.C.G.; Soberón, J.M.G. Dynamic properties variation by irregular superstructure and substructure common bridges. *Procedia Eng.* **2017**, *199*, 2961–2966. [CrossRef]
13. Soleimani, F.; Mangalathu, S.; DesRoches, R. Seismic Resilience of Concrete Bridges with Flared Columns. *Procedia Eng.* **2017**, *199*, 3065–3070. [CrossRef]
14. Eom, J.; Nowak, A.S. Live load distribution for steel girder bridges. *J. Bridge Eng.* **2001**, *6*, 489–497. [CrossRef]
15. *ANSYS*, Version 14.0; ANSYS Inc.: Canonsburg, PA, USA, 2011.

16. Chan, T.H.; Chan, J.H. The use of eccentric beam elements in the analysis of slab-on-girder bridges. *Struct. Eng. Mech.* **1999**, *8*, 85–102. [CrossRef]
17. Barr, P.J.; Eberhard, M.O.; Stanton, J.F. Live-load distribution factors in prestressed concrete girder bridges. *J. Bridge Eng.* **2001**, *6*, 298–306. [CrossRef]
18. Hays, C., Jr.; Sessions, L.; Berry, A. Further studies on lateral load distribution using a finite element method. *Transp. Res. Rec.* **1986**, *1072*, 6–14.
19. Ventura, C.; Felber, A.; Stiemer, S. Determination of the dynamic characteristics of the Colquitz River Bridge by full-scale testing. *Can. J. Civ. Eng.* **1996**, *23*, 536–548. [CrossRef]
20. Chen, Y. Distribution of vehicular loads on bridge girders by the FEA using ADINA: Modeling, simulation, and comparison. *Comput. Struct.* **1999**, *72*, 127–139. [CrossRef]
21. Tabsh, S.W.; Tabatabai, M. Live load distribution in girder bridges subject to oversized trucks. *J. Bridge Eng.* **2001**, *6*, 9–16. [CrossRef]
22. Bishara, A.G.; Liu, M.C.; El-Ali, N.D. Wheel load distribution on simply supported skew I-beam composite bridges. *J. Struct. Eng.* **1993**, *119*, 399–419. [CrossRef]
23. Machado, M.A.; Sotelino, E.D.; Liu, J. Modeling technique for honeycomb FRP deck bridges via finite elements. *J. Struct. Eng.* **2008**, *134*, 572–580. [CrossRef]
24. Mabsout, M.E.; Tarhini, K.M.; Frederick, G.R.; Tayar, C. Finite-element analysis of steel girder highway bridges. *J. Bridge Eng.* **1997**, *2*, 83–87. [CrossRef]
25. Wu, H. *Influence of Live-Load Deflections on Superstructure Performance of Slab on Steel Stringer Bridges*; West Virginia University: Morgantown, WV, USA, 2003.
26. Eamon, C.D.; Nowak, A.S. Effects of edge-stiffening elements and diaphragms on bridge resistance and load distribution. *J. Bridge Eng.* **2002**, *7*, 258–266. [CrossRef]
27. Hraib, F. Finite Element Analysis of Splayed Girder Bridges. Master's Thesis, Civil Engineering Department, American University of Sharjah, Sharjah, United Arab Emirates, June 2015. 162 pages. Available online: https://repository.aus.edu/entities/publication/92989414-e0e1-4937-b64c-20150599b82b (accessed on 1 April 2025).
28. Fang, K.I. Behavior of Ontario-Type Bridge Deck on Steel Girders. Ph.D. Thesis, Department of Civil Engineering, University of Texas at Austin, Austin, TX, USA, 1985.
29. Schonwetter, P. Field Testing and Load Rating of a Steel-Girder Highway Bridge. Master's Thesis, Department of Civil Engineering, University of Texas at Austin, Austin, TX, USA, 1999.
30. Cai, C. Discussion on AASHTO LRFD load distribution factors for slab-on-girder bridges. *Pract. Period. Struct. Des. Constr.* **2005**, *10*, 171–176. [CrossRef]

applied sciences

MDPI

Article

Neural Network-Based Prediction of Amplification Factors for Nonlinear Soil Behaviour: Insights into Site Proxies

Ahmed Boudghene Stambouli [1,2] and Lotfi Guizani [1,*]

1 Department of Construction Engineering, Ecole de Technologie Superieure, University of Quebec, 1100 Notre-Dame St. West, Montreal, QC H3C 1K3, Canada; ahmed.boudghene-stambouli.1@ens.etsmtl.ca or bstambouliahmed@gmail.com

2 Risk Assessment and Management Laboratory (RISAM), Faculty of Technology, Abou Bekr Belkaïd University, Tlemcen P.O. Box 230-13048, Algeria

* Correspondence: lotfi.guizani@etsmtl.ca

Featured Application: The results offer the possibility to better understand the main factors for the prediction of the site effect for the seismic analysis and design of infrastructures. They could also help enhance the codes' provisions in this regard. The methodology offers an insight into further possibilities for the integration of artificial intelligence within the domain of structural and geotechnical engineering.

Abstract: The identification of the most pertinent site parameters to classify soils in terms of their amplification of seismic ground motions is still of prime interest to earthquake engineering and codes. This study investigates many options for improving soil classifications in order to reduce the deviation between "exact" predictions using wave propagation and the method used in seismic codes based on amplification (site) factors. To this end, an exhaustive parametric study is carried out to obtain nonlinear responses of sets of 324 clay and sand columns and to constitute the database for neuronal network methods used to predict the regression equations of the amplification factors in terms of seismic and site parameters. A wide variety of parameters and their combinations are considered in the study, namely, soil depth, shear wave velocity, the stiffness of the underlaying bedrock, and the intensity and frequency content of the seismic excitation. A database of *AF*s for 324 nonlinear soil profiles of sand and clay under multiple records with different intensities and frequency contents is obtained by wave propagation, where soil nonlinearity is accounted for through the equivalent linear model and an iterative procedure. Then, a Generalized Regression Neural Network (GRNN) is used on the obtained database to determine the most significant parameters affecting the *AF*s. A second neural network, the Radial Basis Function (RBF) network, is used to develop simple and practical prediction equations. Both the whole period range and specific short-, mid-, and long-period ranges associated with the *AF*s, F_a, F_v, and F_l, respectively, are considered. The results indicate that the amplification factor of an arbitrary soil profile can be satisfactorily approximated with a limited number of sites and the seismic record parameters (two to six). The best parameter pair is (*PGA*; resonance frequency, f_0), which leads to a standard deviation reduction of at least 65%. For improved performance, we propose the practical triplet $(PGA; V_{s30}; f_0)$ with V_{s30} being the average shear wave velocity within the upper 30 m of soil below the foundation. Most other relevant results include the fact that the *AF*s for long periods (F_l) can be significantly higher than those for short or mid periods for soft soils. Finally, it is recommended to further refine this study by including additional soil parameters such as spatial configuration and by adopting more refined soil models.

Keywords: nonlinear site response; site proxies; neural network; seismic response; site factor

1. Introduction

Local site conditions play a crucial role in determining the level of damage experienced during earthquakes, as demonstrated from past experiences in the Mexico City (1985), Loma Prieta (1989), Northridge (1994), Kobe (1995), and, more recently, Turkey (2023) earthquakes. In many cases, the geological and geotechnical characteristics of an area can amplify seismic waves, leading to greater ground shaking and structural damage. Soil type, liquefaction conditions, site topography, building design, and historical context are some of the key factors affecting the severity of shaking and damage [1–9]. Addressing these factors through better engineering practices and building codes can mitigate damage in future events. To this end, for many decades, using local elastic response spectra based on soil categorization and seismic hazard has been the common way to evaluate seismic input for the design and evaluation of structures. Soil categorization and characterization are also crucial when applying Ground Motion Prediction Equations (GMPEs) for accurate predictions of local ground motion and/or for developing site-specific spectra while taking into account local site conditions.

Numerous researchers have put forward different site parameters to classify and characterize soils. Borcherdt (1994) was among the first to recommend using the average shear wave velocity over the upper 30 m (V_{s30}) as a fundamental criterion for soil characterization [10]. This parameter has since gained widespread acceptance in seismic assessments. However, there are concerns about V_{s30}'s effectiveness in adequately representing site effects on its own [11–20]. More recent GMPEs that utilize various databases continue to rely on V_{s30} to characterize site conditions [21–24]. However, V_{s30} is frequently complemented or replaced by other site parameters, including the fundamental frequency, f_0, the average shear wave velocity at different depths within the soil profile, or the depth to hard bedrock [19,22,25–29]. Recent and current regulatory codes use peak ground acceleration (*PGA*) or spectral values for reference soil, along with V_{s30}, to modify the design spectra's characteristics—specifically the plateau bandwidth, level, and long-period decay—to suit local site conditions. These regulations focus on the free surface response to ascending waves, which governs how seismic waves interact with structures. This methodology is fundamental to major international building codes, including the 1997 NEHRP Provisions, the 1997 Uniform Building Code, Eurocode 8 (ENV 1998), the International Building Code (IBC 2012), and the National Building Code of Canada (NBC) (2015a, b, 2020) [29–34]. Recent studies have incorporated multiple soil parameters using Generalized Regression Neural Network models. These models are founded on the wave propagation theory applied to linear viscoelastic soil profiles sourced from the Japanese Kiban Kyoshin Network (KiK-net) and the US Geological Survey (USGS) [7,35].

Site amplification is usually captured by relying on only few straightforward site proxies' values (like V_{s30} and contrast velocity). However, this simplification can lead to various challenges, including V_{s30}'s inability to reflect the shear wave velocity profile across the entire soil depth, which can result in a skewed seismic response for short, mid, and/or long periods. Given the complex nature of the problem, the selected limited set of site proxies used to predict site amplification may not necessarily be the best combination, especially because the literature lacks a systematic evaluation of the performance of a limited number of different site proxy combinations.

On the other hand, artificial neural networks (ANNs) are particularly effective for prediction tasks due to their capacity to model complex nonlinear relationships within high-dimensional datasets. They perform well across various applications, such as time series forecasting and image recognition. Additionally, techniques like regularization and optimization enhance their performance, improving accuracy and reducing the risk of overfitting. Overall, neural networks are powerful tools for generating reliable predictions in many fields [36–43]. Consequently, the use of ANNs to identify the most prominent selections of site proxies and how they correlate to site amplification factors offers a recent perspective that can help fill the existing gap.

This study aims to identify the most effective site parameters for predicting the site amplification of the seismic response at the free surface of a soil deposit. More specifically, it focuses on establishing relationships between the amplification factor of the spectral response and a limited set of site proxies that describe the characteristics of soil profiles based on predictions of selected ANN methods.

To achieve this goal, this study examines the 1D nonlinear responses of monolayer soil profiles with varying thicknesses and shear wave velocities, situated above a semi-infinite bedrock layer, subject to different seismic records, with varying intensities and frequency contents. More specifically, a total of 324 soil profiles of sand and clay are analyzed, each represented by a single layer with depths varying from 5 to 200 m and a shear wave velocity ranging from 100 to 600 m/s, underlain by a bedrock layer with a shear wave velocity varying between 750 and 1500 m/s. The nonlinear site responses of these profiles to the vertical propagation of shear (S) waves are calculated, including responses at free surfaces, which are used to determine site amplification factors. Fourteen (14) input waveforms are selected based on the recent studies by Boudghene Stambouli et al. (2017) and Dif and Boudghene Stambouli (2023) [8,35]. Each of the considered profiles is subjected to these fourteen input motions, normalized to eleven different *PGA* levels: 0.01, 0.05, 0.1, 0.2, 0.3, 0.4, 0.5, 0.6, 0.75, 0.9, and 1.05 g. This approach allows for the calculation of the mean amplification for each *PGA* value, covering all relevant seismic frequency ranges. In total, 3564 geometric average amplification factors for each soil type are derived (324 soil profiles × 11 *PGA*s).

Several methods exist to calculate ground response with nonlinear behaviour. Nonlinear time history analyses are the common method for investigating the dynamic nonlinear behaviour of structures but are less widely used for soils. They are typically conducted using models discretizing the space domain, such as finite element models. While these methods generally yield accurate results and can model complex problems, they can be time-consuming due to their step-by-step integration procedure in the time domain combined with the iterative process for achieving a direct representation of the nonlinear soil response [44]. Instead, engineers often resort to a viscoelastic equivalent linear model combined with an iterative process, called equivalent linear analysis, to obtain the nonlinear solution for soils. This method, which is typically more computationally efficient, yields satisfactory results for engineering applications, especially within the relevant frequency range and for simple and smooth nonlinear behaviour [44–49]. Johari and Momeni (2015) have shown that nonlinear calculations of soil responses provided only slight performance improvements over equivalent linear methods while demanding considerably more time and complexity [47]. Consequently, given the large number of analyses to carry out (about 100,000), the equivalent linear approach is adopted in this study for efficiency.

Furthermore, numerical methods, based on the finite element method, can solve one-, two-, and three-dimensional nonlinear soil problems, where amplifications can differ significantly from 1D scenarios, especially in the presence of valleys or rugged topography. The effects of such geometric and topographic parameters had been considered in previous studies, such as those of Paolucci and Morstabilini (2006) and Boudghene Stambouli et al. (2018) [50,51]. However, such specific site effects are beyond the scope of this study, and they are not included in basic code specifications. Consequently, for the sake of efficiency, simplification, and direct comparison with codes and the main body of literature, 1D modelling of soil and wave propagation is considered in this study. Furthermore, this avenue offers the possibility of obtaining a simple analytical solution to the wave propagation problem.

Soil nonlinearity is incorporated using shear modulus degradation curves that show the variations in shear modulus (G) with respect to shear strain (γ) levels and loading history (cycling), as detailed hereafter. Shear modulus degradation under cyclic loading has been studied by several researchers using resonant column or enhanced triaxial tests. Seed and Idriss (1970) were pioneers in developing and widely applying shear modulus reduction and damping curves specifically for sand [52]. In 1991, Vucetic and Dobry introduced the plasticity index (IP) to define shear modulus degradation curves for clay [53,54]. In 1993, Ishibashi and Zhang added another parameter, the effective confining pressure, used alongside the IP to better capture the behaviour of soils with low plasticity indices [55]. That same year, the Electric Power Research Institute (EPRI) created its own soil degradation curves by combining shear modulus reduction and damping curves, allowing for broader applicability across cohesionless soils, from gravelly sands to low plasticity silts and sandy clays [56]. More recently, Darendeli (2001) proposed a new set of degradation curves, indicating that the degree of linearity increases with the plasticity index (IP), mean effective stress, and overconsolidation ratio (OCR) [57]. Such a model, while considered more appropriate for representing soil nonlinearity, requires a set of additional soil parameters to those used in our study which will extend the number of required nonlinear analyses. In order to streamline our analysis, limit the number of analyses, and focus on identifying effective proxies for soil profiles, we selected simpler degradation models proposed by Sun et al. (1988) [58] for clay and by Seed and Idriss (1970) [52] for sand. These models require minimal soil properties while still representing the nonlinear behaviour of soils under cyclic loading accurately enough [52,58]. Additionally, we incorporate the damping curves from the study by Idriss (1990), which provide insights into how damping ratios vary with strain levels [59]. This combination allows us to simplify our approach while retaining the critical features for predicting the seismic response of clay and sand, which ultimately should facilitate the identification of the most suitable proxies for soil characterization in seismic assessments.

To explore the correlation between the average amplification factor and various sets of soil characteristics, we use an artificial neural network (ANN) approach. This method effectively captures complex relationships and nonlinear interactions among input variables. To enhance our analysis, a sensitivity analysis is conducted using the neural network model. This analysis aims to identify which soil characteristics serve as the best site proxies for predicting the amplification factor. By examining how changes in these variables (proxies) impact the output, we can pinpoint the most influential parameters, thereby improving our understanding of soil behaviour in seismic contexts and facilitating better site characterization in future studies.

2. Derivation of Amplification Factors (*AF*s)

2.1. Introduction

For a given signal, the amplification factor at given period (T), in terms of spectral acceleration, can be expressed as the ratio of the response spectrum at the surface to the response spectrum at the outcropping reference rock as follows:

$$AF(\mathrm{T}) = \frac{\mathrm{Sa(T)_s}}{\mathrm{Sa(T)_b}} \quad (1)$$

where $\mathrm{Sa(T)_s}$ and $\mathrm{Sa(T)_b}$ are the 5% response acceleration spectra at the free surface and at the reference rock (bedrock), respectively, while T is the period of the structure.

Because the response of soils under strong excitations is nonlinear by nature, and in order to apply the theory of wave propagation in an elastic medium, the problem needs to be linearized. The equivalent linear method, first introduced by Jacobson (1930) [60] and extended by Hudson (1965) [61] and Seed and Idriss (1970), is largely used in nonlinear problems of soils and structures [49,52]. This method replaces the nonlinear behaviour with an equivalent linear spring in parallel with a viscoelastic damper to include the stiffness and damping associated with the hysteresis of the soil or structure. However, given the nonlinear nature of the problem, the stiffness and damping vary with the displacement or deformation level, which is unknown. Consequently, this linearization method adopts an iterative scheme in order to converge at the target displacement, typically the maximum or design displacement. For a wave propagation problem, therefore, the shear modulus (G) and the damping ratio (ξ), which are functions of the deformation level in the middle of each layer, are adjusted at each iteration in order to capture the hysteresis of the soil at the target deformation level. A wave equation can be used to solve the shear wave propagation in soils [62–66]. Schnabel et al. (1972) proposed an algorithm based on the continuous solution of the wave equation, the main steps of which are summarized in the Supplementary File, section (a) [67,68].

2.2. Input Waveforms $Sa(T)_b$

This section discusses and describes the selected input accelerograms, b(t), used in the above algorithm for the calculation of the amplification factors in this study.

From the RESORCE database [69], sets of 14 input waveforms (S1 to S14), recorded on outcropping rock, are selected for this study, and their characteristics are listed in the Supplementary File, section (b). The selected accelerograms are drawn from real earthquakes and satisfy certain conditions to ensure a representative average amplification factor that is not biassed by a spectral content that is too rich in either short, medium, or long frequency [35]. The main characteristics of the 14 reference acceleration time histories are summarized in the Supplementary File. Figure 1 presents the spectra of the selected signals normalized to a *PGA* of 1 g. Each normalized seismic signal is scaled by the appropriate factor to achieve the chosen eleven *PGA* values (0.01 g, 0.05 g, 0.1 g, to 0.9 g, with a step of 0.1 g and 1.05 g). We then compute 49 896 time-history seismic responses (14 records × 11 *PGA* levels × 324 soil profiles) using the shear modulus degradation curve of the clay, and the same process and number of seismic responses are also carried out for sand.

2.3. Site Model, Wave Propagation Solution, and Transfer Function T(f)

In this study, the soil is idealized by one horizontally layered soil deposit resting on a bedrock substratum (see Figure 2).

The layer is fully defined by its shear modulus, G, or shear wave velocity, V; thickness, h; damping ratio ζ; and mass density ρ. The underlying half-space has shear wave velocity $V_{bedrock}$. The vertical z-axis is oriented downwards. The response of the soil column to harmonic, vertically incident plane shear waves is governed by differential Equation (2) [68]:

$$G_i \frac{\partial^2 u_i}{\partial z_i^2} + \eta_i \frac{\partial^3 u_i}{\partial z^2{}_i \partial t} = \rho_i \frac{\partial^2 u_i}{\partial t^2} \quad (2)$$

where u_i is the horizontal displacement in the ith layer (in this study, $i = 1$ for the monolayer soil deposit).

with:

$$\eta_i = 2G_i\zeta_i/\omega \tag{3}$$

where ω is the angular frequency of the exciting harmonic, and ζ_i is the damping ratio.

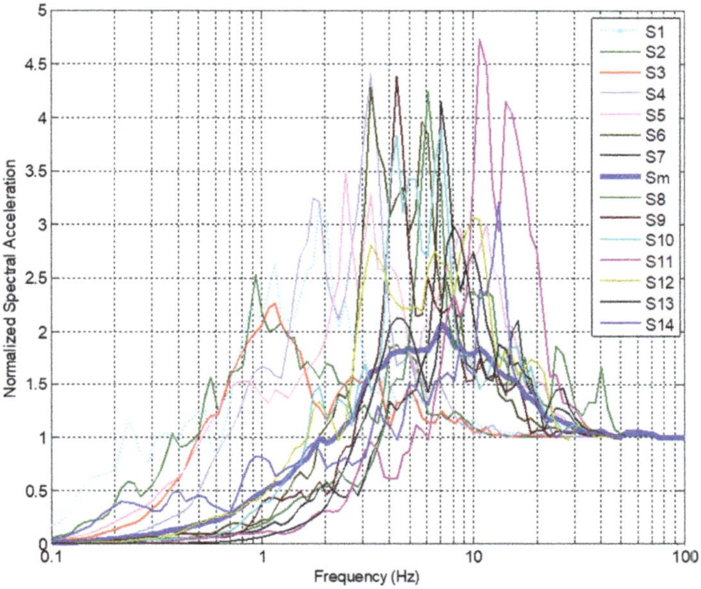

Figure 1. Normalized (to *PGA* = 1 g) acceleration spectra of signals used.

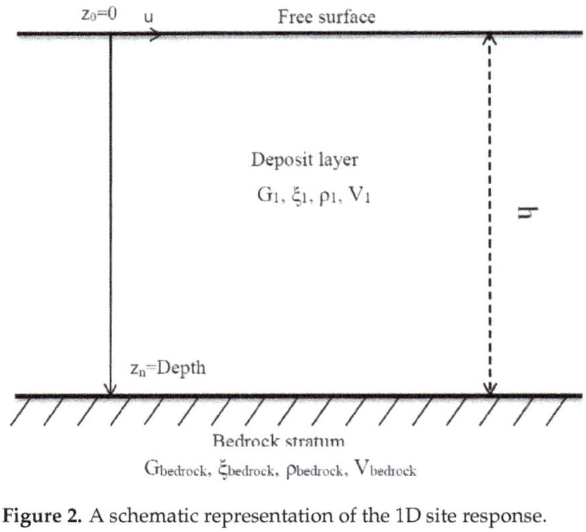

Figure 2. A schematic representation of the 1D site response.

The general solution of governing differential Equation (2) is a summation of up-going and down-going plane waves with unknown amplitudes, A_i and B_i, for each layer, given in Equation (4).

$$u_i(z_i, f) = A_i e^{(i2\pi ft)} e^{(ik_i^* z_i)} + B_i e^{(i2\pi ft)} e^{(-ik_i^* z_i)} \tag{4}$$

The transfer function is a complex function defined as the ratio of the layer surface amplitude to the layer bottom amplitude, and the procedure to drive such a function can be found in the classical literature [68,70,71]. For a single layer of soil, it can be written, using Euler's Law, as follows:

$$T(f) = \frac{1}{\cos K_s^* h + i\alpha_z^* \sin K_s^* h} \quad (5)$$

where K_s^* is a complex wave number, defined as

$$K_s^* = \frac{2\pi f}{V_1(1 + i\zeta_1)} \quad (6)$$

and α_z^* is the complex impedance ratio, defined as

$$\propto_s^* = \frac{\rho_1 V_1(1 + i\zeta_1)}{\rho_{bedrock} V_{bedrock}(1 + i\zeta_{bedrock})} \quad (7)$$

where $\zeta_{bedrock}$ is the damping ratio in the bedrock, which generally has a negligible effect on the results, and is taken as null in this study.

For a particular soil profile, *AF*(*T*) is computed once the transfer function *T*(*f*) is known using the equivalent linear model procedure extensively described in the geotechnical earthquake engineering literature [35,67,68].

As mentioned earlier, we used the simple shear modulus degradation model proposed by Sun et al. (1988) [58] for clay and the model proposed by Seed and Idriss (1970) [52] for sand (G/Gmax is a function of the shear strain). For damping variation with shear strain, the curves of Idriss (1990) are used for both clay and sand [59]. These selected models are illustrated in Figure 3, and their key parameter values are presented in Supplementary File, section (c).

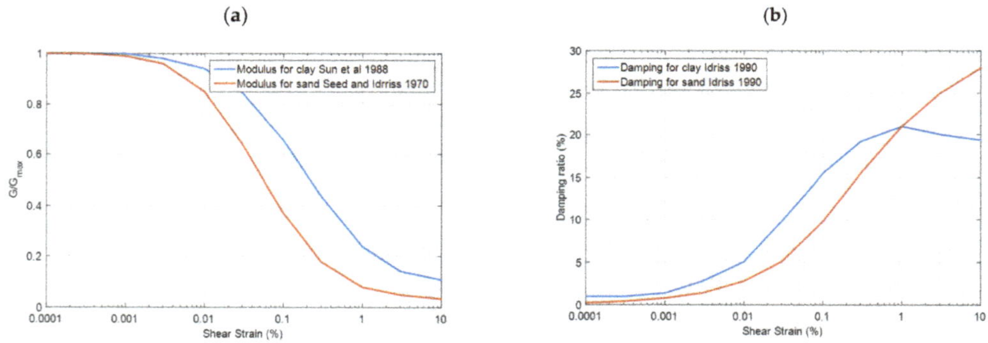

Figure 3. Degradation curves for clay and sand. (**a**) Shear modules. (**b**) Damping ratios. Refs. [52,58,59].

Based on these models, the equivalent shear modulus and damping ratio are obtained by an iterative process to ensure that the values used are consistent with the strain level obtained according to the procedure described above. At the interface of adjacent layers, the stress and displacement continuity equations are solved, and the relationships between these amplitudes for two adjacent layers are established. Afterward, the waves are propagated from the bottom (unit up-going amplitude) to the top layer using the free surface condition (shear stress = 0). The wave amplitudes and the transfer function can be derived with respect to the motion at the outcropping bedrock.

2.4. Database

2.4.1. Descriptions of Soil Profiles Studied

In this study, a monolayer soil profile is characterized by one of six different shear wave velocity values, 100, 150, 200, 300, 400, and 600 m/s, which overcomes a semi-infinite bedrock. The latter, for its part, is characterized by a shear wave velocity equal to 750, 800, 900, 1000, 1200, or 1500 m/s. The depth of the soil layer varies from 5 to 10, 20, 30, 50, 75, 100, 150, and 200 m. In total, three hundred twenty-four (324) soil profiles are generated and considered for this study. Furthermore, all 324 soil profiles are considered in two sets of profiles for a total of 648 different soil conditions: 1—clay/cohesive soil; 2—sand/granular soil. Although certain rare exceptions may exist, the considered variants of simplistic soil profiles thus generated by the aforementioned combinations cover virtually all practical cases encountered in nature and in engineering practice.

2.4.2. Selection of Site Parameters Studied

Each soil profile can be partially described by a few site parameters. In this paper, we investigate seven of them. Extensive studies of seismic site response have been performed in recent decades. In 1994, Borcherdt developed intensity-dependent, short- and long-period amplification factors based on the average shear wave velocity measured over the upper 30 m of a site [10]. Seed et al. (1988) developed a geotechnical site classification system based on the shear wave velocity, the depth to bedrock, and general geotechnical descriptions of soil deposits [72]. Seed et al. (1991) [73] then developed intensity-dependent site amplification factors that modify the baseline "rock" peak ground acceleration (*PGA*) to account for site effects. With such a site, *PGA* value, and site-dependent normalized acceleration response spectra, Martin and Dobry (1994) derived site-dependent design spectra (primarily based on the site classification system and amplification factors) [11], which were incorporated in the 1997 Uniform Building Code (UBC) [30]. The National Building Code of Canada (NBC) (2005, 2015a, b) and a later version (2020) adopt the same philosophy as the UBC, where the soil classification as well as site design spectra are based on the shear wave velocity in the upper 30 m supporting the foundations [32–34]. However, two important limitations are associated with such an approach.

First, it requires a relatively extensive field investigation, and second, it overlooks the potential importance of other site parameters, such as the fundamental frequency or depth to bedrock, among others, which have been identified by many researchers as pertinent for site classification [26–28,35,74].

To overcome the above limitation, our study considers the following parameters in addition to the Vs_{30}: the depth of deposit to bedrock (depth); the average shear velocity (V_{sm}) over that depth; the velocity contrast, that is, the ratio between the shear wave velocity in the bedrock and at the surface (Cv); the second velocity contrast, that is, the ratio between the shear wave velocity in the bedrock and the average shear wave velocity over the upper 30 m (Cv2); the soil profile's fundamental frequency (f_0); and the *PGA* in the bedrock. The inclusion of the *PGA* as a parameter of the study is justified by the well-established fact that site amplification for nonlinear soil is fundamentally dependent on the intensity of the input wave. The site proxies considered are calculated using Relations (8) to (13) as follows:

$$\text{Depth} = \text{h}_1 \tag{8}$$

which is the thickness of the first layer (deposit layer), h_1, with the bedrock being considered as a half-infinite space (see Figure 2).

$$V_{sm} = V_1 \tag{9}$$

where V_1 is the shear wave velocity in the first (deposit) layer:

$$V_{s30} = \sum_{i=1}^{l_{30}} \frac{h_i\ V_i}{30} \tag{10}$$

where l_{30} is the number of distinct layers found in the top 30 m:

$$\mathrm{Cv} = \frac{V_{Bedrock}}{V_1} \tag{11}$$

$$\mathrm{Cv2} = \frac{V_{Bedrock}}{V_{s30}} \tag{12}$$

The fundamental soil frequency f_0 is determined by using the simplified Rayleigh procedure, described by Dobry et al. (1976) [75], based on Equations (13) and (14):

$$f_0 = \frac{\sqrt{\left(4\left(\sum_{i=1}^{n} \frac{(z_i+z_{i+1})^2}{V_i^2}\, h_i\right) / \left(\sum_{i=1}^{n} (X_i + X_{i+1})^2 h_i\right)\right)}}{2\pi} \tag{13}$$

where $\frac{z_{i+1}+z_i}{2}$ is the depth of the midpoint of layer (i), and the X_i values correspond to the estimated deformations due to the fundamental mode shape at the top of each layer (i), derived according to Dobry et al. (1976) as follows [75]:

$$X_n = 0;\ X_{i-1} = X_i + \frac{z_i + z_{i-1}}{V_i^2}\, h_i \tag{14}$$

To better illustrate the domain of validity of this study, the covered site parameters domain is summarized in Table 1 through 10%, 50%, and 90% fractiles for all site parameters.

Table 1. The 10%, 50%, and 90% fractiles of the studied parameters.

		10% Fractile	50% Fractile	90% Fractile
Depth (m)		5	50	200
V_{sm} (m/s)	V_{sm} (m/s)	100	250	600
V_{s30} (m/s)		100	300	642
Cv		1.66	4	10
Cv2		1.33	3	8
f_0 (Hz)		0.31	1.64	12.73

2.4.3. Correlation Between Site Parameters

All the parameters considered are not fully independent, as shown by the coefficient of determination (R^2) presented in Table 2, calculated for each pair of parameters for all soil profiles considered. There exist strong correlations between the pairs (Cv, Cv2), (V_{sm}, Cv), (V_{sm}, V_{s30}), (V_{s30}, Cv), and (V_{s30}, Cv2), as indicated by the coefficients of determination (R^2) generally exceeding 0.62. However, very weak correlations, notably for the pairs (Cv, Depth) and (V_{sm}, Depth), and weak correlations for the pairs (Cv, f_0), (f_0, V_{sm}), and (Depth, V_{s30}), with the R^2 value ranging between 0.25 and 0.37, are observed. Finally, a mitigated correlation, with R^2 = 0.51, is observed for the couple (f_0, Depth). These correlation indicators are useful for selecting independent site parameters for the models relating site amplification to site characteristics.

Table 2. Correlations (R^2) between various site parameters.

	Depth	f_0	Cv	Cv2	V_{sm}	V_{s30}
Depth	1	0.5119	0.0001	0.2584	0.0001	0.3311
f_0		1	0.2926	0.4445	0.3744	0.7496
Cv			1	0.8493	0.7815	0.6208
Cv2				1	0.6638	0.7529
V_{sm}					1	0.7954
V_{s30}						1

3. Computed Amplification Factors: Main Statistical Characteristics

3.1. Validation of Adopted Methodology for Computation of AFs

Earlier studies confirmed that the equivalent linear method combined with the wave propagation solution in the frequency domain, as adopted in this study, instead of the non-linear time history analysis demanding much more computational effort, typically yields satisfactory results for engineering applications [47]. Many examples of the validation of this approach are documented in DEEPSOIL [70]. Furthermore, using the DEEPSOIL 7.0 software, we carry out a nonlinear time history analysis on a specific site among the soil profiles used in this study (a deposit of 5 m in depth with a shear wave velocity of 300 m/s on a bedrock with a shear velocity of 800/s), submitted to the Kobe (1995) record (PGA = 0.82 g). The results are compared to those obtained using DEEPSOIL with the equivalent linear model, and practically identical free surface response time histories and free surface response acceleration spectra are obtained. A difference of 0.005, less than 0.2%, is observed in the amplification factor.

For more flexibility, ease of automation, and to efficiently carry out the parametric study, the above procedure is implemented in Matlab R2023a© and used to compute the ground time-history responses. To validate the developed code, the results are carefully and successfully checked for many validation examples against those provided by other codes, namely, DEEPSOIL [70] and EERA [71]. A relative difference of less than 0.3% is obtained in terms of the maximum acceleration at the free surface.

3.2. General Background of Computed AFs

This section presents on overview of the computed sets of frequency-dependent *AF*s and their short- to mid- and long-period average values. These data are essential and constitute the learning set needed to identify the key parameters controlling the site response characteristics.

In total, 99,792 *AF*(*T*) curves are computed using (Equation (1)) for the 324 soil profiles under the 14 seismic excitations normalized by 11 *PGA* levels for the two types of soils (clay and sand). The *AF* curves for clay are shown in Figure 4, and those for sand are shown in Figure 5. They may be written in the general form of $AF(P_k, \theta, S_l, T_i, PGA(lo))$, where

- P_k, $k = 1, \ldots n_P$ is introduced to identify the soil profile, $n_P = 324$;
- $\theta = 0$ for clay (using the shear modulus degradation curve of clay) and $\theta = 1$ for sand (using the shear modulus degradation curve of sand);
- S_l, $l = 1$, 14 is the lth excitation, where the geometrical average of the 14 amplification factors has been computed for each site and for each *PGA* for the results of Figures 4 and 5;
- T_i, $(i = 1, ..100)$ is the ith structural period, and *AF* values are systematically computed for 100 values, equally spaced between 0.01 and 10 s on a logarithmic period axis;
- $PGA(lo)$, for the identification of the *PGA* level, and lo, vary from 1 (for *PGA* = 0.01 g) to 11 (for *PGA* = 1.05 g).

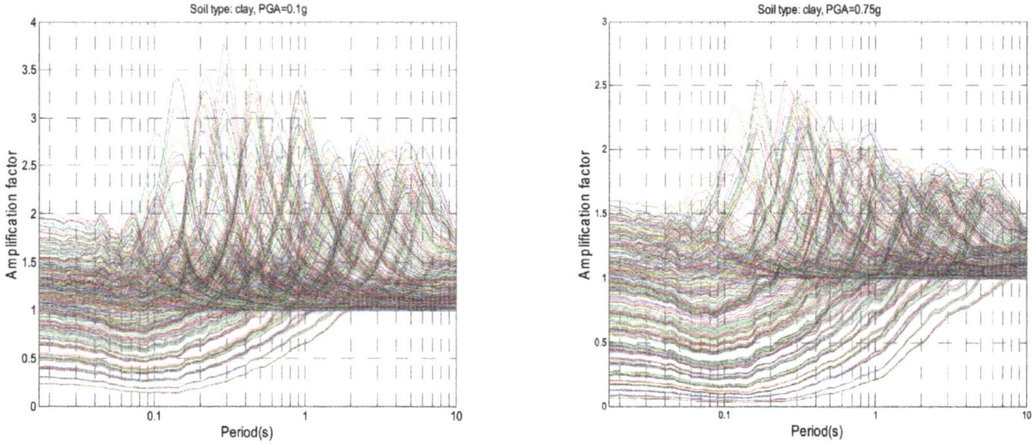

Figure 4. Average amplification factors for 324 sets of clay soil profiles at 0.1 g and 0.75 g *PGA* levels.

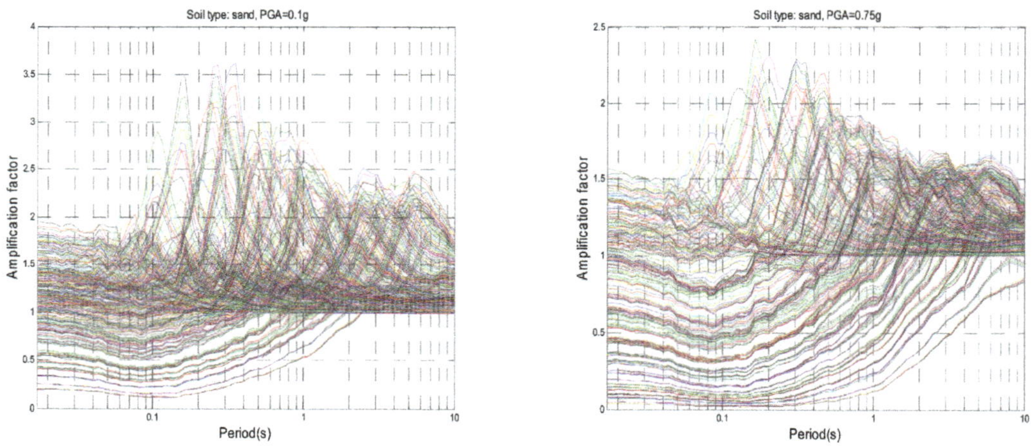

Figure 5. Average amplification factors for 324 sets of sand soil profiles at 0.1 g and 0.75 g *PGA* levels.

For instance, $AF(P_{20}, 0, S_8, T_{50}, 0.1\text{ g})$ stands for the *AF* obtained at the soil profile P_{20} of clay type, subjected to the eighth seismic excitation, S_8, for the 50th period ($T = \text{Log}^{-1}$ $(-2 + (50 - 1) \times (\text{Log } 10 - \text{Log } (0.01)) = 0.295$ s), and normalized to a *PGA* level of 0.1 g (lo = 3).

After the *AF* is calculated for a particular profile, *k*, and for 14 seismic excitations, the average site amplification factor AF_m is computed for the soil profile so that

$$\text{Log}[AF_m(P_k, \theta, T_i, PGA(lo))] = \left(\frac{1}{14}\right) \sum_{l0=1}^{14} \text{Log}[AF(P_k, \theta, S_l, T_i, PGA(lo))] \quad (15)$$

Then, the average amplification factor for each profile, for all intensity levels, is obtained by the following:

$$\text{Log}(AF_0(\theta, T_i)) = \frac{1}{\text{n}_\text{p}} \sum_{k=1}^{\text{n}_\text{p}} [\text{Log}(AF_m(P_k, \theta, T_i))] \quad (16)$$

Consequently, 3564 geometric mean values of the amplification factor are calculated, and stand for each soil profile and type.

Simultaneously, for each profile P_k, the *AF* variability derived from the 11 different average amplification factors is quantified using the corresponding standard deviation:

$$\sigma_{AF}(P_k,\theta,T_i) = \sqrt{\frac{\sum_{lo=1}^{11}[\log(AF_m(P_k,\theta,,T_i,PGA(lo))) - \log(AF_m(P_k,\theta,T_i))]^2}{11}} \quad (17)$$

The σ_{AF} values are displayed in Table 3 for all soil profiles. They exhibit a significant period dependence. The maximum variability is observed at around 0.1 s. It hardly decreases at shorter periods as short as ~0.03 s, but it decreases significantly at intermediate and long periods. These values are greater, especially at short to intermediate periods with a lower degree. It would thus be meaningless to aim to obtain extremely precise models with residuals between observations and predictions much below these values.

Table 3. Initial variability values for clay and sand.

Total Initial Variability (Soil Type: Clay)	**0.2631**	**Total Initial Variability (Soil Type: Sand)**	**0.3241**
Maximum initial variability σ_{max}	0.3905	Maximum initial variability σ_{max}	0.4696
σ (θ = 0, T = 0.01 s)	0.3061	σ (θ = 1, T = 0.01 s)	0.3794
σ (θ = 0, T = 0.02 s)	0.3045	σ (θ = 1, T = 0.02 s)	0.3766
σ (θ = 0, T = 0.04 s)	0.3236	σ (θ = 1, T = 0.04 s)	0.3954
σ (θ = 0, T = 0.07 s)	0.3649	σ (θ = 1, T = 0.07 s)	0.4384
σ (θ = 0, T = 0.1 s)	0.3828	σ (θ = 1, T = 0.1 s)	0.4577
σ (θ = 0, T = 0.2 s)	0.3555	σ (θ = 1, T = 0.2 s)	0.4369
σ (θ = 0, T = 0.4 s)	0.273	σ (θ = 1, T = 0.4 s)	0.3509
σ (θ = 0, T = 0.7 s)	0.2071	σ (θ = 1, T = 0.7 s)	0.2783
σ (θ = 0, T = 1.0 s)	0.1768	σ (θ = 1, T = 1.0 s)	0.2385
σ (θ = 0, T = 2.0 s)	0.1106	σ (θ = 1, T = 2.0 s)	0.1387
σ (θ = 1, T = 4.0 s)	0.0961	σ (θ = 1, T = 4.0 s)	0.0998
σ (θ = 1, T = 7.0 s)	0.0804	σ (θ = 1, T = 7.0 s)	0.081
σ (θ = 0, T = 10.0 s)	0.0652	σ (θ = 1, T = 10.0 s)	0.067

Note that a few additional parameters are introduced to measure the variability of the results, as summarized in Table 3.

The average *AF* for all profiles, noted $AF_0(\theta,T_i)$, and defined as the geometrical average of the n_p average *AF* ($AF(P_k,\theta,T_i)$), is noted for simplicity as AF_0:

$$\text{Log}(AF_0(\theta,T_i)) = \frac{1}{n_p}\sum_{k=1}^{n_p}[\text{Log}(AF_m(P_k,\theta,T_i))] \quad (18)$$

The initial variability, defined as the initial standard deviation of the site average amplification factor over all profiles, is

$$\sigma_0(\theta,T_i) = \sqrt{\frac{1}{11*n_p}\sum_{k=1}^{n_p*11}[\log(AF_m(P_k,\theta,T_i,PGA(lo))) - \log(AF_0(\theta,T_i))]^2} \quad (19)$$

The maximum initial variability, defined as the peak value of the initial variability, σ_0, over the whole period range, is

$$\sigma_{0\max}(\theta) = Max_{T_i}[\sigma_0(\theta, T_i)] \tag{20}$$

The overall initial variability, defined as the average of the initial variability over the whole period range, is

$$\sigma_{0m}(\theta) = \frac{1}{n_T}\sum_{i=1}^{n_T} \sigma_0(\theta, T_i) \tag{21}$$

where n_T is the number of structural periods used, i.e., 100.

3.3. Means and Variability of AFs

- For each profile set, we compute the $n_P \times 14 \times 11$ AF : $AF(P_k, \theta, S_l, T_I)$, the n_P average amplification factors (AF_m), to constitute, for each soil type, a database of 3564 $AF_m(P_k, \theta, T_I, PGA(lo))$ values and their associated variability $\sigma_{AF}(P_k, \theta, T_I)$, the mean amplification factor $AF_0(\theta, T_I)$, and the associated initial variability $\sigma_0(\theta, T_i)$. The results are displayed in Figure 6a,b for the clay and sand sets, respectively. The following main observations are derived:

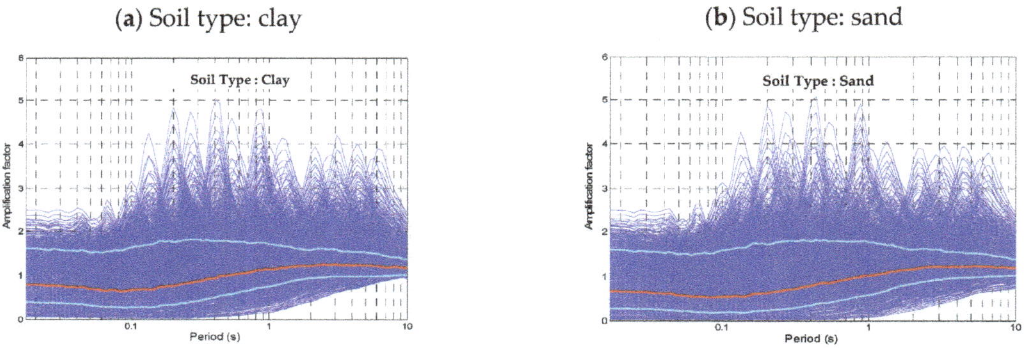

Figure 6. Average amplification factors as functions of real period for each set of soil profiles, namely (**a**) clay soil profiles and (**b**) sand soil profiles. Thin blue lines correspond to every site profile (3564 results), thick red line is geometrical average over whole profile sets, and thick light blue lines are averages ± one standard deviation.

- The peak period, i.e., the period with the peak amplification factor, covers a very broad range from 0.08 s to about 6–7 s for clay and sand soil profiles, which explains the richness of the database.
- As shown in Figures 4 and 5, the corresponding peak amplification ranges from less than 1.0 to 4.0 and up to 5.0 for low levels of *PGA* (0.01 g to 0.05 g). However, increasing the *PGA* level results in a decrease in the peak amplification factors to values of around 2.0 to 2.5 for a *PGA* ranging from 0.75 g to 1.05 g. The average amplification factors for clay are generally higher than those for sand at the mean (2 Hz $\leq f <$ 5 Hz) and high ($f \geq$ 5 Hz) frequency ranges.
- Some amplification factors exhibit a short period of de-amplification. A careful look at the corresponding soil profiles indicates that they correspond to deep soft soils, with low velocity, which act as seismic isolators.
- The overall average amplification factor (Figure 6a,b) is higher than unity for periods greater than 0.5 s for clay soil profiles, but it is greater than 0.9 s for sand soil profiles. The lowest overall average amplification factor is observed in the 0.05 to

0.15 s period range for clay and sand soil profiles. In this period range, the overall average amplification factor is less than 0.7 and 0.6 for clay and sand, respectively. The overall average amplification factors are significantly smaller than the peak values for individual profiles, which highlights the need to identify relevant site parameters that may explain this site-to-site variability.
- The "initial variability" $\sigma(\theta, T_I)$ associated with the average *AF*s (Table 3) has a maximum value at low to intermediate periods (0.01 to 0.4 s), reaching up to 0.39 for clay and 0.47 for sand soil. It then gradually vanishes with the period's increase, reaching a value of around 0.065 at T = 10 s.

3.4. The Division of the Period Range: Short, Intermediate, and Long Periods

The UBC and the National Building Codes are based mainly on research works dating back to the 1990s [10,76]. These earlier studies recognized the dependency of the *AF* with the period and defined three representative amplification factors for three distinct zones: 1- F_a for the short period range (acceleration plateau); 2- F_v for the intermediate to long period range (velocity zone); and 3- F_l for the long period range. In this study, we adopt the same methodology and classification, and in the absence of any consensus regarding the limits of these zones, we use the following ranges: F_a is obtained by the mean *AF* value for periods in the [0.1 s, 0.2 s] range; F_v for periods in the [0.75 s, 1.5 s] range; and F_l, for periods in the [2.82 s, 5.65 s] range.

4. Description and Implementation of Neural Network Method

The principal objective of an artificial neural method is to predict or establish relationships between input and output parameters. It is particularly useful for developing simpler forms of relations between input and output parameters for cases where the interrelations between these parameters are complex and no obvious function exists to describe them. This method is essentially based on a training phase with a database, composed of input and output parameters, with the database being randomly selected. After the training phase is completed, the neuron network can basically be used to predict new input and associated output values for the parameters. In the field of seismology, this method has been used by many researchers to derive new GMPEs [35–39,77]. Two types of ANNs are used in this study: the Generalized Regression Neural Network (GRNN) and the Radial Basis Function (RBF), which are of the same family but with slightly different architecture. Both ANNs have three layers: an input layer, a hidden layer, and an output layer.

In this study, we used a GRNN for the identification of site proxies (input parameters) for both soil types, sand and clay. To this end, we used the 324 soil profiles (soil parameters) of each type for which 11 average amplification factors (i.e., $AF_m(P_k, \theta, T_i, PGA(lo))(P_k, \theta, T_i, PGA(lo))$ were computed from the different seismic signal databases with different *PGA* values ranging from 0.01 g to 1.05 g. That is, there was a database of 3564 (324 × 11) cases for each soil type. The output consisted of the calculated *AF* values for the selected 100 periods of the structure and the amplification factors at short, intermediate, and long periods, named F_a, F_v, and F_l, respectively). Half of the database was used for training, and the second half for testing. Elements of these two subgroups of the database were selected and swapped randomly from one to the other until the network parameter, which is the Gaussian width for the GRNN, was determined. The main advantage of this method is the rapidity of the training phase. Note that many possible combinations of input site parameters were considered. The performance of the GRNN model was measured with various non-independent indicators, namely, the standard deviation of residuals, the reduction in variance with respect to the initial variability, the coefficient of correlation, and the physical tendency. Based on the results obtained with the GRNN, the combination of input site parameters giving the best performance while being handy for engineering practice were selected. For the selected combination of parameters, the RBF neural network was used to obtain a simple prediction equation. However, the training phase needs many adjustments, such as the number and the choice of key neurons in the hidden layer, and it is time-consuming. In the training phase of the RBF, the number and the choice of key neurons can either be set randomly from the training data, or they are iteratively trained or derived using techniques such as K-means, Max–Min algorithms, or Kohonen self-organizing maps [78–93]. After this unsupervised training was carried out and the number and key neurons in the hidden layer were chosen, the weights between this layer and the output layer neurons were determined by multiple regressions in a supervised manner. We used cross-validation, with 50% of the data used for training and the remaining 50% for testing, with each half of the database randomly being swapped from one to other until the RBF network parameters were obtained. More details about the used neural networks and their implementation are given in the Supplementary File, section (d).

5. Results

This section summarizes the main findings and results from the above process, combining a GRNN and RBF.

5.1. Determination of Site Proxies Using GRNN

A total of 200 GRNN models using different combinations of input parameters were derived, and their results were analyzed and compared in terms of the standard deviation of residuals (predicted–actual values) to the initial standard deviation values for each period, i.e., $\sigma_0(\theta, T_i)$, and the overall variability $\sigma_{0m}(\theta)$ defined earlier.

Equations (22) and (23) are used to compute the error between predicted and actual values, which in turn are compared with the initial variabilities found using Equations (18)–(21).

The standard deviation of residuals for each period and each neural network model, for comparison with the initial variability term $\sigma_0(\theta, T_i)$, is given by the following:

$$\varepsilon_{\text{ANN}}(\theta, T_i) = \sqrt{\frac{1}{11 * n_P}\sum_{lo=1}^{11}\sum_{k=1}^{n_P}[\log(AF_{\text{ANN}}(P_k, \theta, T_i, PGA(lo))) - \log(AF_m(P_k, \theta, T_i, PGA(lo)))]^2} \quad (22)$$

where the ANN is GRNN or RBF.

The maximum standard deviation over all periods is defined as

$$\varepsilon_{\text{ANN,max}}(\theta) = Max_{T_i}[\varepsilon_{\text{ANN}}(\theta, T_i)] \quad (23)$$

The overall error is defined as the average over all periods of the error term, that is,

$$\varepsilon_{\text{ANN,m}}(\theta) = \frac{1}{n_T}\sum_{i=1}^{n_T}\varepsilon_{\text{ANN}}(\theta, T_i) \quad (24)$$

To obtain a statistically meaningful insight into the relative performances of each site proxy considered in controlling the amplification factor, the standard deviation for each soil type and period error term $\varepsilon_{\text{GRNN}}(\theta, T_i)$ is plotted, together with the initial variability, $\sigma(\theta, T_i)$, in Figure 7a,b for clay and sand soil profiles, respectively.

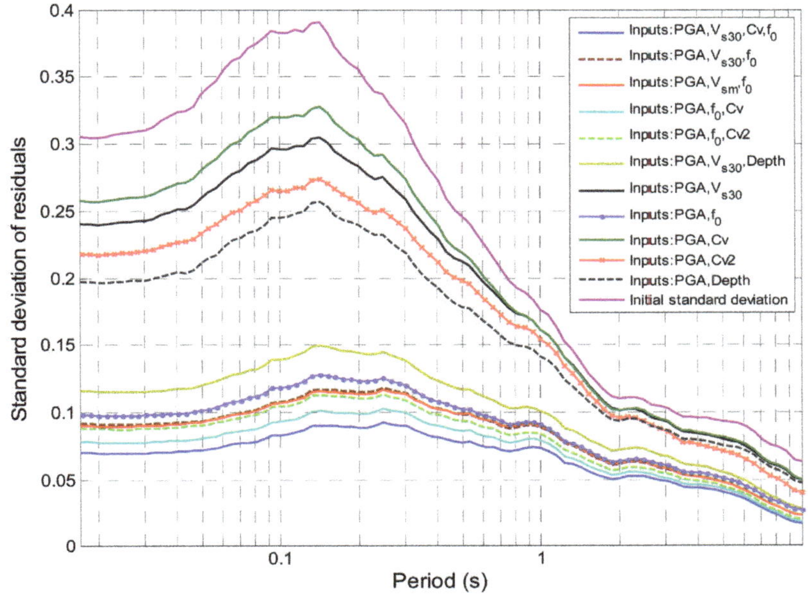

(**a**)

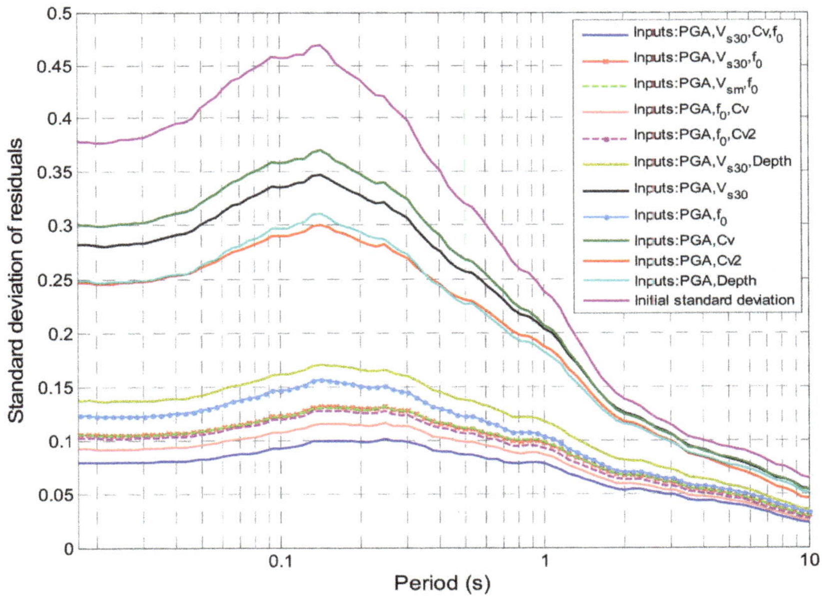

(**b**)

Figure 7. Variations in root mean square error and standard deviation of residuals $\varepsilon_{\text{GRNN}}(\theta, T_i)$ for various GRNN models with various sets of input site parameters compared to initial variability $\sigma_0(\theta, T_i)$ (**a**) for clay soil profiles and (**b**) for sand soil profiles. Data are displayed as function of real period.

In order to better identify the importance of the site proxy, the reduction in the overall standard deviation of residuals is defined by:

$$RS_{ANN,m}(\theta) = 1 - \frac{\varepsilon_{ANN,m}(\theta)}{\sigma_{ANN,m}(\theta)} \quad (25)$$

Table 4 presents the overall standard deviation of residuals as well as the reduction in the standard deviation of residuals for different GRNN models using different input parameter combinations.

Table 4. Standard deviations of model residuals and reduction in standard deviation for various GRNN models involving initial actual frequency amplification factors (clay and sand soil profiles) for various site parameter combinations.

Combination of Parameters	Standard Deviation for Clay $\varepsilon_{GRNN,m}(\theta=0)$	Reduction in Standard Deviation $RS_m(\theta=0)$	Standard Deviation for Sand $\varepsilon_{GRNN,m}(\theta=1)$	Reduction in Standard Deviation $RS_m(\theta=1)$
$PGA + V_{s30} + f_0$ + Cv	**0.0695**	**0.73**	**0.0772**	**0.76**
$PGA + V_{s30}$ + depth + Cv	0.0919	0.65	0.1047	0.68
$PGA + V_{s30} + f_0$ + Cv2	0.0774	0.70	0.0859	0.73
$PGA + V_{s30}$ + depth + Cv2	0.0987	0.62	0.1129	0.65
$PGA + V_{s30} + f_0$	0.0882	0.66	0.0998	0.69
$PGA + V_{sm} + f_0$	0.0877	0.67	0.0997	0.69
PGA + Cv + f_0	**0.0767**	**0.71**	**0.0881**	**0.73**
PGA + Cv2 + f_0	0.0842	0.68	0.0964	0.70
$PGA + V_{s30}$ + Cv	0.2013	0.23	0.2313	0.29
$PGA + V_{s30}$ + Cv2	0.1927	0.27	0.2197	0.32
PGA + Cv2 + Cv	0.1867	0.29	0.2097	0.35
$PGA + V_{s30}$ + depth	0.1086	0.59	0.1265	0.61
PGA + Cv + depth	0.1197	0.54	0.1414	0.56
PGA + Cv2 + depth	0.1255	0.52	0.1482	0.54
$PGA + V_{s30}$	0.2112	0.20	0.2470	0.24
$PGA + V_{sm}$	0.2204	0.16	0.2554	0.21
$PGA + f_0$	**0.0939**	**0.64**	**0.1140**	**0.65**
PGA + Cv	0.2242	0.15	0.2609	0.19
PGA + Cv2	0.1921	0.27	0.2173	0.33
PGA + depth	0.1770	0.33	0.2192	0.32
Overall Initial variability term $\sigma_m(\theta)$	0.2631		0.3241	

Figure 7a,b exhibit several noticeable features:

1. The *PGA* is common to all input parameter combinations. The model did not converge when not considering the *PGA* nor when considering only a single parameter. This means that the *PGA* is a predominant input parameter, and at the very least, a couple of *PGA*s with another parameter are needed to achieve convergence.
2. The *PGA* and f_0 constitute the best couple for the prediction of the amplification factor, producing 64% to 65% reductions in the standard deviation for clay and sand

soil profiles, respectively. The *PGA* performs well for all periods and soil types. In comparison, all other couples of parameters offer a lower reduction in variability, capped at 33%.

3. The triplet (*PGA*, Cv, f_0) is the most pertinent triplet for predicting the *AF*, with a standard deviation reduction of more than 71% for clay and 73% for sand profiles. The other triplets (*PGA*, Cv2, f_0), (*PGA*, V_{sm}, f_0), and (*PGA*, V_{s30}, f_0), which present interesting but slightly lower performances, are also worthy of consideration. However, because parameters such as Cv and Cv2 are difficult to measure in practice and have less physical meaning, the triplet (*PGA*, V_{s30}, f_0) is the triplet retained for predicting the *AF*. Considering more than three parameters will lead to better predictions, but for practical reasons, we decided not to go further.
4. The largest root mean square errors are systematically found in short to intermediate period ranges (see Figure 7).

5.2. Variation in Amplification Factors for Specific Period Ranges Using RBF

Design codes take into account site effects via multiplying the design spectra by amplification factors (i.e., *F*(*T*) in the CNBC). These amplification factors vary with the period of the structure and soil properties. For the sake of simplicity and with respect to earlier versions of codes (i.e., CNB2005), we consider averaged site factors for three specific ranges: F_a for short periods; F_v for intermediate periods; and F_l for long periods.

In this section, we compute these amplification factors, F_a, F_v, and F_l, from the RBF model based on the triplet (PGA, f_0, V_{S30}), which proved to be efficient and accurate. Figures 8–11 display the dependence of these three factors, *PGA*, f_0, and V_{S30}. The [0.15, 40 Hz] interval is considered for f_0 in all cases, covering the full range of the database. In Figure 8, the results are computed for a V_{S30} value varying within the frequency range studied, while in the other figures, typical low (150 m/s), mid (270 m/s), high (450 m/s), and very high (640 m/s) values of V_{S30} are considered. The intensity of the seismic excitation is displayed for three (03) discrete levels: *PGA* = 0.05 g, 0.25 g, and 0.75 g.

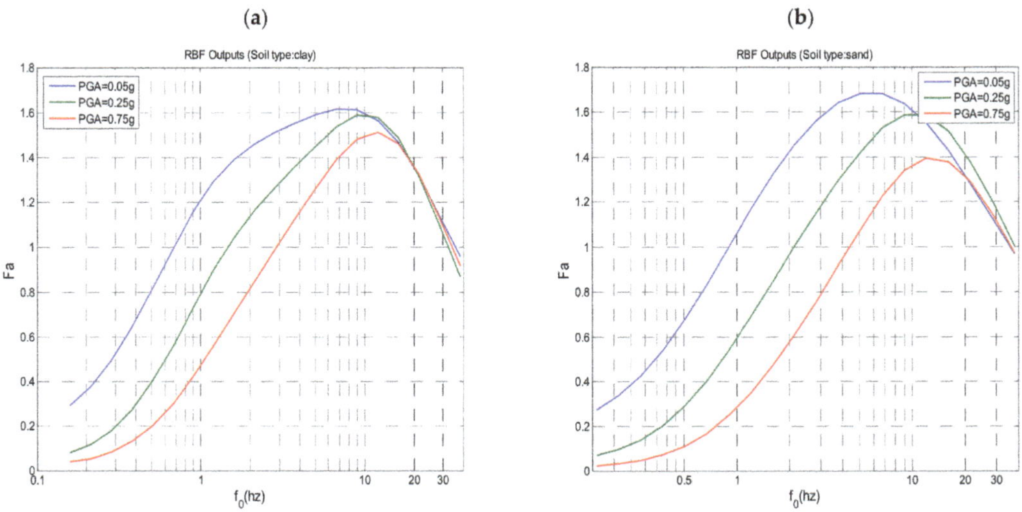

Figure 8. Variations in the short period amplification factors, F_a, with f_0 for different *PGA* values: 0.05 g, 0.25 g, and 0.75 g. (**a**) Clay soil profiles; (**b**) sand soil profiles.

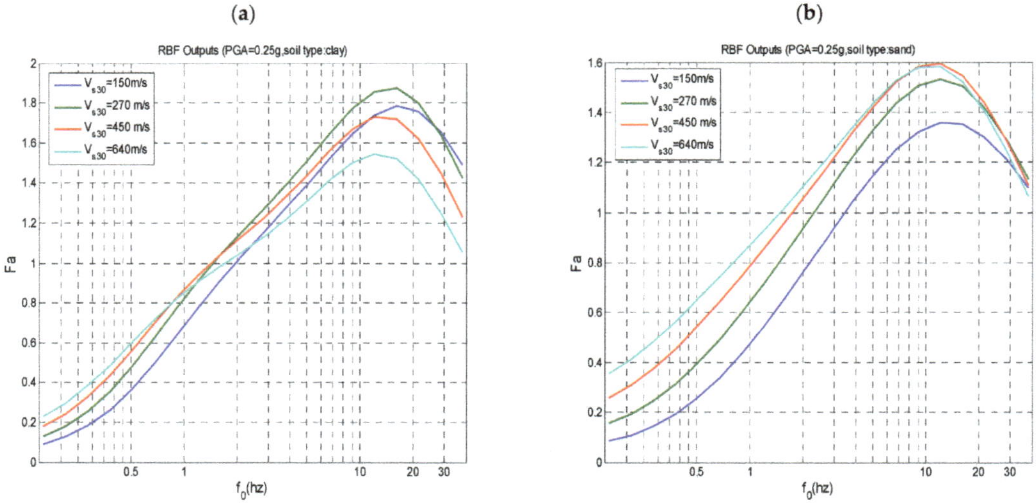

Figure 9. Variation in amplification factor at short period Fa with f_0 and PGA for four values of V_{s30} (150, 270, 450, and 640 m/s) (**a**) for clay soil profiles and (**b**) for sand soil profiles.

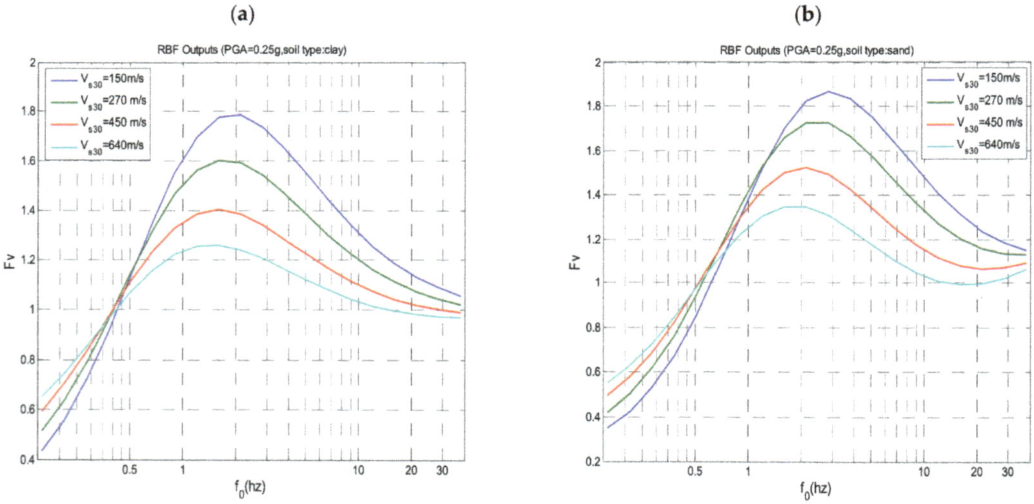

Figure 10. Variation in amplification factor at mid period F_v with f_0 and PGA for four values of V_{s30} (150, 270, 450, and 640 m/s) (**a**) for clay soil profiles and (**b**) for sand soil profiles.

Table 5 presents a summary of the statistical parameters obtained by the RBF triplet predictions. As shown, the initial standard deviation of the residuals is greatly reduced by a factor ranging roughly from 3 to 30. Excellent coefficients of determination, R^2, are obtained and are greater than 0.91, ranging from 0.96 to 0.99 for F_a and F_v for both soil types.

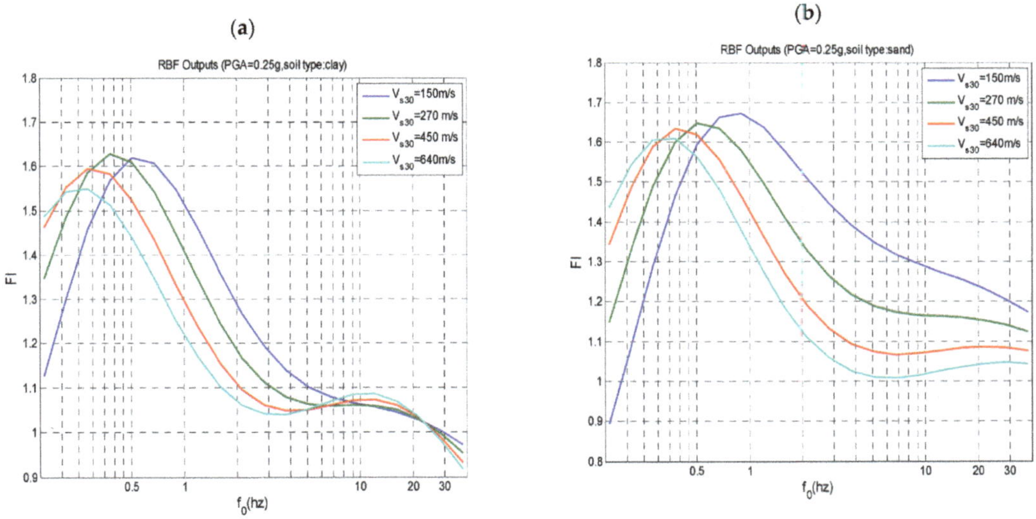

Figure 11. Variation in amplification factor at long period F_l with f_0 and PGA for four values of V_{s30} (150, 270, 450, and 640 m/s) and PGA (**a**) for clay soil profiles and (**b**) for sand soil profiles.

Table 5. Standard deviation of model residuals, coefficient of determination R^2, and numbers of key neurons in hidden layer determined after training phase for various RBF models at short, mid, and long periods (F_a, F_v, and F_l) for clay and sand soil profiles for combination of three soil parameters (PGA, f_0, and V_{s30}).

Statistical Summary for Amplification at Specific Period Ranges	F_a	F_v	F_l
Overall standard deviation $\varepsilon_{RBF,m}$ (Clay, $\theta = 0$)	0.0441	0.0436	0.0363
Initial standard deviation ($\theta = 0$)	0.3759	0.1632	0.0917
R^2 (all database) ($\theta = 0$)	0.9931	0.9637	0.9184
Number of key neurons	16	10	10
Standard deviation for all databases $\varepsilon_{RBF,m}(\theta = 1)$ (soil type: sand)	0.0504	0.0437	0.0341
Initial standard deviation ($\theta = 1$)	0.4551	0.2228	0.0967
R^2 (all database) ($\theta = 1$)	0.9939	0.9805	0.9358
Number of key neurons ($\theta = 1$)	12	10	9

The values of F_a, F_v, and F_l may be predicted as a function of f_0, V_{S30}, and PGA using the following explicit equation based on RBF models:

$$\log(\mathrm{Fa}) = LW^T a1 + b2 \tag{26}$$

where $a1$ is a vector of t lines output by the hidden layer, with t being the number of key neurons in the hidden layer, determined from the training phase. It is estimated from their Euclidian distance (see Section 3 of Supplementary File) using the following equation:

$$a1 = \exp\left(-\left[b1\sqrt{\left[(we1)^2 + (we2)^2 + (we3)^2\right].}\right]^2\right) \tag{27}$$

where $we1$, $we2$, and $we3$ can be expressed by

$$we1 = \left(\log(PGA) - \log(IW_{t,1})\right) \tag{28}$$

$$we2 = \left(\log(f_0) - \log(IW_{t,2})\right) \tag{29}$$

$$we3 = \left(\log(V_{s30}) - \log(IW_{t,3})\right) \tag{30}$$

$we1$, $we2$, and $we3$ are vectors of t lines, while $b1$ and $b2$ are the biases determined after the training phase.

To investigate the performance of the RBF model, we consider only 50% of the data for training and the other for testing. The main results obtained are summarized in Table 6 in terms of the standard deviation $\varepsilon_{RBF}(\theta)$ and coefficient of determination (R^2). They demonstrate the good performance of the model.

Table 6. Performance of RBF model on interpolation, results on terms. Standard deviation of model residuals and coefficient of determination R^2 for database test at F_a, F_v. and F_l for clay and sand.

Main Result for Amplification at Specific Period (Test Database)	F_a	F_v	F_l
Standard deviation for database (50% test database) $\varepsilon_{RBF,m}(\theta = 0)$ (soil type: clay)	0.0479	0.004	0.0369
R^2(50% test database) $(\theta = 0)$	0.9914	0.9667	0.9142
Standard deviation for database (50% test database) $\varepsilon_{RBF,m}(\theta = 1)$ (soil type: sand)	0.0514	0.0436	0.0365
R^2(50% test database) $(\theta = 1)$	0.9933	0.9794	0.9227

The following regression equation between V_{s30} and f_0 is derived from the obtained predictions of the RBF model:

$$(\text{Log}(V_{s30})) = \left(2.37 + 0.385\ \text{Log}(f_0)\right) \pm 0.18 \tag{31}$$

From the results of Figures 8–11 (and similar results for F_v and F_l), we note the following findings:

- Generally, the amplification factors F_a, F_v, and F_l, are higher for clay-type soil than for sand-type soil. This is particularly true for the range of low frequencies f_0 up to 1 Hz. However, for f_0 values higher than 3 Hz and for relatively stiff soils, with a V_{s30} value exceeding 350 m/s, F_a and F_v values are slightly higher for sand than for clay.
- The amplification factors are higher for low PGA values. This holds true for the factor F_l and soft soils with low V_{s30} values.

Furthermore, from the results of Figure 9, we can note that the correlations between f_0 and V_{s30} given by Equation (31) are not always respected as the frequency f_0 does not increase systematically with the V_{s30} value at the peak value of the amplification factors. This is explained by the following two factors:

- Some combinations are not possible in real cases, such as having a soil profile with a fundamental frequency f_0 greater than 10 Hz while its V_{s30} value is lower than 500 m/s. In such situations, the predictions of Equation (26) are somehow extrapolations that are very likely erroneous and meaningless.

- The predictions of Equation (26) shall be considered for explaining the global tendency regarding the interactions between different parameters such as the change in the amplification factors with the *PGA* level, V_{s30} (about 600 m/s). We can observe, for instance, that stiff soils with high V_{s30} values have higher Fa but lower Fl than soft soils with low values of V_{s30} (about 150 m/s). Similarly, the tendencies of F_v and F_l with V_{s30} and *PGA* can be deduced from the results shown in Figures 10 and 11, respectively.

Based on the RBF neural network, we can compute the synaptic weight, which characterizes the respective impacts of each of the three input parameters considered, namely, *PGA*, V_{s30}, and f_0 (see Supplementary File, section (e)). The resulting synaptic weights and the main results of these computed weights are reported in Table 7. We observe that generally, the fundamental frequency f_0 has the highest synaptic weight, which is typically more than 45% for clay soil profiles, and there are even higher values of the synaptic weight for F_a and F_v values of sand soil profiles. The *PGA* synaptic weight is generally of secondary importance, except in the case of the amplification factor F_l in a sand soil profile, where it is the highest.

Table 7. Participation of synaptic weights for different site parameters used in RBF model.

Participation of Synaptic Weights	(%) *PGA*	(%) f_0	(%) V_{s30}
For F_a ($\theta = 0$) (soil type: clay)	30.51	46.82	22.67
For F_v ($\theta = 0$) (soil type: clay)	32.73	45.81	21.45
For F_l ($\theta = 0$) (soil type: clay)	33.44	45.26	21.30
For F_a ($\theta = 1$) (soil type: sand)	30.76	51.10	18.14
For F_v ($\theta = 1$) (soil type: sand)	31.75	47.11	21.14
For F_l ($\theta = 1$) (soil type: sand)	41.52	37.62	20.86

6. Conclusions

The main objective of this study was to identify the key soil parameters influencing 1D site seismic response and amplification factors (*AF*s) at the free surface. To achieve this, we computed the nonlinear site response using a linear equivalent approach under vertically incident plane waves, utilizing a representative set of real input accelerograms that cover a broad range of peak frequencies and intensities.

We focused on monolayer soil profiles with variable thicknesses and shear wave velocities, situated above a semi-infinite bedrock with varying shear wave velocities. Soil nonlinearity was modelled using degradation curves proposed by Sun et al. (1988) [58] for clay and Seed and Idriss (1970) [52] for sand, which are simple and require minimal soil parameters and are effective enough. In total, 324 soil profiles of sand and an equal number of profiles of clay were analyzed under 14 records at 11 *PGA* levels each for a total of 99,792 time-history analyses.

For each soil profile, we calculated geometric average amplification factors for short to mid and long period ranges, denoted as F_a, F_v, and F_l for periods of [0.1 s, 0.2 s], [0.75 s, 1.5 s], and [2.82 s, 5.65 s], respectively.

Two types of neural networks were utilized in our analysis: (1) the Generalized Regression Neural Network (GRNN) was used to identify the most effective combination of soil proxies in predicting *AF*s, and (2) the Radial Basis Function neural network (RBF) was used to develop regression equations between these proxies and *AF*s. This combination of methods facilitated a comprehensive investigation into the factors that govern site response. The main findings can be summarized as follows:

- The pair (PGA, f_0) was identified as effective for predicting *AF*s, achieving reductions in the standard deviation of 64% and 65% for clay and sand profiles, respectively.
- The triplet (PGA, Cv, f_0) proved to be particularly powerful in predicting actual *AF*s, resulting in standard deviation reductions of over 71% for both clay and sand. Other combinations, such as (PGA, Cv2, f_0), (PGA, V_{sm}, f_0), and (PGA, V_{s30}, f_0), also yielded promising results and can be utilized.
- Because parameters Cv and Cv2 are often difficult and costly to measure in engineering practices, we recommend using the combination (PGA, V_{s30}, f_0), which, while offering slightly lower performance, still provides a significant reduction in the standard deviation, making it a practical alternative for field applications.

Furthermore, the prediction equations derived from the GRNN were found to be quite complex due to the high number of neurons in the hidden layer, making them challenging to implement in practice. As a result, we opted for the RBF neural network, which offers simpler prediction equations, albeit at the cost of a more intensive and prolonged training phase. We focused on the most effective site proxies identified through the GRNN analysis: PGA, f_0, and V_{s30}. Using the RBF, we established correlation relationships between these proxies and site-specific average *AF*s, considering the variability in initial *AF*s across the 3564 values for each soil type. Both the overall period range and specific short- and mid- to long-period ranges associated with the amplification factors F_a, F_v, and F_l were analyzed.

Using 50% of the database for training and the remaining 50% for testing, the RBF results demonstrated a lower standard deviation of the error term (ε_{RBF}) for F_a, F_v, and F_l, generally with a reduction in the standard deviation exceeding 61%, along with a strong coefficient of determination exceeding 91%. These findings highlight the effectiveness of the RBF neural network method in accurately predicting amplification factors.

The main findings of this study will aid in enhancing site classification by establishing a physical relationship between the *AF* and site proxies. However, several recommendations regarding regulatory codes can be made:

- The Inclusion of the Fundamental Frequency: The fundamental frequency (f_0) should be considered alongside the peak ground acceleration (PGA) and shear wave velocity (V_{s30}) to improve predictions of amplification factors.
- The Distinction of Period Ranges: It is crucial to differentiate between amplification factors for short (F_a), medium (F_v), and long (F_l) period ranges. The long-period amplification (F_l) is particularly significant, often exceeding F_a and F_v values for soil profiles with low V_{s30} values, which typically correspond to site classes C, D, and E in Eurocode 8 (EC8) and site classes D and E in UBC/CNBC codes.

Finally, we propose new equations for predicting the amplification factor across specific short-, mid-, and long-period ranges.

It is important to acknowledge the limitations of this study, primarily related to the use of simplistic soil profiles (homogeneous, monolayer, and 1D) and basic degradation curves, specifically those used by Sun et al. (1988) [58] for clay and Seed and Idriss (1970) [52] for sand. Adopting more recent degradation curves and realistic soil profiles and site geometry could yield more accurate and site-specific amplification factors. Nevertheless, this study provides straightforward and effective prediction equations for amplification factors, utilizing a limited number of parameters that could enhance current design codes.

Supplementary Materials: The following supporting information can be downloaded at: https://www.mdpi.com/article/10.3390/app15073618/s1.

Author Contributions: Conceptualization, A.B.S.; Methodology, A.B.S.; Validation, L.G.; Investigation, A.B.S.; Writing—original draft, A.B.S.; Writing—review & editing, L.G.; Funding acquisition, L.G. All authors have read and agreed to the published version of the manuscript.

Funding: This work was partially supported by the Ecole de technologie superieure, University of Québec, 1100 Notre-Dame Ouest, Montréal, QC H3C 1K3, Canada: https://www.etsmtl.ca/ and by the Fonds de recherche du Quebec: https://frq.gouv.qc.ca/.

Institutional Review Board Statement: Not applicable.

Informed Consent Statement: Not applicable.

Data Availability Statement: The original contributions presented in this study are included in the article/Supplementary Material. Further inquiries can be directed to the corresponding author.

Acknowledgments: The department of Construction Engineering of Ecole de technologie superieure and its staff for their support and material resources used along the project.

Conflicts of Interest: The authors declare no conflicts of interest.

References

1. Bard, P.-Y.; Campillo, M.; Chavez-Garcia, F.J.; Sanchez-Sesma, F.J. The Mexico Earthquake of September 19, 1985—A Theoretical Investigation of Large- and Small-scale Amplification Effects in the Mexico City Valley. *Earthq. Spectra* **1988**, *4*, 609–633. [CrossRef]
2. Kawase, H.; Aki, K. A study on the response of a soft soil basin for incident S, P, and Rayleigh waves with special reference to the long duration observed in Mexico City. *Bull. Seismol. Soc. Am.* **1989**, *79*, 1361–1382.
3. Chávez-García, F.J.; Bard, P.Y. Site effects in Mexico City eight years after the September 1985 Michoacan earthquakes. Soil Dyn. *Earthq. Eng.* **1994**, *13*, 229–247.
4. Del Gaudio, V.; Pierri, P.; Rajabi, A.M. An approach to identify site response directivity of accelerometer sites and application to the Iranian Area. *Pure Appl. Geophys.* **2015**, *172*, 1471–1490. [CrossRef]
5. Cherid, D.; Hammoutene, M.; Tiliouine, B.; Berrah, M.K. Local seismic site amplification: Effects of obliquely incident antiplane motions. *J. Seismol.* **2017**, *21*, 509–524. [CrossRef]
6. Sandıkkaya, M.A.; Akkar, S. Cumulative absolute velocity, Arias intensity and significant duration predictive model from a pan-European strong-motion dataset. *Bull. Earthq. Eng.* **2017**, *15*, 1881–1898. [CrossRef]
7. Boudghene Stambouli, A.; Zendagui, D.; Bard, P.Y.; Dif, H. Influence of site parameters on fourier amplification application for 1D linear viscoelastic method. *Period. Polytech. Civ. Eng.* **2021**, *65*, 229–241. [CrossRef]
8. Dif, H.; Boudghene Stambouli, A. Impact of site condition on Arias intensity and Cumulative absolute velocity: Application for low seismicity areas. *Asian J. Civ. Eng.* **2023**, *24*, 1411–1424. [CrossRef]
9. Buckle, I.G.; Friedland, I.; Mander, J.; Martin, G.; Nutt, R.; Power, M. *Seismic Retrofitting Manual for Highway Structures. Part 1, Bridges (No. FHWA-HRT-06-032)*; Turner-Fairbank Highway Research Center: McLean, VA, USA, 2006.
10. Borcherdt, R.D. Estimates of site dependent response spectra for design (methodology and justification). *Earthq. Spectra* **1994**, *10*, 617–653. [CrossRef]
11. Martin, G.R.; Dobry, R. Earthquake site response and seismic code provisions. *NCEER Bull.* **1994**, *8*, 1–6.
12. Dickenson, S.E.; Seed, R.B. Nonlinear Dynamic Response of Soft and Deep Cohesive Soil Deposits. In Proceedings of the International Workshop on Site Response Subjected to Strong Earthquake Motions, Yokosuka, Japan, 16–17 January 1996; Volume 2, pp. 67–81.
13. Dobry, R.; Borcherdt, R.D.; Crouse, C.B.; Idriss, I.M.; Joyner, W.N.; Martin, G.R.; Power, M.S.; Rinne, E.E.; Seed, R.B. New Site coefficients and site classification system used in recent building seismic code provisions. *Earthq. Spectra* **2000**, *16*, 41–67. [CrossRef]
14. Steidl, J.H. Site response in southern California for probabilistic seismic hazard analysis. *Bull. Seismol. Soc. Am.* **2000**, *90*, S149–S169. [CrossRef]
15. Rodriguez-Marek, A.; Bray, J.D.; Abrahamson, N.A. An empirical geotechnical seismic site response procedure. *Earthq. Spectra* **2001**, *17*, 65–87. [CrossRef]
16. Pitilakis, K.D.; Makra, K.A.; Raptakis, D.G. 2D vs 3D Site Effects with Potential Applications to Seismic Norms: The case of EUROSEISTEST and Thessaloniki. In Proceedings of the XVth ICSMGE, Istanbul, Turkey, 27–31 August 2001; pp. 123–133.
17. Stewart, J.P.; Liu, A.H.; Choi, Y. Amplification Factors for Spectral Acceleration in Tectonically Active Regions. *Bull. Seismol. Soc. Am.* **2003**, *93*, 332–352. [CrossRef]
18. Choi, Y.; Stewart, J.P. Nonlinear site amplification as function of 30 m shear wave velocity. *Earthq. Spectra* **2005**, *21*, 1–30. [CrossRef]
19. Castellaro, S.; Mulargia, F.; Rossi, P.M. Vs30: Proxy for seismic amplification? *Seismol. Res. Lett.* **2008**, *79*, 540–543. [CrossRef]

20. Lee, V.W.; Trifunac, M.D. Should average shear-wave velocity in the top 30m of soil be used to describe seismic amplification? *Soil Dyn. Earthq. Eng.* **2010**, *30*, 1250–1258. [CrossRef]
21. Abrahamson, N.; Atkinson, G.; Boore, D.; Bozorgnia, Y.; Campbell, K.; Chiou, B.; Idriss, I.; Silva, W.; Youngs, R. Comparisons of the NGA ground-motion relations. *Earthq. Spectra* **2008**, *24*, 45–66. [CrossRef]
22. Gregor, N.; Abrahamson, N.A.; Atkinson, G.M.; Boore, D.M.; Bozorgnia, Y.; Campbell, K.W.; Chiou, B.S.-J.; Idriss, I.M.; Kamai, R.; Seyhan, E.; et al. Comparison of NGA-West2 GMPEs. *Earthq. Spectra* **2014**, *30*, 1179–1197. [CrossRef]
23. Ancheta, T.D.; Darragh, R.B.; Stewart, J.P.; Seyhan, E.; Silva, W.J.; Chiou, B.S.-J.; Wooddell, K.E.; Graves, R.W.; Kottke, A.R.; Boore, D.M.; et al. NGA-West 2 database. *Earthq. Spectra* **2014**, *30*, 989–1005. [CrossRef]
24. Douglas, J.; Akkar, S.; Ameri, G.; Bard, P.-Y.; Bindi, D.; Bommer, J.J.; Singh Bora, S.; Cotton, F.; Derras, B.; Hermkes, M.; et al. Comparisons among the five ground-motion models developed using RESORCE for the prediction of response spectral accelerations due to earthquakes in Europe and the Middle East. Bull. *Earthq. Eng.* **2014**, *12*, 341–358. [CrossRef]
25. Luzi, L.; Puglia, R.; Pacor, F.; Gallipoli, M.; Bindi, D.; Mucciarelli, M. Proposal for a soil classification based on parameters alternative or complementary to Vs, 30. *Bull. Earthq. Eng.* **2011**, *9*, 1877–1898. [CrossRef]
26. Cadet, H.; Bard, P.-Y.; Duval, A.-M.; Bertrand, E. Site effect assessment using KiK-net data—Part 2—Site amplification prediction equation (SAPE) based on f_0 and Vsz. *Bull. Earthq. Eng.* **2012**, *10*, 451–489. [CrossRef]
27. Pitilakis, K.; Riga, E.; Anastasiadis, A. Design spectra and amplification factors for Eurocode 8. *Bull. Earthq. Eng.* **2012**, *10*, 1377–1400. [CrossRef]
28. Pitilakis, K.; Riga, E.; Anastasiadis, A. New code site classification, amplification factors and normalized response spectra based on a worldwide ground-motion database. *Bull. Earthq. Eng.* **2013**, *11*, 925–966. [CrossRef]
29. *NEHRP Guidelines for the Seismic Rehabilitation of Buildings*; FEMA-273; Federal Emergency Management Agency: Washington, DC, USA, 1997. Available online: https://courses.washington.edu/cee518/fema273.pdf (accessed on 10 March 2025).
30. Uniform Building Code. Structural engineering design provisions. In Proceedings of the International Conference of Building Officials, Whittier, CA, USA, 14 July 1997; Volume 2.
31. *EN 1998-1*; EC8 Eurocode 8. Design of Structures for Earthquake Resistance—Part 1: General Rules, Seismic Actions and Rules for Buildings. European Committee for Standardization (CEN): Brussels, Belgium, 2004. Available online: https://eurocodes.jrc.ec.europa.eu (accessed on 1 February 2016).
32. IBC2012. *International Building Code*, 2012 ed.; International Code Council: Washington, DC, USA, 2012; ISBN 978-1-60983-039-7. Available online: https://archive.org/details/gov.law.icc.ibc.2012 (accessed on 1 November 2016).
33. Canadian Commission on Building and Fire Codes, National Research Council of Canada. *National Building Code of Canada (NBC) (2015) National Building Code of Canada*; Canadian Commission on Building and Fire Codes, National Research Council of Canada: Ottawa, ON, Canada, 2015.
34. Canadian Commission on Building and Fire Codes, National Research Council of Canada. *National Building Code of Canada (NBC) (2015b) User's Guide—NBC 2015 Structural Commentaries (Part 4 of Division B)*; Canadian Commission on Building and Fire Codes, National Research Council of Canada: Ottawa, ON, Canada, 2015.
35. Boudghene Stambouli, A.; Zendagui, D.; Bard, P.-Y.; Derras, B. Deriving amplification factors from simple site parameters using generalized regression neural networks: Implications for relevant site proxies. *Earth Planets Space* **2017**, *69*, 99. [CrossRef]
36. Ghaboussi, J.; Garrett, J., Jr.; Wu, X. Knowledge-based modeling of material behavior with neural networks. *J. Eng. Mech.* **1991**, *117*, 132–153. [CrossRef]
37. Ghaboussi, J.; Sidarta, D. New nested adaptive neural networks (NANN) for constitutive modeling. *Comput. Geotech.* **1998**, *22*, 29–52. [CrossRef]
38. Paolucci, R.; Colli, P.; Giacinto, G. Assessment of seismic site effect in 2-D alluvial Valleys Using Neural Networks. *Earthq. Spectra* **2000**, *16*, 661–680. [CrossRef]
39. Hurtado, J.; Londono, J.; Meza, M. On the applicability of neural networks for soil dynamic amplification analysis. *Soil Dyn. Earthq. Eng.* **2001**, *21*, 579–591. [CrossRef]
40. Derras, B.; Bard, P.Y.; Cotton, F.; Bekkouche, A. Adapting the neural network approach to PGA prediction: An example based on the KiK-net data. *Bull. Seismol. Soc. Am.* **2012**, *102*, 1446–1461. [CrossRef]
41. Poggio, T.; Girosi, F. Networks for approximation and learning. *Proc. IEEE* **1990**, *78*, 1481–1497. [CrossRef]
42. Liu, S.; Wang, L.; Zhang, W.; Sun, W.; Wang, Y.; Liu, J. Physics-informed optimization for a data-driven approach in landslide susceptibility evaluation. *J. Rock Mech. Geotech. Eng.* **2024**, *16*, 3192–3205. [CrossRef]
43. Zhang, Z.; Pan, Q.; Yang, Z.; Yang, X. Physics-informed deep learning method for predicting tunnelling-induced ground deformations. *Acta Geotechnica* **2023**, *18*, 4957–4972. [CrossRef]
44. Constantopoulos, I.V.; Roesset, J.M.; Christian, J.T. A comparison of linear and exact nonlinear analysis of soil amplification. In Proceedings of the 5th WCEE, Rome, Italy, 25–29 June 1973; pp. 1806–1815.
45. Yoshida, N.; Kobayashi, S.; Suetomi, I.; Miura, K. Equivalent linear method considering frequency dependent characteristics of stiffness and damping. *Soil Dyn. Earthq. Eng.* **2002**, *22*, 205–222. [CrossRef]

46. Assimaki, D.; Kausel, E. An equivalent linear algorithm with frequency-and pressure-dependent moduli and damping for the seismic analysis of deep sites. *Soil Dyn. Earthq. Eng.* **2002**, *22*, 959–965. [CrossRef]
47. Johari, A.; Momeni, M. Stochastic analysis of ground response using non-recursive algorithm. *Soil Dyn. Earthq. Eng.* **2015**, *69*, 57–82. [CrossRef]
48. Trifunac, M.D.; Todorovska, M.I. Can aftershock studies predict site amplification factors? Northridge, CA, earthquake of 17 January 1994. *Soil Dyn. Earthq. Eng.* **2000**, *19*, 233–251. [CrossRef]
49. Nguyen, X.D.; Guizani, L. On the application limits and performance of the single-mode spectral analysis for seismic analysis of isolated bridges in Canada. *Can. J. Civ. Eng.* **2022**, *49*, 1747–1763. [CrossRef]
50. Boudghene Stambouli, A.; Bard, P.Y.; Chaljub, E.; Moczo, P.; Kristek, J.; Stripajova, S.; Durand, C.; Zendagui, D.; Derras, B. 2D/1D aggravation factors: From a comprehensive study to estimation with a neural network model. In Proceedings of the 16Ith European Conference of Earthquake Engineering, Thessaloniki, Greece, 18–21 June 2018.
51. Paolucci, R.; Morstabilini, L. Non-dimensional site amplification functions for basin edge effects on seismic ground motion. In Proceedings of the Third International Symposium on the Effects of Surface Geology on Seismic Motion, Grenoble, France, 30 August–1 September 2006; Volume 30, pp. 823–831.
52. Seed, H.B.; Idriss, I.M. *Soil Moduli and Damping Factors for Dynamic Response Analyses*; EERC-70-10; University of California Berkeley Structural Engineers and Mechanics: Berkeley, CA, USA, 1970.
53. Vucetic, M.; Dobry, R. Cyclic triaxial strain-controlled testing of liquefiable sands. In *Advanced Triaxial Testing of Soil and Rock, ASTM STP 977*; ASTM International: West Conshohocken, PA, USA, 1988.
54. Vucetic, M.; Dobry, R. Effect of soil plasticity on cyclic response. *J. Geotech. Eng.* **1991**, *117*, 89–107. [CrossRef]
55. Ishibashi, I.; Zhang, X. Unified dynamic shear moduli and damping ratios of sand and clay. *Soils Found.* **1993**, *33*, 182–191. [CrossRef]
56. EPRI. *Guidelines for Determining Design Basis Ground Motions*; EPRI Tr-102293; Electric Power Research Institute: Palo Alto, CA, USA, 1993.
57. Darendeli, M.B. *Development of a New Family of Normalized Modulus Reduction and Material Damping Curves*; The University of Texas at Austin: Austin, TX, USA, 2001.
58. Sun, J.I.; Golesorkhi, R.; Seed, H.B. *Dynamic Moduli and Damping Ratios for Cohesive Soils*; Report No. UCB/EERC-88/15, EERC; College of Engineering, University of California: Berkeley, CA, USA, 1988.
59. Idriss, I. Response of soft soil sites during earthquakes. In *Proc. H. Bolton Seed Memorial Symposium*; BiTech Publishers: Berkeley, CA, USA, 1990; pp. 273–289.
60. Jacobsen, L.S. Steady forced vibration as influenced by damping: An approximate solution of the steady forced vibration of a system of one degree of freedom under the influence of various types of damping. *Trans. Am. Soc. Mech. Eng.* **1930**, *52*, 169–178. [CrossRef]
61. Hudson, D.E. Equivalent Viscous Friction for Hysteretic Systems with Earthquake-Like Excitations. In Proceedings of the 3rd World Conference on Earthquake Engineering, Auckland, New Zealand, 22 January–1 February 1965; pp. II-185/II-201.
62. Kanai, K. *Relation Between the Nature of Surface Layer and the Amplitude of Earthquake Motions*; Bulletin, Tokyo Earthquake Research Institute: Tokyo, Japan, 1951.
63. Matthiesen, R.B.; Duke, C.M.; Leeds, D.J.; Fraser, J.C. *Site Characteristics of Southern California Strong-Motion Earthquake Stations, Part Two*; Report No. 64-15; Department of Engineering, University of California: Los Angeles, CA, USA, 1964.
64. Roesset, J.M.; Whitman, R.V. *Theoretical Background for Amplification Studies*; Massachusetts Institute of Technology: Cambridge, MA, USA, 1969.
65. Tsai, N.; Housner, G. Calculation of surface motions of a layered half-space. *Bull. Seismol. Soc. Am.* **1970**, *60*, 1625–1651. [CrossRef]
66. Lysmer, J.; Seed, H.B.; Schnabel, P.B. Influence of base-rock characteristics on ground response. *Bull. Seismol. Soc. Am.* **1971**, *61*, 1213–1231. [CrossRef]
67. Schnabel, P.B.; Lysmer, J.L.; Seed, H.B. *SHAKE: A Computer Program for Earthquake Response Analysis of Horizontally Layered Sites*; Report EERC-72-12; University of California: Berkeley, CA, USA, 1972.
68. Kramer, S.L.; Stewart, J.P. *Geotechnical Earthquake Engineering*; CRC Press: Boca Raton, FL, USA, 2024.
69. Akkar, S.; Sandıkkaya, M.A.; Şenyurt, M.; Azari Sisi, A.; Ay, B.Ö.; Traversa, P.; Douglas, J.; Cotton, F.; Luzi, L.; Hernandez, B.; et al. Reference Database for Seismic Ground-Motion in Europe (RESORCE). *Bull. Earthq. Eng.* **2014**, *12*, 311–339. [CrossRef]
70. Hashash, Y.M.A.; Groholski, D.R.; Phillips, C.A.; Park, D.; Musgrove, M. *DEEPSOIL 5.1, User Manual and Tutorial*; Department of Civil and Environmental Engineering, University of Illinois at Urbana-Champaign: Urbana, IL, USA, 2012.
71. Bardet, J.P.; Ichii, K.; Lin, C.H. *EERA: A Computer Program for Equivalent-Linear Earthquake Site Response Analyses of Layered Soil Deposits*; Department of Civil Engineering, University of Southern California: Los Angeles, USA, USA, 2000.
72. Seed, H.B.; Romo, M.P.; Sun, J.I.; Jaime, A.; Lysmer, J. Relationships between soil conditions and earthquake ground motions. *Earthq. Spectra* **1988**, *4*, 687–729. [CrossRef]

73. Seed, R.B.; Dickenson, S.E.; Mok, C.M. Seismic response analysis of soft and deep cohesive sites: A brief summary of recent findings. In Proceedings of the CALTRANS First Annual Seismic Response Workshop, Sacramento, CA, USA, 3–4 December 1991.
74. Chang, S.W.; Bray, J.D.; Gookin, W.B. *Seismic Response of Deep Stiff Soil Deposits in the Los Angeles, California Area During the 1994 Northridge Earthquake*; University of California at Berkeley: Barkeley, CA, USA, 1997.
75. Dobry, R.; Oweis, I.; Urzua, A. Simplified Procedures for Estimating the Fundamental Period of a Soil Profile. *Bull. Seismol. Soc. Am.* **1976**, *66*, 1293–1321.
76. Borcherdt, R.D.; Glassmoyer, G. On the Characteristics of Local Geology and Their Influence on Ground Motions Generated by the Loma Prieta Earthquake in the San Francisco Bay Region, California. *Bull. Seismol. Soc. Am.* **1992**, *82*, 603–641. [CrossRef]
77. Salameh, C.; Bard, P.-Y.; Guillier, B.; Harb, J.; Cornou, C.; Almakari, M. Using ambient vibration measurements for risk assessment at an urban scale: From numerical proof of concept to a case study in Beirut (Lebanon). In Proceedings of the 5th IAEE International Symposium: Effects of Surface Geology on Seismic Motion, Taipei, Taiwan, 15–17 August 2016.
78. Yilmaz, A.S.; Özer, Z. Pitch angle control in wind turbines above the rated wind speed by multi-layer perceptron and radial basis function neural networks. *Expert Syst. Appl.* **2009**, *36*, 9767–9775. [CrossRef]
79. Barrile, V.; Cacciola, M.; D'Amico, S.; Greco, A.; Morabito, F.C.; Parrillo, F. Radial basis function neural networks to foresee aftershocks in seismic sequences related to large earthquakes. In *Neural Information Processing, Proceedings of the 13th International Conference, ICONIP 2006, Hong Kong, China, 3–6 October 2006*; Proceedings, Part II 13, 909-916; Springer: Berlin/Heidelberg, Germany, 2006.
80. Baddari, K.; Aïfa, T.; Djarfour, N.; Ferahtia, J. Application of a radial basis function artificial neural network to seismic data inversion. *Comput. Geosci.* **2009**, *35*, 2338–2344. [CrossRef]
81. Günaydın, K.; Günaydın, A. Peak ground acceleration prediction by artificial neural networks for northwestern Turkey. *Math. Probl. Eng.* **2008**, *2008*, 919420. [CrossRef]
82. Broomhead, D.S.; Lowe, D. Radial basis functions, multi-variable functional interpolation and adaptive networks: Royal Signals and Radar Establishment Malvern (United Kingdom). *Complex Syst.* **1988**, *2*, 321–355.
83. Moody, J.; Darken, C.J. Fast learning in networks of locally-tuned processing units. *Neural Comput.* **1988**, *1*, 281–294. [CrossRef]
84. Borgi, A. Apprentissage Supervisé par Génération de Règles: Le Système SUCRAGE. Ph.D. Thesis, Université de Paris, Paris, France, 1999. (In French).
85. Seghouane, A.K.; Fleury, G. Apprentissage de réseaux de neurones à fonctions radiales de base avec un jeu de données à entrée-sortie bruitées. *Trait. Signal* **2003**, *20*, 414.
86. Chen, S.; Cowan, C.F.; Grant, P.M. Orthogonal least squares learning algorithm for radial basis function networks. *IEEE Trans. Neural Netw.* **1991**, *2*, 302–309. [CrossRef]
87. Specht, D.F. A general regression neural network. *IEEE Trans. Neural Netw.* **1991**, *2*, 568–576. [CrossRef]
88. Cigizoglu, H.K.; Alp, M. Generalized regression neural network in modelling river sediment yield. *Adv. Eng. Softw.* **2005**, *37*, 63–68. [CrossRef]
89. Hannan, S.A.; Manza, R.R.; Ramteke, R.J. Generalized regression neural network and radial basis function for heart disease diagnosis. *Int. J. Comput. Appl.* **2010**, *7*, 7–13. [CrossRef]
90. Wasserman, P.D. *Advanced Methods in Neural Computing*; John Wiley & Sons Inc.: Hoboken, NJ, USA, 1993.
91. Kim, B.; Lee, D.W.; Parka, K.Y.; Choi, S.R.; Choi, S. Prediction of plasma etching using a randomized generalized regression neural network. *Vacuum* **2004**, *76*, 37–43. [CrossRef]
92. Kohonen, T. Learning vector quantization. *Self-Organ. Maps* **1997**, 203–217.
93. Haykin, S. *Neural Networks: A Comprehensive Foundation*, 2nd ed.; Prentice Hall: Hoboken, NJ, USA, 1999; p. 842.

applied sciences

MDPI

Article

Construction Control Technology and Monitoring Analysis of Walking Incremental Launching Construction of Small-Curvature Steel Box Girder Bridges Across Expressways

Hai Huang [1,2], Xiaomin Xue [1], Haitao Ma [2], Zhijun Zhou [2], Jianbo Dou [2,*], Heng Yang [3], Erwei Guo [3] and Jianjun Ma [3,4]

[1] School of Himan Settlements and Civil Engineering, Xi'an Jiaotong University, Xi'an 710049, China; 13164309209@163.com (H.H.); xuexm@mail.xjtu.edu.cn (X.X.)

[2] China Construction Seventh Engineering Division Co., Ltd., Zhengzhou 450003, China; c18839212258@163.com (H.M.); zhouzhijun526@aliyun.com (Z.Z.)

[3] School of Civil Engineering and Architecture, Henan University of Science and Technology, Luoyang 471023, China; yh17516234788@163.com (H.Y.); guollwei@126.com (E.G.); majianjun@haust.edu.cn (J.M.)

[4] Henan Province Engineering Technology Research Center of Building Safety and Protection, Luoyang 471023, China

* Correspondence: doudou_19860228@126.com

Abstract: As urban areas keep growing, there are higher requirements for the carrying capacity of traffic operations, and there are more and more curve incremental launching projects with complex construction conditions. This paper builds upon the walking incremental launching project of a small-curvature steel box girder in Xuchang City and has developed a detailed construction method and monitoring technology. Due to the bridge's longitudinal gradient being designed as a two-way slope and falling under the category of a small-radius steel box girder, the front end of the main beam exhibits significant lateral deviation, and linear control is difficult. It is necessary to carry out stress and displacement monitoring of the whole process of construction of the curved steel box girder and the guide girder to guide the construction process. The stress conditions of the incremental launching pier and the settlement of the concrete substructure are also studied, and we analyze the stress and displacement characteristics. Firstly, the finite element tool MIDAS Civil is adopted to build a model for the construction. The five most unfavorable working conditions are selected from the entire incremental launch process to analyze the internal force and displacement state of the steel box girder bridge, which is and then compared with the site monitoring value. It is demonstrated by the outcomes that the internal force and deflection of the steel box girder and the guide girder are within the safe construction range, which ensures the security of the incremental launching construction. In the maximum cantilever condition, the guide girder experiences significant stress, but the maximum value is not observed during the maximum cantilever condition of the guide girder. Therefore, whole-process monitoring should be carried during construction to maintain safety measures and quality management.

Keywords: construction control; small-curvature steel box girder bridge; walking incremental launching; numerical simulation; site monitoring

1. Introduction

Bridges are closely related to our lives. Over the past few years, as the social economy has rapidly developed and technology has gradually improved, new demands have been

https://doi.org/10.3390/app15020585

raised for bridge construction methods, and the demand for green technology is increasing. Particularly in the construction of bridges across expressways, subways, and rivers, it is necessary to maintain the smooth flow of existing traffic lines. In order to meet this requirement, the incremental launching technique has been developed and implemented, which is a technology representing the prospect of bridge construction. Practice has proven that the incremental launching technique a cost-effective and secure method for completing the aerial construction of a beam body. Incremental launching construction has the following benefits: the field equipment can be reused and the construction expenses are reduced; it enables the prefabrication of steel box girder engineering, on-site assembly and welding, and fast construction speed; it has a negligible effect on the surrounding environment and does not affect the normal operation of traffic lines [1,2]. Considering the above advantages, the incremental launching technique is crucial in contemporary bridge building.

Limited research has been carried out on walking incremental launching construction. The walking incremental launching method controls the stress state in the incremental launching process through the control system. Walking incremental launching integrates jacking, pushing, falling, and lateral adjustment. Through these, timely rectification in the incremental launching process can be carried out to guarantee that the geometric shape of the incremental launching construction girder section is consistent with the design. The use of walking incremental launching can ensure accuracy and efficiency, and the construction quality can be better guaranteed.

The modern incremental launching technique was introduced with the Ager Bridge constructed in 1959. The Caroni River Bridge built in Venezuela in 1962 was the first to feature a steel launching nose at the leading edge of the girder and a temporary pier at the midpoint of the incremental launching span. After construction, it was found that the characteristics of the girder section during the incremental launching process were improved through these improvement measures. Therefore, this project is considered to be a real bridge incremental launching construction [3]. With the increasing number of long-span suspension bridges, their complex construction conditions make it difficult for traditional construction methods to meet the requirements. Therefore, the advancement of incremental launching engineering technology is an inescapable trend of current development [4,5].

Incremental launching construction and its construction monitoring have garnered significant attention from scholars and resulted in substantial research. The Jiubao Bridge in Hangzhou is 1855 m long. This is the inaugural large river-crossing project with an entirely composite structure in China. The primary navigation span and the approach span use the continuous hybrid arch–girder structure and the steel–concrete composite structure, respectively. Both of them adopt the construction method of multi-point incremental launching. This method prevents the risk associated with assembling the arch over water and improves the construction quality [6]. Aiming at the problems encountered in the traditional construction method, Zhang et al. [7] developed a novel incremental launching approach composed of the girder curvature adaptation technology of the main girder and an intelligent surveillance system, which made it possible to incrementally launch small-radius, variable-cross-section steel box girders. Xie et al. [8] introduced the mechanical properties throughout the incremental advancement process and analyzed the influence of the design parameters of Guihua Bridge throughout the process of incremental launching. Aiming at the characteristics of engineered curve shape, multiple segments, and road complexity, Feng et al. [9] adopted the multi-point unidirectional double-launching method. Through real-time monitoring, issues like detachment between the steel girder and the incremental launching platform during the incremental advancement were resolved. Zhang et al. [10] examined the comprehensive and specific mechanical behaviors of a wide-span cable-stayed bridge with dual towers and dual cable planes during incremental launching construction. Ding et al. [11] designed the incremental launching construction scheme of suspension stiffening chords through a comprehensive analysis of Jinan Huanghe rail-cum-road Bridge, and effectively implemented it in the construction of huge steel truss girders. Wang et al. [12] monitored the spatial configuration and internal loading of the steel box girder in the whole process of incremental launching, modified the relevant parameters based on the on-site construction conditions and initial stress condition of steel box girder, and compared the modified theoretical value and monitoring value. Du et al. [13] investigated the impact of local stress and deviation error in steel box girders while performing skew incremental launching under two working conditions.

In terms of finite element analysis, Wu et al. [14] utilized the ABAQUS software to model the incremental launching process of steel box girders, aiming to better understand and reduce the complex stresses generated during the process, and analyzed the mechanical properties and local buckling of steel box girders. Wang et al. [15] studied the structural characteristics of steel U-shaped girders during incremental launching and proposed local stress control measures based on finite element analysis. Hu et al. [16] studied the feasibility of using incremental launching for the construction and assembly of continuous hybrid arch–girder bridges by combining experimental and numerical methods. Chacón et al. [17] conducted finite element numerical modeling of the incremental launching process of a steel bridge and verified it by a laboratory scale-reduced test. Chai et al. [18] utilized the finite element method to investigate the feasibility of incremental launching construction for a 180 m main span long-span bridge without auxiliary supports. Wang et al. [19] advanced a construction approach for the bidirectional incremental launching and mid-span closure of long cantilever steel box girders. On this basis, a finite element model for pre-camber calculation was established. The study results demonstrate that the mid-span closure method substantially boosts construction efficiency. Li et al. [20] created a finite element model for the incremental launching of steel box girders, evaluated the local stress under the maximum cantilever condition and the stress distribution during vertical jacking asynchrony and lateral deflection, and formulated a dual-parameter control formula for incremental launching.

Presently, few studies have been conducted on the control and monitoring technology of incremental launching construction for small-curvature-radius longitudinally variable-slope steel box girder bridges. Based on this, this paper studies the incremental launching process and full-process monitoring of a steel box girder bridge with a two-way slope, small curvature radius, large lateral deviation of the front end of the main girder, and difficulty in correcting the deviation, so as to guide the incremental launching design and construction of other ramp bridges in the project and offer guidance for the incremental launching construction of other small-curvature steel box girders.

2. Project Overview

This project involves the existing expressway line, and the span of the expressway is 28 m. In order not to affect the passage of the existing line, the incremental launching construction method is adopted. Considering the large span of the permanent piers on both sides of the expressway, with the aim of reducing the span of the incremental launching construction, on the one hand, temporary piers are set on both sides of the expressway to ensure the orderly progress of the project. On the other hand, the guide girder is added to guide the construction, and the span of the guide girder is 23 m. During the construction process, the steel box girder will appear in a cantilever state, and the end part will display downward deflection. The setting of the guide girder can ensure that the downward deflection is within a reasonable range. To guarantee the safety of construction, it is essential to monitor the comprehensive stages of the incremental launching process.

The first contract section of the Xuchang–Xinyang expressway is located in Jianan District of Xuchang City. The Yulin Hub Interchange of the project is situated at the intersection of the Xuchang–Guangzhou expressway and Yongdeng expressway. It is an interchange reconstruction project. The plan can be seen in Figure 1.

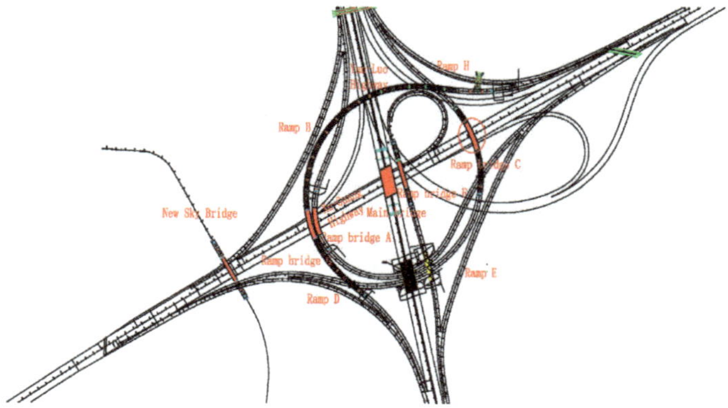

Figure 1. Plane position diagram of steel box girder bridge.

Among them, the longitudinal slope of ramp bridge C is changed from 3.15% to −3.2%. The bridge plane is located on a circular curve with a curve radius of 230 m, which involves a small-curvature-radius steel box girder. The leading edge of the primary girder exhibits significant lateral displacement, making it challenging to correct. Therefore, ramp bridge C is designated as the research focus to guide the incremental launching construction of other ramp bridges. In the middle span of ramp bridge C, the steel box girder has a length of 60 m, a height of 2.6 m, and a weight of 435.8 t. Ramp bridge C is the second full bridge (3 × 20 m prestressed concrete cast in place in a 140 (40 + 60 + 40) m steel box girder); the bridge length is 207 m, consisting of a single design. Among the bridges, the second one across the Xuguang Expressway (subgrade width of 28 m, for the two-way four-lane expressway) consists of 61.4 m of steel box girders, composed of four girders; the length of each girder is 15.9 m, 17.9 m, 17.4 m, 10.2 m. The weight of each segment girder is 98.4 t, 110.8 t, 112 t, and 56.2 t, respectively. The crossing angle is 86°; the width of the steel box girder is 10.06 m; the height of the girder is 2.6 m; and the cantilever length is 2316 mm. The transverse slope of the steel box girder is 6%, and a typical section of the steel box girder is depicted in Figure 2. The net span of incremental launching is 33.64 m, and the mid-span pre-camber of incremental launching of a steel box girder is 7.4 cm.

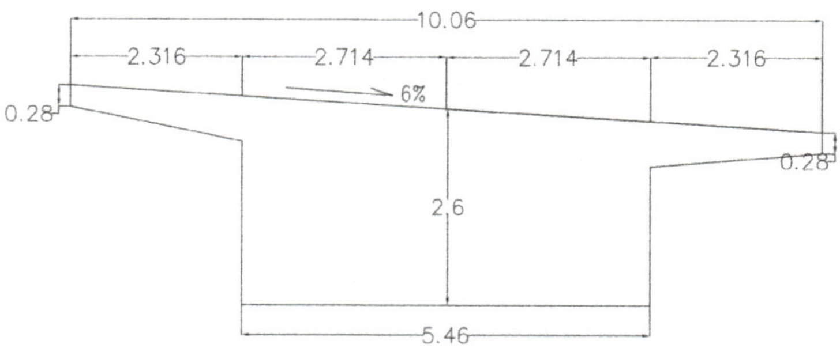

Figure 2. General sectional drawing of steel box girder bridge.

3. Incremental Launching Construction and Monitoring Technology

In light of the structural properties of the bridge, construction is carried out by means of curve incremental launching. The curve incremental launching equipment is a walking-type incremental launching system. The walking-type incremental launching system is a construction method that combines the jacks with three functions of jacking, translation, and transverse movement in a support frame. When the console instructs the starting equipment to carry out incremental launching, the jacks of the walking pre-jacking equipment are cyclically repeated according to the four steps of jacking, pushing, falling, and pulling back, so as to push the girder to the design position.

3.1. Incremental Launching Construction Steps

Step 1: Site preparation, temporary pier foundation concrete construction on the basis of hardening, erection of assembly support, establishment of the incremental launching pier and emergency temporary pier in the central separation belt, and then installation of the incremental launching equipment are carried out. Before incremental launching, site inspection of the incremental launching equipment such as the jack, high-pressure oil pump, and control device is carried out to guarantee the proper functioning of the incremental launching equipment.

Step 2: Steel box girder assembly is carried out using a crane and other equipment for steel box girder hoisting, positioning, alignment matching, and temporary connection installation; on the splicing platform, assembly welding is performed on 5-section and 6-section steel box girders and 23 m guide girders and on both sides of the steel box girder to install a guardrail, and the axis and elevation are reviewed.

Step 3: Steel box girder incremental launching is initiated. The vertical jack of the walking-type translation incremental launching device lifts the steel girder The horizontal jack completes the forward pushing. After the girder is dropped, it is placed on the supporting beam, and then the horizontal jack returns to the original state to complete the stroke incremental launching. The steel box girder advances 17.4 m through cyclic incremental launching, and the incremental launching is stopped. The axis and elevation of the steel box girder are checked. After confirmation, the 4-section steel box girder is hoisted and welded, followed by the 5-section steel box girder. Subsequently, the guardrail of the 4-section steel box girder is installed.

Step 4: The girder is pushed forward 15.7 m and incremental launching is stopped to check the axis and elevation of the steel box girder, confirm the correctness, hoist and weld the 3- and 4-section steel box girders, and install the 3-section steel box girder's guardrail.

Step 5: The equipment continues pushing forward and stops incremental launching when reaching the falling girder position; it then dismantles the guide girder, adjusts the alignment, and then carries out the overall falling girder maneuver. The elevation is checked by the total station, and the other steel box girders are lifted in situ. The assembly support, incremental launching pier, and emergency pier of the central isolation belt of the expressway are removed.

3.2. Overview of Temporary Facilities

The steel box girder is in a curved shape and it crosses over the expressway. In order to accurately control the alignment of the steel box girder throughout the incremental launching process, two incremental launching piers are set between permanent piers 3 and 4 and permanent piers 4 and 5, and each incremental launching pier and the assembly support are arranged along the arc of the bridge to maintain the alignment of the bridge, while permanent piers 4 and 5 are, respectively, on both sides of the existing expressway. When temporary piers are set, they are also set near both sides of the expressway. At the same time, an assembly support is set near permanent pier 3, as shown in Figure 3. In order to avoid incremental launching suspension or an emergency situation when the steel box girder is pushed to the upper part of the existing expressway, a row of emergency temporary piers is set in the central isolation belt of the expressway to ensure the stability of the structure, as depicted in Figure 4.

Figure 3. Assembly support and Nikon XS total station.

Figure 4. Central isolation belt emergency temporary pier.

3.3. Alignment Control

Alignment control is an important part of the bridge incremental launching construction process. In this project, the alignment of the ramp guide girder and the arrangement of the incremental launching piers are consistent with the alignment of the steel box girder. Considering that the alignment of the steel box girder in this project is on a curve with a radius of 230 m, the actual alignment deviation is too large. It will cause secondary stress when the bearing is disengaged, which will cause irreversible collapse and overturning throughout the incremental launching construction, cause uneven stress between the pier and the steel box girder, and reduce the service life of the bridge. Therefore, in splicing within construction, it is necessary to control the actual alignment to ensure that it is basically consistent with the theoretical alignment. Considering the influence of welding, summer temperature, construction environment, and other aspects in practical engineering, deviation in alignment cannot be avoided. A Nikon XS total station (manufactured by Nikon Corporation in Tokyo, Japan) is used to collect alignment data in the process of incremental launching in real time and then adjust the actual alignment according to that data.

Alignment control includes horizontal alignment control and vertical alignment control. Horizontal alignment control is adjusted by rectification measures during incremental launching, and vertical alignment is adjusted by the coordination of the total station and the incremental launching rectification device. By arranging the prism elevation control points on the top of the steel box girder, the actual alignment of the steel box girder is obtained according to the measured elevation data.

In order to carry out total station observation, four displacement reference points are set up at the front end and the back end of the steel box girder. The incremental launching trajectory of the steel box girder is plotted with AutoCAD. According to the four displacement measuring points on the upper part of the steel box girder, horizontal alignment control of incremental launching is carried out to advance along the incremental launching trajectory. The incremental launching trajectory of the whole section of the steel box girder is illustrated in Figure 5.

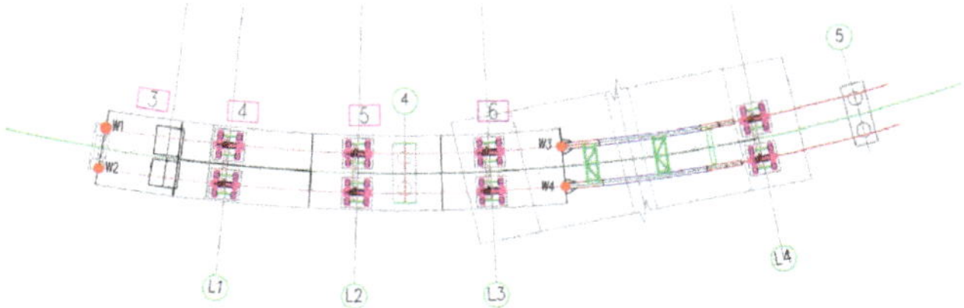

Figure 5. Incremental launching curve diagram.

The incremental launching displacement control steps are as follows:

(1) When carrying out incremental launching on each one-meter section, the lateral movement of the control point is tracked in real time;

(2) Comparing the calculated displacement with the monitored displacement;

(3) Using the correction function of the incremental launching equipment to correct the deviation.

3.4. Monitoring Conditions

Based on the specific safety plan for the incremental launching of steel box girders on the Xuchang–Xinyang Expressway, incremental launching monitoring of this project is carried out from the assembly state, and each 1 m is set as a working condition in the incremental launching process. The construction stage is analyzed through the finite element tool MIDAS Civil. Provided that the calculation outcomes adhere to the specification criteria, five critical working conditions are selected from the complete incremental launching sequence for analysis. The specific conditions of the five most unfavorable working conditions are illustrated in Figure 6 and presented in Table 1.

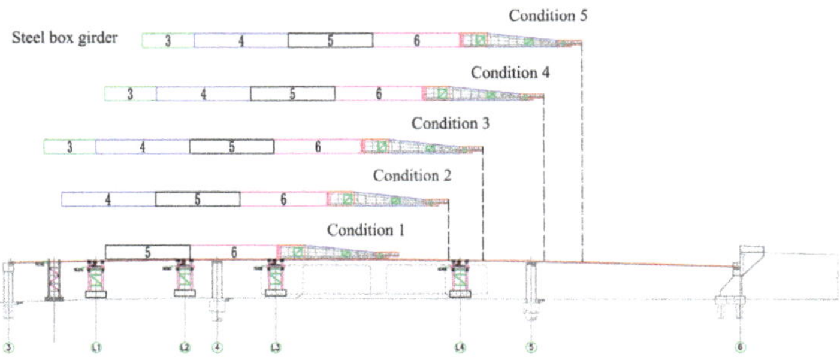

Figure 6. Five most unfavorable conditions in the incremental launching process.

Table 1. The most unfavorable working conditions.

Working Condition	Statement
Condition 1	The cumulative incremental launching is 15 m, and the tail of the steel box girder is cantilevered by 17 m (the most unfavorable condition of steel box girder tail cantilever).
Condition 2	The cumulative incremental launching is 30 m, with the leading edge of the guide girder extending 35.5 m (the worst-case operating condition of the front cantilever of the guide girder).
Condition 3	The cumulative incremental launching is 36 m.
Condition 4	The cumulative incremental launching is 42 m, and the tail of the steel box girder is cantilevered by 17 m (the most unfavorable condition of steel box girder tail cantilever).
Condition 5	Cumulative incremental launching of 53 m, pushing to the design position.

The single incremental launching range of the walking jack for incremental launching is 500 mm. Due to the progress of the project and the need for correction, during the incremental launching procedure, the single incremental launching distance is arranged according to the need. When incremental launching is pushed to the most unfavorable working condition specified in the project, the regular incremental launching process is carried out to better analyze its stress and displacement state. The incremental launching steps before and after the most adverse conditions can be seen in Figure 7.

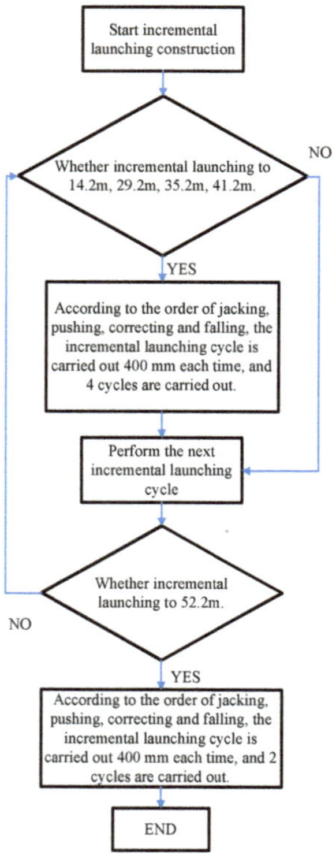

Figure 7. The incremental launching steps before and after the most unfavorable working conditions.

4. Monitoring System Design

For this segment, the finite element analysis application MIDAS Civil is applied to simulate the whole process of incremental launching, and the stress and deformation of the structure under the most unfavorable conditions in Table 1 are obtained, including the displacement and stress data for the most unfavorable position of the steel box girder. The finite element model of the steel box girder is presented in Figure 8.

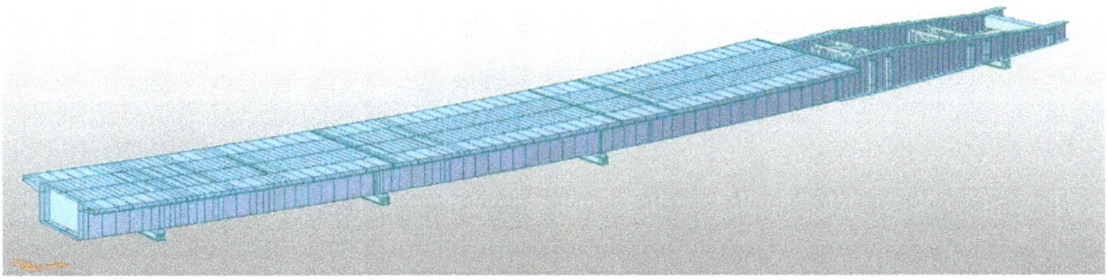

Figure 8. Finite element model of steel box girder.

To safeguard the structural integrity of incremental launching, the real-time alignment is adjusted, ensuring the geometric positioning and stress levels of the steel box girder are managed during incremental launching. According to reference [12], the error limits of internal force and deflection in the incremental launching construction process are stipulated to guide the stability of the incremental launching construction structure. In order to analyze the stress for the most adverse scenarios in the incremental launching process and ensure construction safety, the real-time monitoring values of the most unfavorable working conditions are contrasted with the finite element simulation outcomes.

With the advancement of incremental launching construction progress, the 3-section and 4-section steel box girders are successively welded to the incremental launching steel box girder section. During this process, the stress state of the overall structure is constantly changing. The monitoring system focuses on controlling the alignment of the steel box girder to make it move forward according to the incremental launching trajectory. At the same time, the stress state of the steel box girder, the guide girder, and the incremental launching piers are elucidated in real time to ensure the structural safety and personnel safety of incremental launching construction.

4.1. Displacement Measuring Points

In the incremental launching process, settlement monitoring points are set at the concrete foundation of the temporary piers, and settlement monitoring points are made on the incremental launching piers. The monitoring points are marked with reflective paste, and this reflective paste is set at the front end of the guide girder, as illustrated in Figure 9, the measurement points are marked with red circles. Four displacement data points are set on the upper part of the steel box girder, and four prisms are set at the front end and the back end of the steel box girder. Throughout the incremental launching process, the displacement changes of the front end and back end of the steel box girder are monitored in real time, and the horizontal alignment of the incremental launching is controlled. The incremental launching trajectory should be followed to compare the monitoring value with the calculated value.

Figure 9. Displacement monitoring points of incremental launching and guide girder.

4.2. Stress Measurement Points

For this construction project, during incremental launching, the strategic placement of stress measurement points is crucial. Initial measurement points are set on the bottom side of the girder sections at temporary piers and at the midspan, with corresponding points on the top of the girder, especially at locations with varying cross-sections. As the steel box girder is incrementally launched, new measurement points are added to the new segments, continuously monitoring stress distributions to ensure safety. Additional points are also arranged on the temporary piers and midspan of the new steel box girder segments as they advance. Furthermore, measurement points at the connection between the guide girder and steel box girder ensure the reliability of this critical node. Comprehensively, the stress measurement points are specifically arranged as follows:

Four stress measuring points are arranged at a height of 6 cm from the bottom of each temporary pier column, and for temperature compensation, the temperature-compensated strain-gauge method is adopted and stress compensation measuring points need to be arranged, with a total of 19 stress measurement points. The 60 m steel box girder spans four segments. A stress test section is set up in the middle of each segment. Five measuring points are arranged in each test section. Meanwhile, four temperature compensation ports are configured in the stress–strain acquisition device. A stress test section is also set up at the junction of the five- and six-section girders, a total of 29 stress measuring points. At the same time, two sections are arranged at the connection between the guide girder and the steel box girder. For each section, four measuring points are placed, and overall, eight stress measuring points are positioned. One interface is on the side of the guide girder, and the other section is on the side of the steel box girder. The layout of the measuring points is illustrated in Figure 10. Z_0 is the compensation strain gauge. Through these stress measuring points, the stress of each unfavorable working condition throughout the incremental launching is monitored, and the data are recorded by a JM3816 static resistance strain gauge. Finally, the stress is determined using the material's stress–strain characteristics.

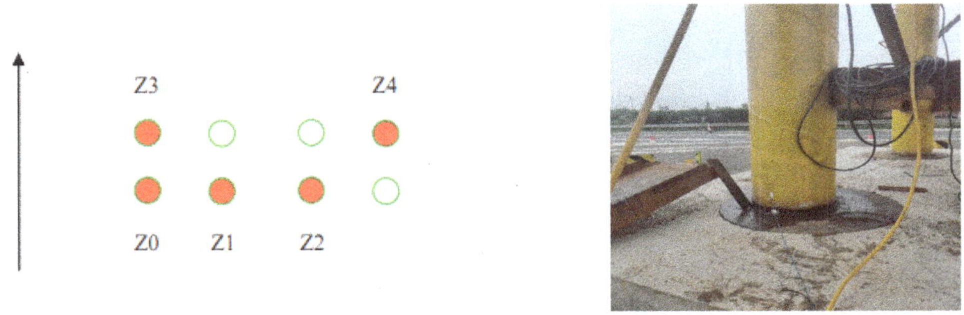

(a)

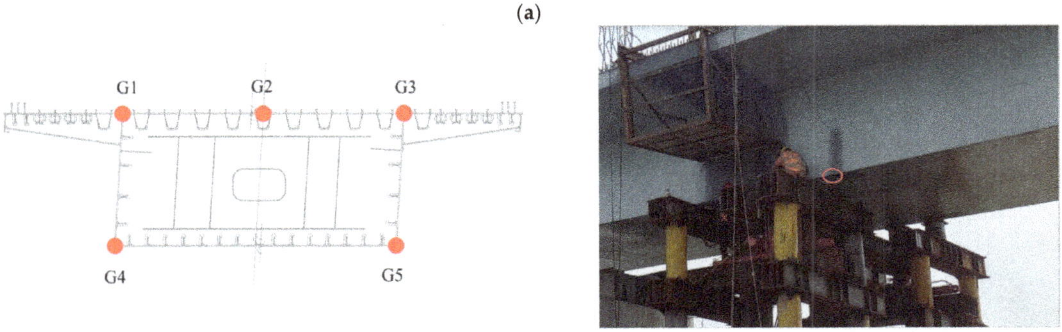

(b)

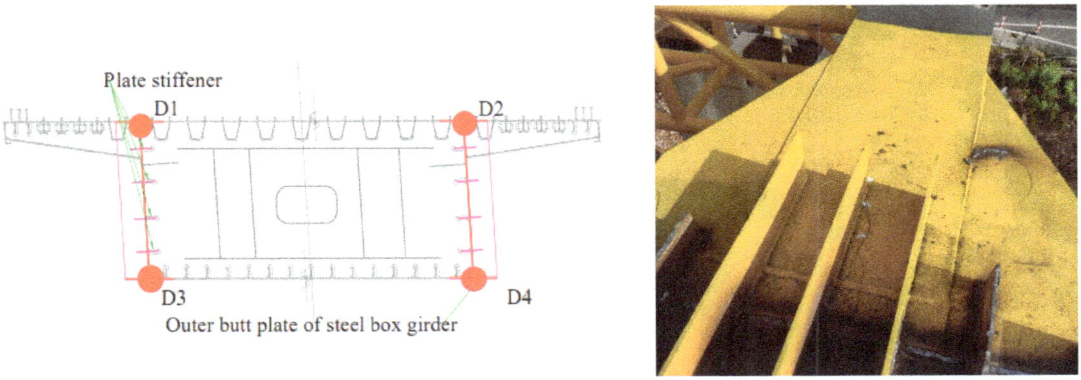

(c)

Figure 10. Monitoring point arrangement. (**a**) Temporary pier measurement point layout; (**b**) arrangement of measuring points of steel box girder; (**c**) arrangement of measuring points at the junction of the guide girder and steel box girder.

5. Comparative Analysis of Incremental Launching Monitoring Results

The stress measuring points of the two sections at the connection between the guide girder and the steel box girder are D_{11}, D_{12}, D_{13}, D_{14}, D_{21}, D_{22}, D_{23}, and D_{24}, respectively, as presented in Figure 10c. The stress measurement points of the section at the junction of the five-section and six-section girders, the six-section girder, the five-section girder, the four-section girder, and the three-section girder are G_{j1}, G_{j2}, G_{j3}, G_{j4}, G_{j5}, respectively, where j = 1, 2, 3, 4, 5; the position is illustrated in Figure 10b.

5.1. Steel Box Girder Displacement Monitoring

The theoretical displacements at the leading and trailing edges of the steel box girder for five operational scenarios are provided by the finite element analysis and contrasted with the monitoring data. Figure 11 shows the overall vertical deformation cloud map of the finite element simulation of the steel box girder. Table 2 shows the calculated and monitored values of the displacement of the steel box girder under five working conditions, and the displacement limit is given according to reference [12]. According to the data in the table, the displacement is within the limit value, indicating that the displacement of the steel box girder meets the safety design requirements.

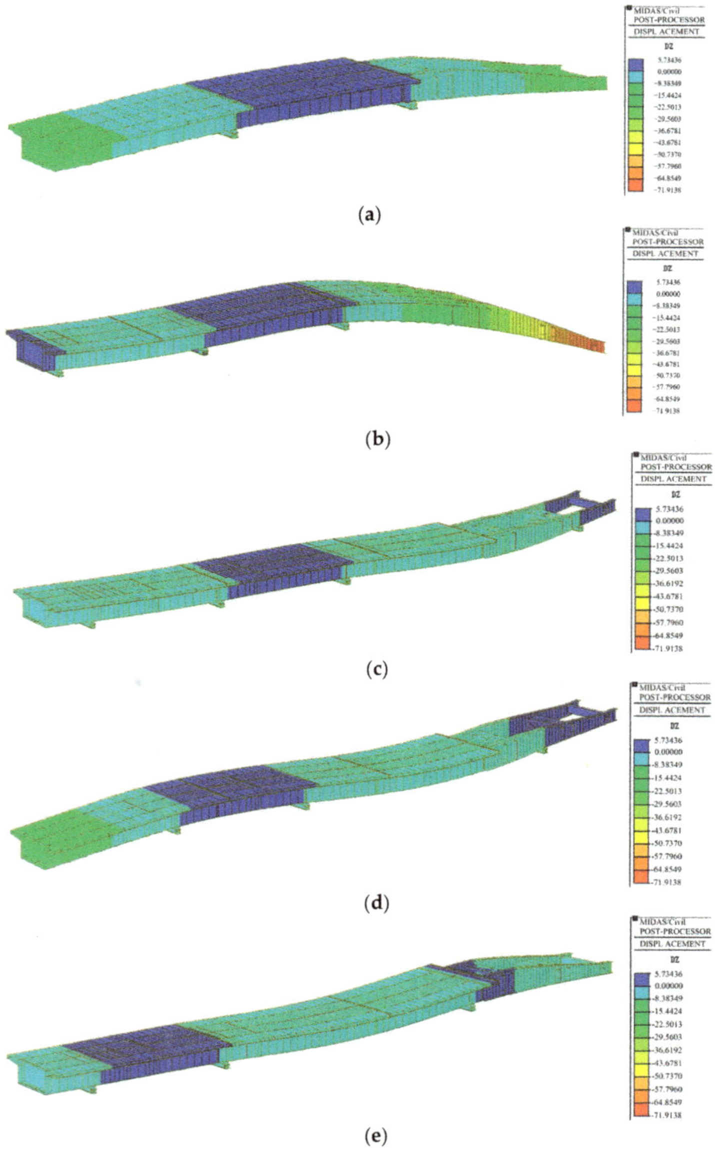

Figure 11. Overall deformation under five working conditions. (**a**–**e**) Corresponding working conditions 1–5.

Table 2. Contrast between measured and calculated displacements of steel box girder.

Condition	Position	Monitoring Data (mm)	Calculated Value (mm)	Limiting Value (mm)
1	Front	0.402	0.267	±20
	End	−14.016	−12.29	
2	Front	−30.581	−19.506	
	End	−0.108	−0.074	
3	Front	−5.245	−7.906	
	End	−0.071	−0.013	
4	Front	−4.558	−6.974	
	End	−19.487	−18.097	
5	Front	0.468	0.279	
	End	−0.319	−0.654	

Note: During the incremental launching process, if the monitoring data exceed the limiting value, an early warning will be issued. At this time, measures need to be taken for the support to correct the deviation and remove the limit warning to ensure the smooth progress of the incremental launching construction.

5.2. Guide Girder Displacement Monitoring

The theoretical deflection of the front end of the guide girder under five working conditions is given by the finite element simulation outcomes. The displacement at the leading edge of the guide girder is related to whether the guide girder can reach the piers smoothly. Therefore, the theoretical deflection is contrasted with the actual monitoring outcomes. Table 3 gives the calculated values and monitoring values of the deflection of the guide girder under five working conditions. In the second working condition, the guide girder experiences the highest theoretical deflection, and the cantilever of the guide girder is 35.5 m; 35.5 m/200 = 177.5 mm. According to Table 3, the deflection monitoring value and calculation value of the guide girder under working condition 2 are less than 177.5 mm. Table 3 indicates that the deflection values of the guide girder are within the control range, and the deflection of the guide girder meets the safety design requirements.

Table 3. Contrast between measured and calculated deflections of guide girder.

Condition	Monitoring Data (mm)	Calculated Value (mm)	Limiting Value (mm)
1	−14.022	−16.068	±115
2	−39.303	−66.683	±177.5
3	4.515	3.435	±30
4	2.596	5.095	±60
5	−9.867	−7.566	±115

5.3. Temporary Pier Concrete Foundation Settlement Monitoring

The settlement of the concrete foundation of incremental launching pier L3 under unfavorable working condition 4 is selected for analysis. At this time, the incremental launching process is relatively regular. There are four incremental launching cycles. The five points of each incremental launching process are, respectively, expressed as the falling of the previous cycle, jacking, pushing, correcting, and falling. The settlement of the concrete foundation of incremental launching pier L3 during the four incremental launching cycles is presented in Figure 12. The diagram indicates that the foundation displacement of each incremental launching cycle exhibits an initial increase followed by a decrease; the variation between the lowest values remains within 4 mm, indicating that the incremental launching process has a negligible influence on the settlement of the concrete foundation at the incremental launching pier.

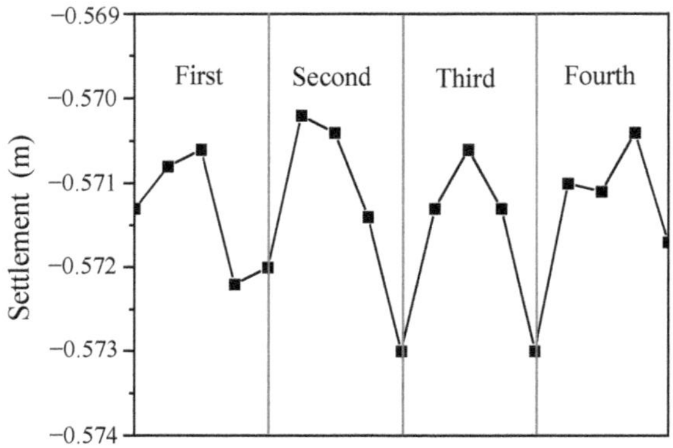

Figure 12. The settlement of incremental launching pier L3's concrete foundation.

5.4. Stress Monitoring of Steel Box Girder

Table 4 shows the key stress monitoring values and calculated values of steel box girders under five working conditions. Table 4 indicates that the stress values are all within the limit range, which meets the monitoring and control requirements, meaning that the incremental launching operation can be carried out smoothly. Additionally, the elevated summer temperatures and the impact of on-site construction conditions contribute to a significant discrepancy between the monitored and calculated values.

Table 4. Contrast between measured and calculated stress in steel box girder.

Condition	Measuring Point	Monitoring Data (MPa)	Calculated Value (MPa)	Limiting Value (MPa)
1	D_{24}	−9.8	−15.2	295
	G_{12}	0.2	1.5	
	G_{15}	−13.8	−36.9	
	G_{32}	−1.6	−0.3	
2	D_{24}	−65.8	−20.2	
	G_{12}	30.8	2.8	
	G_{15}	−52.0	−69.5	
	G_{44}	−17.6	13.2	
3	D_{24}	41.4	17.2	
	G_{12}	−2.8	−0.4	
	G_{15}	−14.4	−18	
	G_{44}	3.4	5.9	
	G_{55}	6.6	−4	
4	D_{24}	44.4	22.2	
	G_{12}	3.6	0.3	
	G_{44}	30.6	−51.8	
	G_{52}	55.4	0.3	
	G_{55}	37.2	−4.4	
5	D_{24}	22	−19.6	
	G_{12}	−3.8	0.2	
	G_{44}	21.4	−13	
	G_{52}	13.8	0.5	
	G_{55}	5.4	−3.8	

5.5. Whole-Process Stress Monitoring of Incremental Launching Pier

Temporary incremental launching pier L3 is selected for the whole-process stress monitoring and analysis of the incremental launching process, and the locations of the measuring points are illustrated in Figure 10a. Because the four-section and three-section steel box girders need to be welded, the incremental launching process is segmented into three distinct rounds: the first, second, and third stages of incremental launching. The monitored stress throughout the incremental launching of temporary pier L3 is illustrated in Figure 13. Figure 13a indicates that column Z1 experiences initial compression, followed by tension, with the stress ultimately reaching a stable state. With the increase in the incremental launching distance, column Z_3 is first compressed and then pulled and finally compressed. When Z_1 and Z_3 are at 1–12 m, the stress trend is roughly the same. After 12 m, Z_3 suddenly changes from the tensile state to the compressive state. The stress of Z_2 and Z_4 changes greatly, and the stress is in an unstable state. When the incremental launching distance reaches about 15 m, that is, when the unfavorable working condition is 1, the tail of the steel box girder is in a cantilever state, and the stress changes abruptly. The stress of Z_4 suddenly increases to 34 MPa; Z_2 changes from compression to tension, and Z_3's stress decreases. In the first stage of incremental launching, the maximum stress at each incremental launching distance almost appears on Z_4 and Z_3, with the maximum tensile stress of 37.4 MPa and the maximum compressive stress of 45.4 MPa.

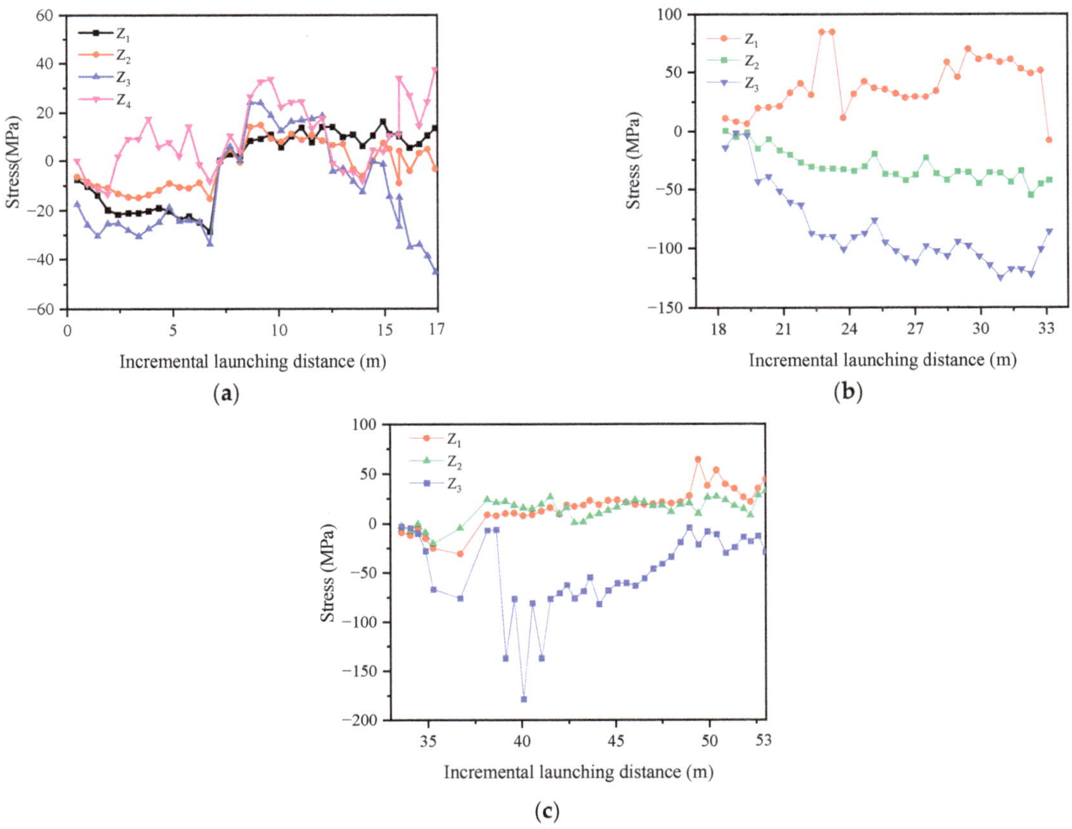

Figure 13. Temporary incremental launching pier L3's stress. (**a**) First stage of incremental launching; (**b**) second stage of incremental launching; (**c**) third stage of incremental launching.

According to Figure 13b, in the second phase of incremental launching, column Z_1 is almost in full tension, and the maximum stress is 85.2 MPa. As the incremental launching distance increases, the stress of column Z3 shows an overall increasing trend, and it remains in a compressed state throughout the process. Z2 is also under compression during the entire process, with a relatively stable stress that generally exhibits an increasing trend. In the second stage of incremental launching, with the increase in incremental launching distance, the stress of Z_1, Z_2, and Z_3 gradually increases; the maximum tensile stress is 85.2 MPa, and the maximum compressive stress is 124.8 MPa.

It can be seen from Figure 13c that columns Z_1 and Z_2 are first compressed, and this gradually changes to tension after the incremental launching distance reaches unfavorable working condition 3, and the stress curve is relatively stable. Z_3 is compressed in the whole process, and the stress increases first and then decreases. When the incremental launching distance reaches 36 m, the stress of Z_1, Z_2, and Z_3 fluctuates. When the incremental launching distance is about 36–42 m, the stress of Z_3 fluctuates greatly and is more dangerous. The maximum stress reaches 178.6 MPa, and the stress gradually decreases after 42 m. In the third stage of incremental launching, the maximum stress at each incremental launching distance almost appears on Z_3, and the maximum compressive stress is 178.6 MPa.

5.6. Whole-Process Expression of Guide Girder Stress

In unfavorable working condition 2, the front end of the guide girder is cantilevered by 35.5 m, and the guide girder is in the most unfavorable condition. Therefore, the whole-process analysis of the second stage of incremental launching is carried out to analyze the change in the stress of the guide girder. Figure 14 shows the stress change diagram of the stress of the guide girder section during the second stage of incremental launching. From the diagram, the general pattern of stress in the guide girder section exhibits an initial rise followed by a decline as the incremental launching distance increases. The trend of the stress curves of measuring points D_1 and D_2 is roughly the same. With the increase in incremental launching distance, it is first pulled and then compressed. When the incremental launching distance is 20 m, it attains its peak. The maximum stress of D_1 is 117.4 MPa. The maximum stress of D_2 is 103.4 MPa, lower than the steel strength limit of 295 MPa. The trend of the stress curve of measuring points D_3 and D_4 is the same, and the whole process is tensioned. As the distance in incremental launching grows, the overall trend increases first and then decreases. When the incremental launching distance is 25 m, the stress of the measuring points is the largest, 98 MPa and 138 MPa, respectively, both less than 295 MPa.

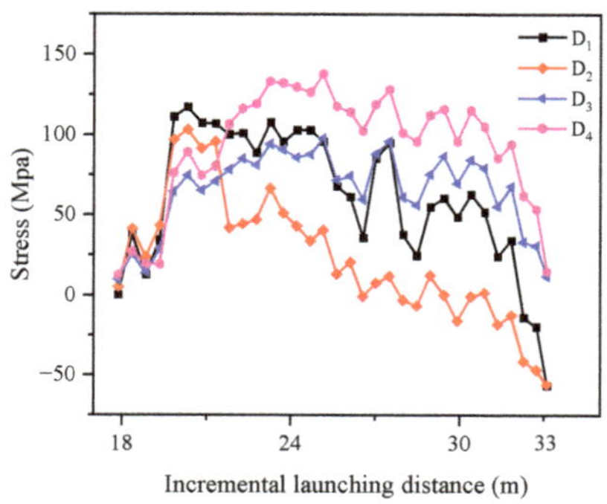

Figure 14. The guide girder section stress of the second stage of incremental launching.

When the incremental launching is 30 m, the guide girder reaches the maximum cantilever state, and the stress of the four measuring points of the guide girder has a sudden downward trend. Meanwhile, the stress begins to decrease gradually once the leading edge of the guide girder arrives at temporary incremental launching pier L4, aligning with the findings from theoretical analysis.

5.7. Construction Results

In this project, the structural stress state of the bridge was comprehensively monitored throughout the entire incremental launching construction process. Through in-depth comparative analysis with the finite element theoretical calculations, the internal force characteristics of the structure during the process were studied, which provided valuable guidance for the construction process. The monitoring results revealed that the alignment of the steel girder is smooth and the installation of each component is precise, and all monitoring data related to the structural stress state met the design requirements. This indicates that the incremental launching construction technology of the small-curvature curve steel box girder bridge has been successfully applied and verified in this project, offering some practical basis for the application of this technology in similar bridge projects and ensuring the smooth acceptance of the bridge, thus guaranteeing the subsequent normal operation of this bridge. Figure 15a shows the scene of the curved steel box girder bridge across the highway after being incrementally launched into position, and Figure 15b shows the scene after the construction is completed and the bridge is open to traffic.

(a) (b)

Figure 15. Steel box girder bridge of span expressway is pushed in place and opened to traffic after construction is completed. (**a**) is the scene of the curved steel box girder bridge across the highway after being incrementally launched into position, (**b**) is the scene after the construction is completed and the bridge is open to traffic.

6. Conclusions

In the presented study, the stress and displacement of curved steel box girders, incremental launching piers, and guide girders during the construction process are monitored in combination with the incremental launching project of small-curvature-radius curved steel box girders at the intersection of Xuchang–Xuguang Expressway and Yongdeng Expressway. For the analysis, five of the most adverse working conditions were selected. At the same time, finite element analysis was performed by altering the location of the temporary support to realize the simulation of the incremental launching process. The states of internal forces and displacements under the five working conditions were evaluated, and the following conclusions were drawn:

(1) The comparison of monitored and theoretical values for the steel box girder and guide girder under key working conditions shows that their internal forces and deflections are within the safe range, ensuring the safety of incremental launching construction.

(2) During the incremental launching process in working condition 4, the foundation displacement of each incremental launching cycle first increased and then decreased, with the variation range of the minimum value within 4 mm. This indicates the relatively small impact of the incremental launching process on the foundation settlement.

(3) Temporary pier L3 is in an ideal low-stress state in the first stage of incremental launching. The stress is generally large in the second stage of incremental launching, but it is within the safe range. In the third stage of the incremental launching process, when the incremental launching distance of Z_3 is about 36–42 m, the stress exhibits significant fluctuations and demands heightened attention. Although it poses a relatively higher risk level compared to other scenarios, it remains within the acceptable safety boundaries. Therefore, temporary pier L3 is in a safe state during the whole pushing process.

(4) The second stage of incremental launching is the most unfavorable working condition of the guide girder (working condition 2). As the incremental launching distance increases, the stress of the guide girder section first rises and then falls. When the guide girder is reaches its maximum cantilever position, the stress on it increases, but the peak stress does not occur at this position. Therefore, the construction process must be monitored throughout to effectively ensure construction safety and quality.

Meanwhile, considering that the construction monitoring is carried out in the summer season, where temperatures are over 35 °C and the surface temperature of the steel box girder is 50–60 °C, the strain value will be affected, which will make the strain larger, but the stress is still in the safe range. In subsequent research, the influence of increased temperatures on the strain of steel box girders should be addressed.

Author Contributions: Conceptualization, H.H.; methodology, H.H. and H.M.; software, X.X.; validation, H.H., X.X. and J.D.; formal analysis, H.M., J.D. and Z.Z.; investigation, Z.Z., H.Y., E.G. and J.M.; resources, H.M.; data curation, Z.Z. and E.G.; writing—original draft preparation, H.H. and H.M.; writing—review and editing, H.H., H.M. and J.D.; funding acquisition, H.M. and J.M. All authors have read and agreed to the published version of the manuscript.

Funding: This work was supported by the Key R&D and Promotion Special Project in Henan Province-Science and Technology Research (joint fund) Project (Grant No. 232103810082) and the Key R&D and Promotion Special Project in Henan Province-Science and Technology Research Project (Grant No. 222102320462). The authors are grateful for the funds provided by China Construction Seventh Engineering Division Corp., Ltd. (Grant No. China Construction 07110120200499020).

Institutional Review Board Statement: Not applicable.

Informed Consent Statement: Not applicable.

Data Availability Statement: The original contributions presented in the study are included in the article, further inquiries can be directed to the corresponding author.

Acknowledgments: We are thankful to the reviewers for their thoughtful and constructive critiques, which have helped to enhance the overall quality of the paper.

Conflicts of Interest: Authors Hai Huang, Haitao Ma, Zhijun Zhou and Jianbo Dou were employed by the company China Construction Seventh Engineering Division Co., Ltd. The remaining authors declare that the research was conducted in the absence of any commercial or financial relationships that could be construed as a potential conflict of interest. The authors declare that this study received funding from China Construction Seventh Engineering Division Corp., Ltd. The funder had the following involvement with the study: Its employees, namely Hai Huang, Haitao Ma,

Zhijun Zhou and Jianbo Dou, participated in the relevant research work as detailed in the Author Contributions section.

References

1. Virlogeux, M. Bridges with multiple Cable-stayed spans. *Struct. Eng. Int.* **2001**, *11*, 61–82. [CrossRef]
2. Jiang, T.Y.; Tian, Z.C.; Xu, J.H. Key technologies of whole incremental launching construction control for inclined continuous box girder with steep longitudinal gradient. *Appl. Mech. Mater.* **2012**, *204–208*, 2034–2039. [CrossRef]
3. Dai, J.; Di, J.; Qin, F.J.; Zhao, M.; Lu, W.R. Finite Element Analysis on Incremental Launching Construction for Steel Box Girder. *Adv. Mater. Res.* **2013**, *671–674*, 974–979. [CrossRef]
4. Xu, K.M. Incremental launching construction method for steel Truss suspension bridge. *Adv. Mater. Res.* **2011**, *204–210*, 842–845. [CrossRef]
5. Zhang, Z.H.; Zhang, J.Y.; Hao, W.S.; Dai, J.G.; Shen, Y. Hangzhou Jiangdong Bridge Designed as a Spatial Self-Anchored Suspension Bridge, China. *Struct. Eng. Int.* **2010**, *20*, 303–307. [CrossRef]
6. Shao, C.Y. Innovative structure design and construction-Jiubao Bridge, China. *Struct. Eng. Int.* **2015**, *25*, 102–105. [CrossRef]
7. Zhang, H.; Guo, Q.; Wang, W.X. A new Technology for Incremental Launching the Small Curvature Radius Variable Cross-Section Steel Box Girder. *Int. Conf. Manuf. Sci. Eng.* **2016**, *32*, 510–516.
8. Xie, F.J. Influence analysis of the incremental launching of steel box girder and local stability control. *Int. Conf. Civil. Archit. Environ. Eng.* **2017**, *1–2*, 275–283.
9. Feng, Z.W.; Que, R.B. Study on Incremental Launching of Lane L8 Steel Box Girder of Main Bridge in Xiamen Xianyue Flyover. *Adv. Eng. Res.* **2017**, *124*, 226–230.
10. Zhang, P.Y.; Jiang, X.W.; Gan, H.L. Research on the Overall and Local Mechanical Behaviors of Steel Box Girder Cable-Stayed Bridge via Incremental Launching Construction. *Insight Civ. Eng.* **2020**, *3*, 35–42. [CrossRef]
11. Ding, S.H.; Fang, J.; Zhang, S.L.; Liang, C.S. A Construction Technique of Incremental Launching for a Continuous Steel Truss Girder Bridge with Suspension Cable Stiffening Chords. *Struct. Eng. Int.* **2020**, *10*, 93–98.
12. Wang, H.P.; Wang, K.; Guo, N. Coordinate Monitoring Technology in the Process of Incremental Launching Construction of Curved Steel Box Girder Bridge. *Adv. Civ. Eng.* **2022**, *2022*, 3641683. [CrossRef]
13. Du, J.B.; Niu, W.; Shi, Y.; Wu, Y.Z.; Tang, Q. Local Mechanical Behaviors of Steel Box Girder During Skew Incremental Launching. *Adv. Geospat. Technol. Min. Earth Sci.* **2023**, 371–382. [CrossRef]
14. Wu, H.S.; Ma, J.M.; Zou, Y.; Wang, H.J. Mechanical Analysis of Steel Box Girder During Incremental Launching. *Int. Conf. Logist. Eng. Manag. Comput. Sci.* **2014**, *101*, 446–449.
15. Wang, J.F.; Lin, J.P.; Fan, X.L. Evaluation of Long Multi-Span Steel U-Shaped Girder During Incremental Launching Construction. *J. Test. Eval.* **2015**, *43*, 425–433. [CrossRef]
16. Hu, Z.J.; Wu, D.J.; Sun, L.Z. Integrated investigation of an incremental launching method for the construction of long-span bridges. *J. Constr. Steel Res.* **2015**, *112*, 130–137. [CrossRef]
17. Chacón, R.; Uribel, N.; Oller, S. Numerical Validation of the Incremental Launching Method of a Steel Bridge Through a Small-Scale Experimental Study. *Exp. Tech.* **2016**, *40*, 333–346. [CrossRef]
18. Chai, H.; Song, Y.M. Mechanical analysis for incremental launching construction of long-span continuous steel truss bridge. *Mech. Archit. Des.* **2017**, 181–188. [CrossRef]
19. Wang, K.; Lu, W.; Zhang, Z.B. Walking-Type Bidirectional Incremental Launching and Mid-Span Closure Technology for Steel Box Girders with Long Cantilevers. *IOP Conf. Ser. Earth Environ. Sci.* **2021**, *632*, 022032. [CrossRef]
20. Li, Q.F.; Guo, H.; Guo, B. The Dual-Parameter Control of Synchronization in Steel Box Girder Incremental Launching Construction. *Appl. Sci.* **2023**, *13*, 12074. [CrossRef]

applied sciences

MDPI

Article

Seismic Design and Ductility Evaluation of Thin-Walled Stiffened Steel Square Box Columns

Mwaura Njiru and Iraj H. P. Mamaghani *

Department of Civil Engineering, University of North Dakota, Grand Forks, ND 58202, USA
* Correspondence: iraj.mamaghani@und.edu; Tel.: +1-701-777-3563

Abstract: This paper investigates the seismic performance of thin-walled stiffened steel square box columns, modeling bridge piers subjected to unidirectional cyclic lateral loading with a constant axial load, focusing on local, global, and local-global interactive buckling phenomena. Initially, the finite element model was validated against existing experimental results. The study further explored the degradation in strength and ductility of both thin-walled and compact columns under cyclic loading. Thin-walled, stiffened steel square box columns exhibited buckling near the base, forming a half-sine wave shape. The research also addresses discrepancies from different material models used to analyze steel tubular bridge piers. Analysis using a modified two-surface plasticity model (2SM) yielded results closer to experimental data than a multi-linear kinematic hardening model, particularly for compact sections. The 2SM, which accounts for cycling within the yield plateau and strain hardening regime, demonstrated enhanced accuracy over the multi-linear kinematic hardening model. Additionally, a parametric study was conducted to assess the impact of key design parameters—such as width-to-thickness ratio (R_f), column slenderness ratio (λ), and magnitude of axial load (P/P_y)—on the performance of thin-walled stiffened steel square box columns. Design equations were then developed to predict the strength and ductility of bridge piers. These equations closely matched experimental results, achieving an accuracy of 95% for ultimate strength and 97% for ductility.

Keywords: thin-walled; stiffened; steel; square box section; multi-linear kinematic-hardening rule; modified two-surface model; cyclic loading; buckling; strength; ductility

1. Introduction

Thin-walled steel box columns, commonly used as bridge piers, are vital to urban transportation infrastructure. Structural hollow sections offer several advantages over open sections, such as enhanced aesthetic appeal, excellent flexural resistance about all axes, and superior torsional rigidity. These features make them ideal for various applications, including columns in buildings, members in lattice girders, and bridge piers.

Typically produced through cold-forming techniques [1–3], thin-walled steel box columns are created by welding steel sheets longitudinally to form the final section profile without prior heating or subsequent heat treatment. This method not only improves structural performance but also provides an economical solution for withstanding instability due to severe environmental loads, such as seismic forces [4–6].

Research on the local [7–13] and global [14–22] buckling of cold-formed hollow sections is extensive. While some studies have explored the interaction between local and global buckling in welded box sections [23–25], research on cold-formed tubular members with slender cross-sections remains limited [26–30].

Steel bridge piers with box sections are generally designed as single cantilever columns or one- to three-story frames. Highway bridge systems frequently utilize thin-walled members with closed-box cross-sections due to their high strength and torsional rigidity. These structures differ markedly from building columns, often failing due to local buckling in thin-walled members, irregular distribution of story mass and stiffness, strong beams and weak columns, low rise (one to three stories), and the need for residual displacement evaluations. Consequently, thin-walled steel box columns in bridge piers are particularly susceptible to damage from local and overall interaction buckling during severe earthquakes [31–34].

For example, Figure 1 depicts the performance of thin-walled steel box section bridge piers during the Hyogoken-Nanbu earthquake near Kobe City, Japan, on 17 January 1995. The steel bridge piers with rectangular box sections experienced collapse, weld seam tear-outs, and severe local buckling damage, as shown in Figure 1 [32,33]. Local buckling was observed near the base of the column (Figure 1c).

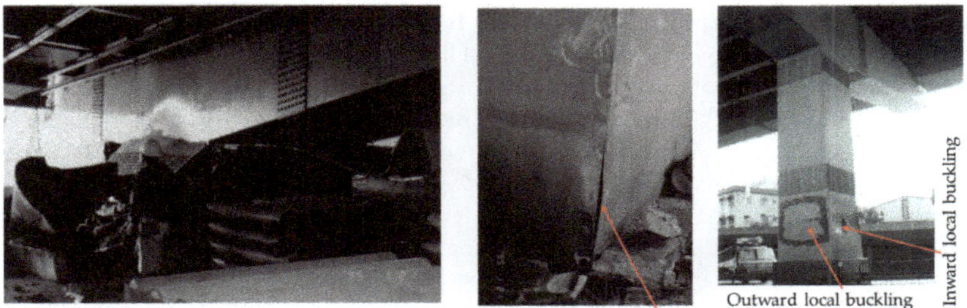

(a) Collapsed thin-walled steel box column (Bruneau, M., 1998)
(b) Tear-out of a weld-seam (Bruneau, M., 1998)
(c) Damaged steel column (Gao, S. B., and Ge., H. B., 2007)

Figure 1. The performance of the thin-walled steel box section bridge pier was during the Hyogoken-Nanbu earthquake near Kobe City, Japan, on 17 January 1995: (**a**) Collapsed thin-walled steel box column [32], (**b**) Tear-out of a weld-seam [32], (**c**) Damaged steel column [33].

Post-Kobe earthquake research has increasingly focused on steel tubular columns with various cross-sections subjected to constant axial loads and cyclic lateral loads in one or multiple directions. To enhance the seismic capacity of steel bridge piers with tubular box sections, researchers have investigated both experimentally and theoretically the stability and plastic ductility of steel tubular columns [31–52]. The strength of these tubes is influenced by local buckling of the tube walls, which depends on the slenderness of the plate elements forming the tube. Local buckling patterns in rectangular or square bare steel tubes can include inward and outward buckles (see Figure 1c), which are considered in major steel design specifications.

Thin-walled steel tubular columns typically experience local buckling near the base at heights corresponding to the diameter and width of circular and square thin-walled columns, respectively [2,32–36]. Research indicates that the ability of these piers to withstand severe earthquakes depends heavily on their strength and ductility [2,15].To improve seismic performance, studies have explored incorporating stiffeners and diaphragms to enhance both strength and ductility [2,6,39,40].

In response to the damage observed in steel bridge piers during the Kobe earthquake, a critical consideration in the seismic design and retrofit of steel tubular columns is to increase ductility while maintaining nearly unchanged ultimate strength [2,31]. Therefore, understanding the cyclic inelastic behavior of thin-walled steel tubular columns is essential for developing rational performance-based seismic design methodologies and evaluating ductility.

Finite element analysis (FEA) of thin-walled steel tubular columns has utilized both isotropic and kinematic hardening material models for cyclic analysis. The kinematic hardening model generally aligns better with test data compared to the isotropic hardening model [47,49,51]. However, it may overestimate strengths under cyclic loading [50]. To improve numerical accuracy, further investigation into additional material models, such as those implemented through ABAQUS-UMAT subroutines in FE ABAQUS 2023 software [52], is necessary. Recent advancements, particularly in Japan, have introduced alternative material models that better replicate test results, such as the fiber-based model with a failure segment [47].

FE analysis of cyclic behavior involves solving a system of non-linear equations derived from applied nodal forces and internal equivalent nodal forces [53–55]. The constitutive material model of structural steel is described using incremental plastic theory, which includes strength rules, yield criteria, flow rules, and criteria for plastic loading and unloading [56–58]. Accurate finite element methods are crucial for predicting the cyclic behavior of these structures. The modified two-surface plasticity model (2SM), developed by Mamaghani et al. [6,58,59], which accounts for cyclic strain hardening and the reduction of the yield plateau and elastic range (Bauschinger effect) due to cyclic loading, was employed in the commercial software ABAQUS 2023 (ABAQUS-2SM).

This study aims to thoroughly investigate the stability behavior of cold-formed hollow steel box sections subjected to cyclic lateral loads combined with axial compression, focusing on local, global, and local-global interactive buckling phenomena. Thin-walled, stiffened steel square box columns are susceptible to damage from local buckling, global buckling, or a combination of both. Local buckling is influenced by the width-to-thickness ratio (R_f) of the flange plate, while flexural buckling is determined by the column's slenderness ratio (λ) [2]. Extreme seismic loads can cause significant damage or collapse due to these buckling modes [32,33]. Large earthquakes generate substantial cyclic lateral loads, necessitating precise design parameters. This study employs ABAQUS-2SM to validate the inelastic cyclic behavior of thin-walled, stiffened steel square box columns [45]. A parametric analysis assessed key design parameters, including steel material modeling, diaphragms, stiffeners, the width-thickness ratio for the flange (R_f), and the column slenderness ratio (λ). The study developed interaction equations for these parameters, facilitating a safe and cost-effective design approach for thin-walled, stiffened steel square box columns used in bridge piers.

2. Characteristics of Thin-Walled Stiffened Steel Square Box Columns

Thin-walled stiffened steel square box columns modeling bridge piers are designed by monitoring specific ranges for various parameters: the width-to-thickness ratio for the flange (R_f), the slenderness parameter (λ), and the axial force parameter (P/P_y). Where P_y is the squash load of the column. The following equations describe these parameters [44]:

$$R_f = \frac{d}{t}\frac{1}{n\pi}\sqrt{3(1-v^2)}\sqrt{\frac{\sigma_y}{E}} \tag{1}$$

$$\lambda = \frac{2h}{r}\frac{1}{\pi}\sqrt{\frac{\sigma_y}{E}} \tag{2}$$

where, d = column depth (square section size), h = column height, t = plate thickness, v = Poisson's ratio, E = modulus of elasticity, σ_y = yield strength, n = the number of subpanels separated by the stiffeners [44,47].

A series of experiments on thin-walled, stiffened steel square box columns has been conducted by [44,48]. Based on these experiments, the practical values for R_f for the thin-walled stiffened steel square box columns have been proposed in the range of 0.2 to 0.5, with the axial load not exceeding 20% of the yield load [39,47]. A schematic diagram of the thin-walled, stiffened steel square box column is illustrated in Figure 2a,c.

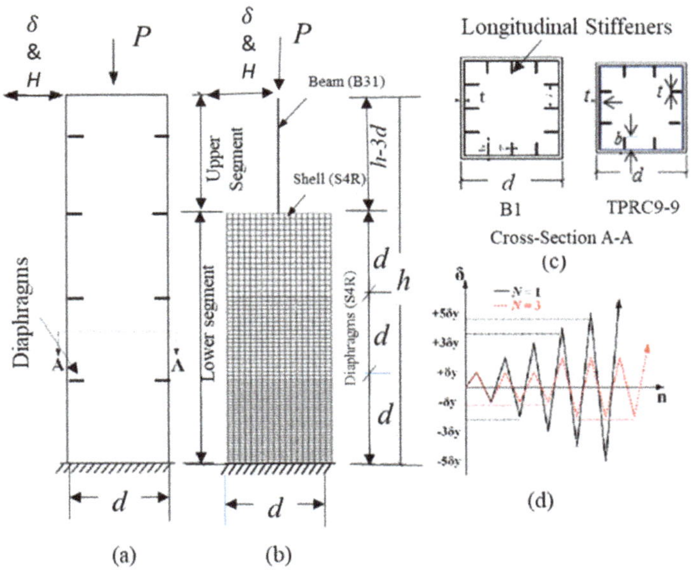

Figure 2. Finite Element Modeling (**a**) Column, (**b**) FE Meshing, (**c**) Stiffened Cross-section, and (**d**) Loading Path Protocol.

3. Finite Element Analysis

In this study, the results from two tested columns available in the literature—the TPRC9-9 column [43] and the B1 column [3]—were used to validate the finite element modeling with the 2SM model for material nonlinearity [58]. The modeling was conducted using ABAQUS [52] employing the 2SM for material nonlinearity [58], hereafter referred to as ABAQUS-2SM.

As shown in Figure 2c, the column features eight longitudinal stiffeners for the TPRC9-9 column and twelve longitudinal stiffeners for the B1 column, each 6 mm thick and 60 mm wide, and two diaphragms, spaced 450 mm apart and 6 mm thick. The cross-sectional plate thickness is 9 mm for both the flange and the web. Table 1 lists the material and geometric properties of the analyzed specimens.

Table 1. Geometric and material parameters for the specimens.

Column	h (mm)	d (mm)	E (MPa)	v	σ_y (MPa)	R_f	λ	P/P_y
TPRC9-9	2141	449	204,000	0.26	266.0	0.30	0.29	0.150
B1	3403	900	206,000	0.30	378.6	0.56	0.26	0.122

3.1. Brief Description of the 2SM Model

The 2SM model can more effectively capture the cyclic behavior of structural steel compared to existing material models, as it incorporates the yield plateau and explicitly accounts for its reduction and eventual disappearance under cyclic loading [6,58]. A detailed description of the 2SM and its application to simulating the cyclic behavior of structural steel is provided by Shen et al. [58]. This model requires more constants than the multi-linear kinematic hardening (MKH) model, totaling 18 parameters: 5 material constants (σ_y, E, ν, $\varepsilon_{st}{}^{p}$, and E_{st}) and 13 model parameters [6,58]. Here, $\varepsilon_{st}{}^{p}$ represents the plastic strain, and E_{st} denotes the initial hardening modulus at the end of the yield plateau. The 13 model parameters were calibrated using experimental data from a simple load test on a steel section [6,58].

The cyclic behavior of structural steel with a yield plateau influences the constitutive model based on the principles of time-independent plasticity theory. This model employs a Von Mises yield surface in conjunction with an associative flow rule [58]. The flow rule is described by the following equations:

$$f = \left[\frac{3}{2}(\zeta - \eta) : (\zeta - \eta)\right]^{\frac{1}{2}} - \kappa \leq 0 \tag{3}$$

$$k = R + \rho \tag{4}$$

$$d\varepsilon^p = \xi \frac{\partial f}{\partial \sigma} = \frac{3}{2}\xi \frac{\zeta - \eta}{K} \tag{5}$$

$$dp = \left[\frac{2}{3} d\varepsilon^p : d\varepsilon^p\right]^{1/2} = \xi \tag{6}$$

In these equations, bold letters represent second-order tensors, $\boldsymbol{\varepsilon}^p$ represents the plastic strain tensor, $\boldsymbol{\sigma}$ represents the stress tensor, $\boldsymbol{\zeta}$ is the stress deviator, $\boldsymbol{\eta}$ is the back stress, ρ is the initial size of the yield surface, k is the size of the yield surface, R is the isotropic hardening/softening variable, ξ is the plastic multiplier, p is the accumulated plastic strain, and the double dot represents the inner product [6,58]. Although the stiffeners and diaphragms have different geometric properties, they share the same material properties as the column's component plate. The 2SM model is also applied to these components since they are integral to the column and affect its buckling behavior.

3.2. Loading Description

Strength and ductility analysis of bridge piers typically involves evaluating their performance under various loading conditions, including axial, monotonic, and cyclic loads. Axial loading applies a force along the axis of the bridge pier, simulating dead loads, live loads, and other service loads within the bridge [54]. Monotonic lateral loading involves the gradual application of force or displacement in a single lateral direction without reversal.

Cyclic lateral loading, which represents the horizontal component of seismic excitation, can be either force-controlled or displacement-controlled. This loading involves increasing the lateral load or displacement by integer multiples of the initial yield load or displacement in alternating directions. Experimentally, cyclic loading is commonly applied using a push-pull hydraulic jack [39]. In this study, a unidirectional displacement-controlled cyclic loading protocol is used, as illustrated in Figure 2d. The solid line represents one-cycle loading (N = 1), while the dashed line denotes three-cycle loading (N = 3). During the loading history, a quasi-static cyclic lateral displacement (δ) is applied to the top of the column, accompanied by a constant axial force (P), see Figure 2a. The amplitude of the cyclic displacement is incrementally increased in multiples of the yield displacement (δ_y), which is defined by Equation (7).

$$\delta_y = \frac{H_y h^3}{3EI} \tag{7}$$

$$H_y = \left(\sigma_y - \frac{P}{A}\right)\frac{s}{h} \tag{8}$$

Equation (8) is used to calculate the lateral yield load (H_y), where A, h, E, I, and S represent the cross-sectional area, height, Young's modulus, the moment of inertia of the cross-section, and the elastic section modulus of the column, respectively [6,44].

3.3. FE Meshing

In thin-walled, stiffened steel square box columns, excessive local deformation can inhibit stress redistribution. Test results indicate that local buckling typically occurs near the column base over a region approximately equal to the cross-sectional size (d) [3,44,46]. To address this, a two-node beam element (B31) is used for the upper segment of the column, with a height of $h - 3d$, while reduced integration four-node shell elements (S4R), which effectively capture localized deformation, are applied to the lower segment, with a height of $3d$, as shown in Figure 2b. The interface between the shell (S4R) and beam (B31) elements is modeled using a multi-point constraint (MPC). The elements used are available in the Abaqus/Standard library [52].

For computational efficiency and accuracy, the bottom and middle portions of the lower segment of the column (equal to the side width, d) are, respectively, divided into 30 and 20 S4R elements, while the remaining height (d) is divided into 15 elements, with an additional 20 elements used in the width direction for the lower segment, see Figure 2b. The upper segment of the column, with a height of $h - 3d$, is divided into beam elements (B31), each sized at 90 mm. Finally, each longitudinal stiffener and subpanel between the stiffeners, respectively, have 3 and 6 columns of S4R. These mesh sizes were optimized through trial and error to ensure efficient and accurate results. A default of five integration points through the thickness of the S4R element is used. The analysis employs a displacement convergence criterion with a tolerance of 10^{-5}. Initial geometric imperfections and residual stresses are not considered, as these were not quantified in the tested columns [3,43]. Additionally, these factors are believed to have a negligible influence on the overall cyclic behavior after the first half-cycle [6,31]. Further details on elastoplastic large displacement analyses are provided in [6].

4. Numerical Results

4.1. FE Model Validation for Column TPRC9-9

Finite element (FE) analysis was performed using both the multi-linear kinematic hardening (MKH) model and the 2SM model, and the results were compared with experimental data for the column TPRC9-9 by Fukumoto et al. [43]. Figure 3a,b compare the test and analysis results for the normalized lateral load versus lateral displacement hysteresis loops and envelope curve, respectively, for the column TPRC9-9. Deformation visualizations are shown in Figure 4a,b, where the buckled shape near the base displays a half-sine wave formation. The ABAQUS-2SM more accurately predicted the strength values compared to the ABAQUS-MKH in the post-buckling range. For thin-walled, stiffened square box columns, the maximum strength predictions were similar for both MKH and 2SM models, but the accuracy diverged beyond this point (in the post-buckling regime). This discrepancy is attributed to the 2SM model's superior ability to predict the cyclic elastoplastic behavior of structural steel [58] used in the analysis.

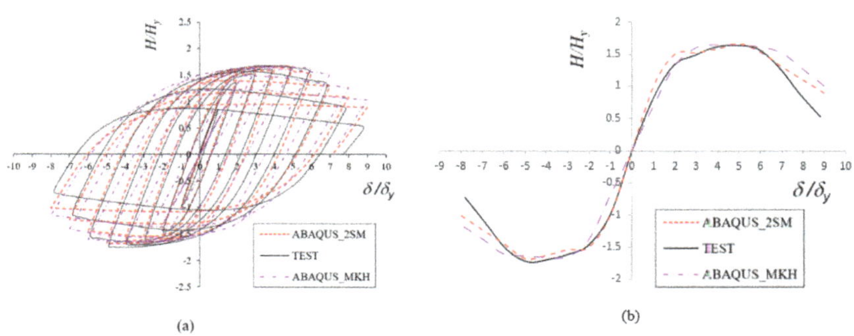

Figure 3. (**a**) Hysteresis curve for column TPRC9-9 (**b**) Envelope curve for column TPRC9-9.

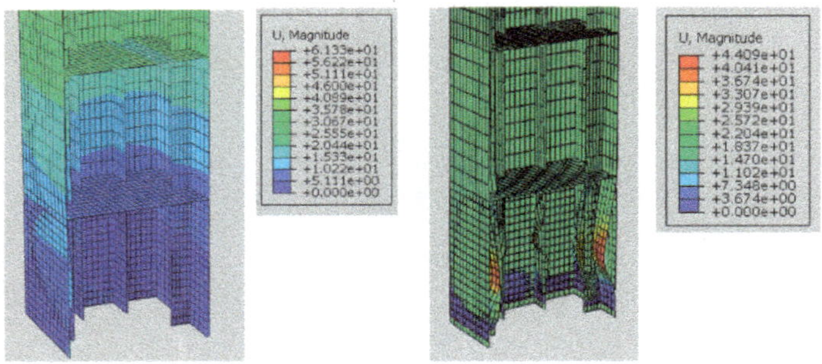

(a) (b)

Figure 4. Buckling of column TPRC9-9: (**a**) ABAQUS-MKH model (**b**) ABAQUS-2SM model.

4.2. Model Validation for Column B1

Column B1, as reported in the literature [3], was also used to validate the 2SM model against experimental results. The comparison of results from the ABAQUS-2SM model and the multi-linear kinematic hardening (ABAQUS-MKH) model is shown in Figure 5a,b. The ABAQUS-MKH model predicted higher strength values than both the experimental data and the ABAQUS-2SM model. However, the ABAQUS-2SM model provided strength values closer to the experimental results compared to the ABAQUS-MKH model. Figure 6a,b show the comparison of buckling shapes for the two material models. The lower portion of the stiffener exhibited a buckling shape resembling a half-sine wave, while the plates buckled alternately inwards and outwards.

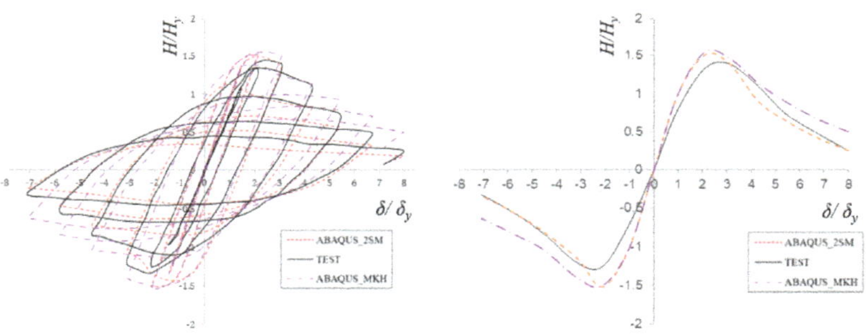

(a) (b)

Figure 5. (**a**) Hysteresis curve for column B1 (**b**) Envelope curve for column B1.

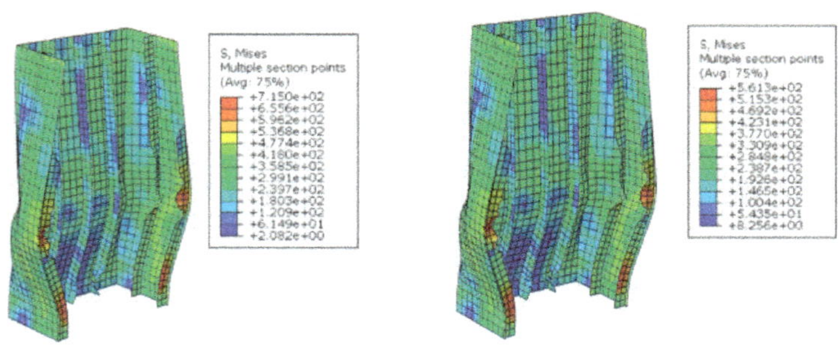

(a) (b)

Figure 6. Buckling of column B1 (**a**) ABAQUS_MKH (**b**) ABAQUS_2SM.

4.3. Comparison of Material Model Performance

To evaluate the performance of the 2SM model, further analysis was conducted using a column with compact section properties. The material and geometric properties of this compact section were similar to those of column B1, as detailed in Table 1, except that the plate thickness of the compact section is 25 mm instead of 9 mm. Figure 7a,b show that the MKH model predicted higher strength values for the compact section compared to the 2SM model. This observation suggests that the 2SM model yields more reliable results for FE analysis of compact sections. For the compact column under cyclic lateral loading, material nonlinearity predominates over geometric nonlinearity. Since the 2SM model accurately represents the material characteristics under cyclic loading, it provides a more reliable analysis.

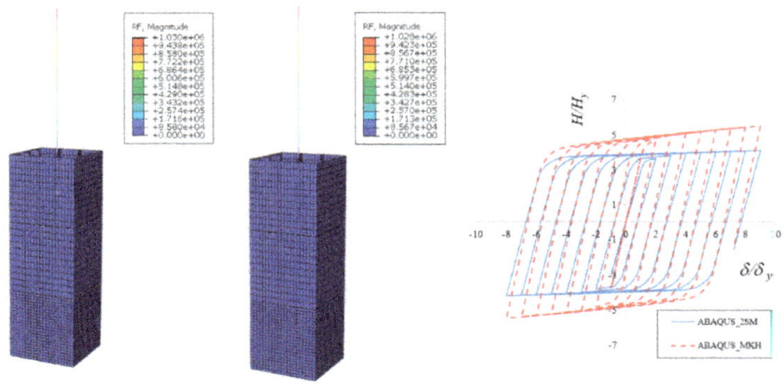

(a) (b) (c)

Figure 7. FE Analysis of a Column with a Compact Section: (**a**) Deformed Configuration from ABAQUS-MKH, (**b**) Deformed Configuration from ABAQUS-2SM, (**c**) Comparison of Hysteresis Loops between ABAQUS-MKH and ABAQUS-2SM.

5. Parametric Study

Using the ABAQUS-2SM, a parametric study was conducted on thin-walled, stiffened steel square box columns to investigate their overall interaction and the impact of key design parameters on their cyclic behavior. The parameters examined include the width-to-thickness ratio (R_f), the column slenderness ratio (λ), and the magnitude of axial load (P/P_y). Based on previous studies, the practical ranges for these parameters are: $0.2 \leq R_f \leq 0.5$, $0.2 \leq \lambda \leq 0.5$, and $P/P_y \leq 0.2$ [39,47].

In this study, the size of diaphragms and stiffeners were kept constant at a thickness of 9 mm, while the web and flange thicknesses were calculated for forty columns by varying R_f and λ values. Details of the column properties used in the parametric study are listed in Table 2. The finite element analysis (FEA) results are summarized in Table 3.

Table 2. Proposed columns for the parametric study.

Column Name	Axial Load	R_f	λ	t_f (mm)	Height (mm)
T-20-29-0.00	0.00	0.20	0.29	14.39	2269
T-30-29-0.00	0.00	0.30	0.29	9.00	
T-40-29-0.00	0.00	0.40	0.29	7.20	
T-50-29-0.00	0.00	0.50	0.29	6.00	
T-20-29-0.05	0.05	0.20	0.29	14.39	
T-30-29-0.05	0.05	0.30	0.29	9.00	
T-40-29-0.05	0.05	0.40	0.29	7.20	
T-50-29-0.05	0.05	0.50	0.29	6.00	

Table 2. *Cont.*

Column Name	Axial Load	R_f	λ	t_f (mm)	Height (mm)
T-20-29-0.10	0.10	0.20	0.29	14.39	2269
T-30-29-0.10	0.10	0.30	0.29	9.00	
T-40-29-0.10	0.10	0.40	0.29	7.20	
T-50-29-0.10	0.10	0.50	0.29	6.00	
T-20-29-0.15	0.15	0.20	0.29	14.39	
T-30-29-0.15	0.15	0.30	0.29	9.00	
T-40-29-0.15	0.15	0.40	0.29	7.20	
T-50-29-0.15	0.15	0.50	0.29	6.00	
T-20-29-0.20	0.20	0.20	0.29	14.39	
T-30-29-0.20	0.20	0.30	0.29	9.00	
T-40-29-0.20	0.20	0.40	0.29	7.20	
T-50-29-0.20	0.20	0.50	0.29	6.00	
T-30-29-0.00	0.00	0.30	0.20	9.00	1565
T-30-29-0.00	0.00	0.30	0.29	9.00	2269
T-30-29-0.00	0.00	0.30	0.50	9.00	3912
T-30-29-0.05	0.05	0.30	0.20	9.00	1565
T-30-29-0.05	0.05	0.30	0.29	9.00	2269
T-30-29-0.05	0.05	0.30	0.40	9.00	3130
T-30-29-0.05	0.05	0.30	0.50	9.00	3912
T-30-29-0.10	0.10	0.30	0.20	9.00	1565
T-30-29-0.10	0.10	0.30	0.29	9.00	2269
T-30-29-0.10	0.10	0.30	0.40	9.00	3130
T-30-29-0.10	0.10	0.30	0.50	9.00	3912
T-30-29-0.15	0.15	0.30	0.20	9.00	1565
T-30-29-0.15	0.15	0.30	0.29	9.00	2269
T-30-29-0.15	0.15	0.30	0.40	9.00	3130
T-30-29-0.15	0.15	0.30	0.50	9.00	3912
T-30-29-0.20	0.20	0.30	0.20	9.00	1565
T-30-29-0.20	0.20	0.30	0.29	9.00	2269
T-30-29-0.20	0.20	0.30	0.40	9.00	3130
T-30-29-0.20	0.20	0.30	0.50	9.00	3912

Table 3. Parametric study: strength and ductility evaluation.

Column	P/P_y	t_f (mm)	R_f	λ	$(1 + P/P_y).\ R_f\lambda$	$R_f\lambda$	H_m/H_y	δ_m/δ_y	$H_{0.95}/H_y$	$\delta_{0.95}/\delta_y$
T-20-29-0.00	0.00	14.39	0.20	0.29	0.058	0.058	3.00	8.00	2.85	9.00
T-30-29-0.00	0.00	9.00	0.30	0.29	0.087	0.087	2.00	8.00	1.90	8.00
T-40-29-0.00	0.00	7.20	0.40	0.29	0.116	0.116	1.48	5.00	1.41	6.00
T-50-29-0.00	0.00	6.00	0.50	0.29	0.145	0.145	1.23	4.00	1.17	4.00
T-20-29-0.05	0.05	14.39	0.20	0.29	0.061	0.058	2.89	7.00	2.75	8.97
T-30-29-0.05	0.05	9.00	0.30	0.29	0.091	0.087	1.84	7.00	1.75	7.94
T-40-29-0.05	0.05	7.20	0.40	0.29	0.122	0.116	1.40	4.00	1.33	6.00
T-50-29-0.05	0.05	6.00	0.50	0.29	0.152	0.145	1.16	3.00	1.10	4.00
T-20-29-0.10	0.10	14.39	0.20	0.29	0.064	0.058	2.89	7.00	2.75	8.44
T-30-29-0.10	0.10	9.00	0.30	0.29	0.096	0.087	1.81	6.00	1.72	7.34
T-40-29-0.10	0.10	7.20	0.40	0.29	0.128	0.116	1.37	4.00	1.30	5.61
T-50-29-0.10	0.10	6.00	0.50	0.29	0.160	0.145	1.10	3.00	1.04	3.82
T-20-29-0.15	0.15	14.39	0.20	0.29	0.067	0.058	2.770	6.00	2.632	7.775
T-30-29-0.15	0.15	9.00	0.30	0.29	0.100	0.087	1.756	6.00	1.668	6.870
T-40-29-0.15	0.15	7.20	0.40	0.29	0.133	0.116	1.329	4.00	1.263	5.444
T-50-29-0.15	0.15	6.00	0.50	0.29	0.167	0.145	1.067	3.00	1.014	3.800

Table 3. *Cont.*

Column	P/P_y	t_f (mm)	R_f	λ	$(1 + P/P_y). R_f\lambda$	$R_f\lambda$	H_m/H_y	δ_m/δ_y	$H_{0.95}/H_y$	$\delta_{0.95}/\delta_y$
T-20-29-0.20	0.20	14.39	0.20	0.29	0.070	0.058	2.570	6.00	2.442	7.500
T-30-29-0.20	0.20	9.00	0.30	0.29	0.104	0.087	1.620	6.00	1.539	6.600
T-40-29-0.20	0.20	7.20	0.40	0.29	0.139	0.116	1.270	3.00	1.207	5.000
T-50-29-0.20	0.20	6.00	0.50	0.29	0.174	0.145	0.960	2.00	0.912	3.500
T-30-29-0.00	0.00	9.00	0.30	0.20	0.060	0.060	2.500	3.00	2.375	4.260
T-30-29-0.00	0.00	9.00	0.30	0.29	0.087	0.087	2.000	8.00	1.900	8.000
T-30-29-0.00	0.00	9.00	0.30	0.40	0.120	0.120	1.175	6.00	1.116	8.148
T-30-29-0.00	0.00	9.00	0.30	0.50	0.150	0.150	0.920	6.00	0.874	8.800
T-30-29-0.05	0.05	9.00	0.30	0.20	0.063	0.060	2.450	3.00	2.328	3.940
T-30-29-0.05	0.05	9.00	0.30	0.29	0.091	0.087	1.844	7.00	1.752	7.944
T-30-29-0.05	0.05	9.00	0.30	0.40	0.126	0.120	1.130	5.00	1.074	5.800
T-30-29-0.05	0.05	9.00	0.30	0.50	0.158	0.150	0.900	6.00	0.855	8.000
T-30-29-0.10	0.10	9.00	0.30	0.20	0.066	0.060	2.400	3.00	2.280	3.420
T-30-29-0.10	0.10	9.00	0.30	0.29	0.096	0.087	1.805	6.60	1.715	7.342
T-30-29-0.10	0.10	9.00	0.30	0.40	0.132	0.120	1.090	4.00	1.036	6.300
T-30-29-0.10	0.10	9.00	0.30	0.50	0.165	0.150	0.870	5.00	0.827	8.000
T-30-29-0.15	0.15	9.00	0.30	0.20	0.069	0.060	2.350	2.00	2.233	3.280
T-30-29-0.15	0.15	9.00	0.30	0.29	0.100	0.087	1.756	6.00	1.668	6.870
T-30-29-0.15	0.15	9.00	0.30	0.40	0.138	0.120	1.080	4.00	1.026	5.630
T-30-29-0.15	0.15	9.00	0.30	0.50	0.173	0.150	0.830	5.00	0.789	7.160
T-30-29-0.20	0.20	9.00	0.30	0.20	0.072	0.060	2.260	2.00	2.147	3.150
T-30-29-0.20	0.20	9.00	0.30	0.29	0.104	0.087	1.620	6.00	1.539	6.600
T-30-29-0.20	0.20	9.00	0.30	0.40	0.144	0.120	1.040	4.00	0.988	5.267
T-30-29-0.20	0.20	9.00	0.30	0.50	0.180	0.150	0.800	4.00	0.760	6.000

5.1. Effect of Width-to-Thickness Ratio Parameter (R_f)

The influence of the R_f parameter on the hysteretic behavior of the analyzed columns was examined. As R_f increased, the thickness of the column decreased. Although the width of the column also affects this parameter, it was kept constant for this study. With $\lambda = 0.29$, both the ultimate strength and maximum displacement decreased by 59% and 50%, respectively, as R_f increased from 0.2 to 0.5, as illustrated in Figure 8a,b.

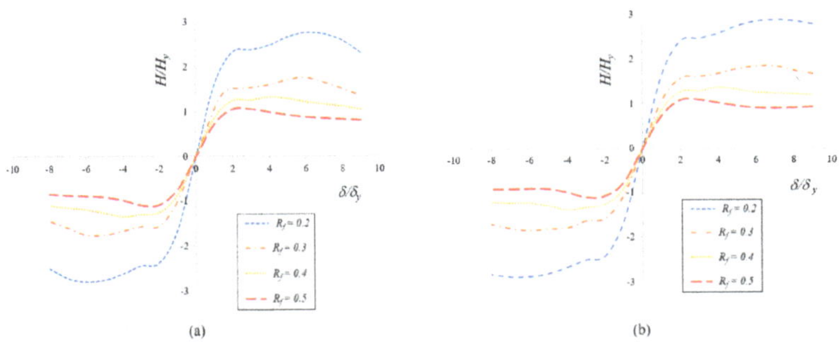

Figure 8. (**a**) Envelope curve ($\lambda = 0.29$, $P/P_y = 0.15$) (**b**) Envelope curve ($\lambda = 0.29$, $P/P_y = 0.05$).

5.2. Effect of Slenderness Ratio Parameter (λ)

The height of the column is directly proportional to the λ parameter. While the plate thickness of the column remained constant as the λ parameter varied from 0.2 to 0.5, with R_f and P/P_y held constant, the height changed from 1563 mm to 3907 mm due to this variation. As shown in Figure 9a,b, both ultimate strength and maximum displacement decreased by 54% and 25%, respectively, when the λ parameter increased from 0.2 to 0.5.

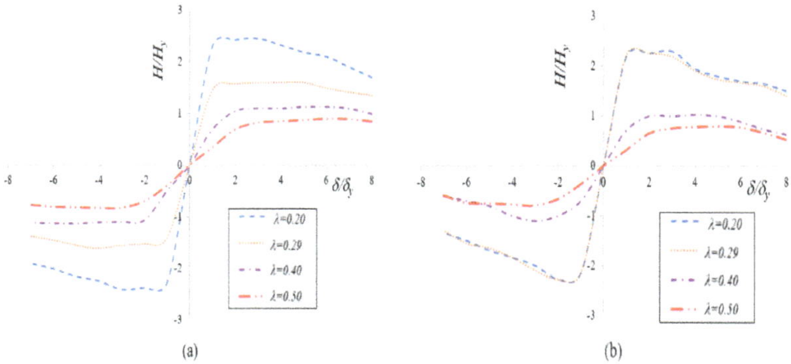

Figure 9. (**a**) Envelope curve (R_f = 0.30, P/P_y = 0.05) (**b**) Envelope curve (R_f = 0.30, P/P_y = 0.20).

5.3. Effect of P/Py

Envelope curves of the proposed columns were examined under varying axial loads ($0 \leq P/P_y \leq 0.2$). As shown in Figure 10, the ratio H_m/H_y decreased with increasing axial load. Although ultimate strength and maximum displacement were only slightly affected as the axial load increased from 0 to $0.2P_y$, lateral strength and lateral deformation were significantly reduced at higher axial loads (e.g., for $P/P_y \geq 0.30$).

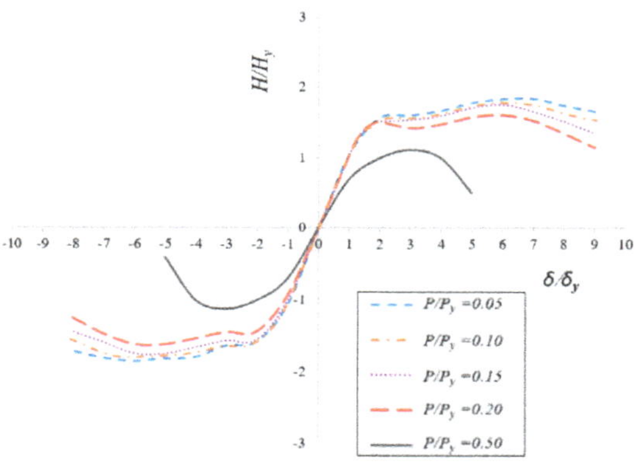

Figure 10. Effect of P/P_y on strength and ductility.

5.4. Post-Buckling Behavior of the Columns

The post-buckling behavior of the columns was analyzed, with Figure 11 illustrating the variation in maximum strength and deformation for each column. The analysis revealed that columns with lower values of R_f and P/P_y exhibited higher maximum lateral strength and lateral deformation. To accurately capture failure and assess critical strength and ductility, a failure criterion corresponding to 95% of the maximum strength was used to define the post-buckling behavior [39,44,46]. This threshold represents the failure point beyond which the column is considered damaged and unable to support additional load, requiring retrofitting or replacement. Post-buckling strength and deformation values decreased as R_f and P/P_y increased for all columns.

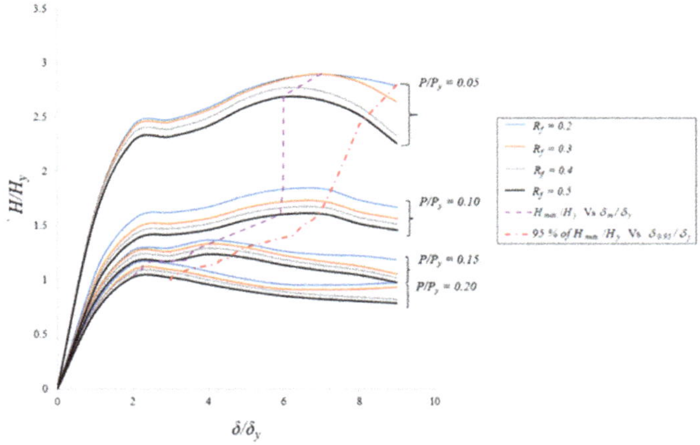

Figure 11. Post buckling strength and ductility of the columns.

6. Proposed Design Equations

The ultimate strengths and ductility of 40 analyzed columns were determined through finite element (FE) analysis using the 2SM for material nonlinearity in ABAQUS. Figure 12 plots H_m/H_y against the integrated parameters (1 + P/P_y)·R_f·λ, considering the interaction between R_f, λ, and P/P_y on column strength. The best-fitting equation relating to the computed ultimate strength is given by:

$$\frac{H_m}{H_y} = \frac{0.1402}{\left[\left(1+\frac{P}{P_y}\right)R_f\lambda\right]^{1.084}} \tag{9}$$

As shown in Figure 12, the ultimate strength improved as the integrated parameters (1 + P/P_y)·R_f·λ, decreased. For this study, failure of the thin-walled stiffened steel box columns was considered to occur at a displacement equal to either δ_m or $\delta_{0.95}$. Here, δ_m represents the maximum displacement corresponding to H_m/H_y, while $\delta_{0.95}$ is the displacement at which post-peak strength drops to 95% of H_m/H_y. These normalized parameters are key to evaluating the ductile behavior of the columns. The $\delta_{0.95}/\delta_y$ parameter was used to analyze ductility, as column strength significantly deteriorates after this point due to local buckling.

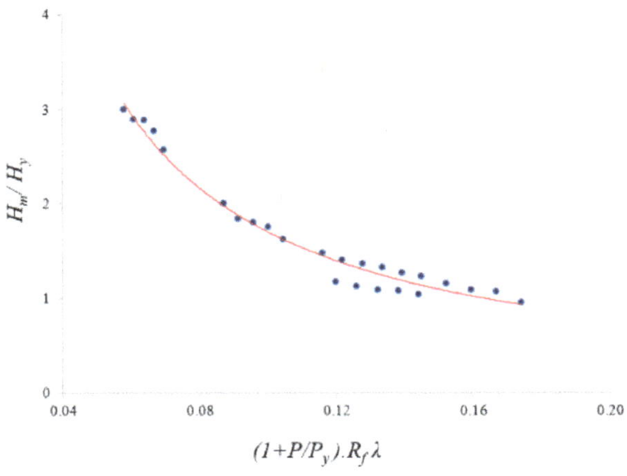

Figure 12. The ultimate strength of the analyzed columns.

Figure 13a,b illustrate a reduction in both maximum deformation and post-buckling deformation as $(1 + P/P_y)\cdot R_f\cdot\lambda$ and $R_f\cdot\lambda$ increased. Additionally, ductility equations were proposed:

$$\frac{\delta_m}{\delta_y} = \frac{0.5654}{\left(R_f\lambda\right)^{0.938}} \tag{10}$$

$$\frac{\delta_{0.95}}{\delta_y} = \frac{1.0877}{\left[\left(1+\frac{P}{P_y}\right)R_f\lambda\right]^{0.783}} \tag{11}$$

Equations (10) and (11) were derived from the design parameters to determine maximum deformation and post-buckling deformation, respectively, within the ranges $0.2 \le R_f \le 0.50$, $0.2 \le \lambda \le 0.50$, and $P/P_y \le 0.2$. These limits for parameters are in the practical range for bridge design [44,46].

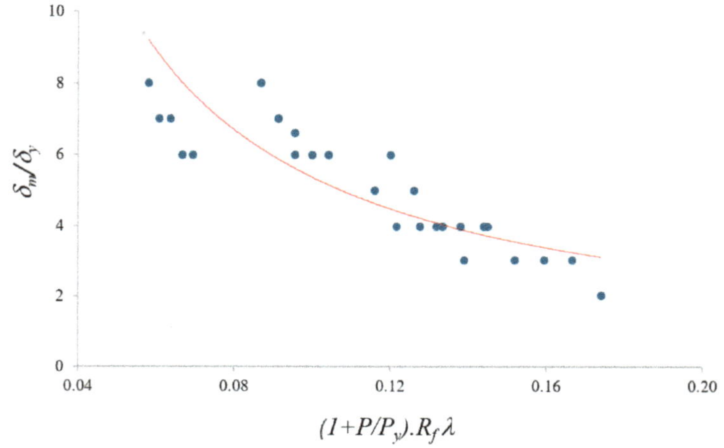

(a)

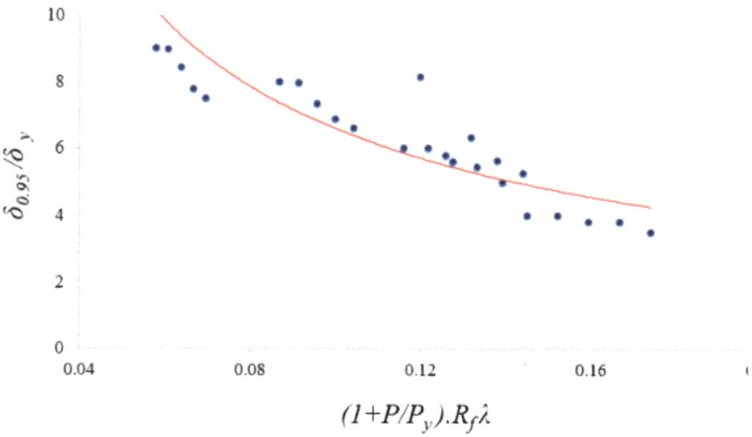

(b)

Figure 13. (**a**) Ductility of the analyzed columns: (**a**) (δ_m/δ_y) vs. $(1 + P/P_y)\cdot R_f\cdot\lambda$, and (**b**) $(\delta_{0.95}/\delta_y)$ vs. $(1 + P/P_y)\cdot R_f\cdot\lambda$.

Application of the Proposed Design Equations

To verify the proposed design equations, two columns were designed and analyzed using ABAQUS-2SM: Column-1 with $P/P_y = 0.15$, $R_f = 0.35$, and $\lambda = 0.25$; and Column-2 with $P/P_y = 0.18$, $R_f = 0.30$, and $\lambda = 0.30$. The verification results are shown in Table 4. Figure 14a,b present the FE analysis results for Columns 1 and 2, respectively. The strength was predicted with 92% accuracy, while maximum deflection and post-buckling deflections were predicted with 94% accuracy. This study thus establishes practical equations for engineers to use in designing and constructing thin-walled, stiffened square box columns for bridge piers.

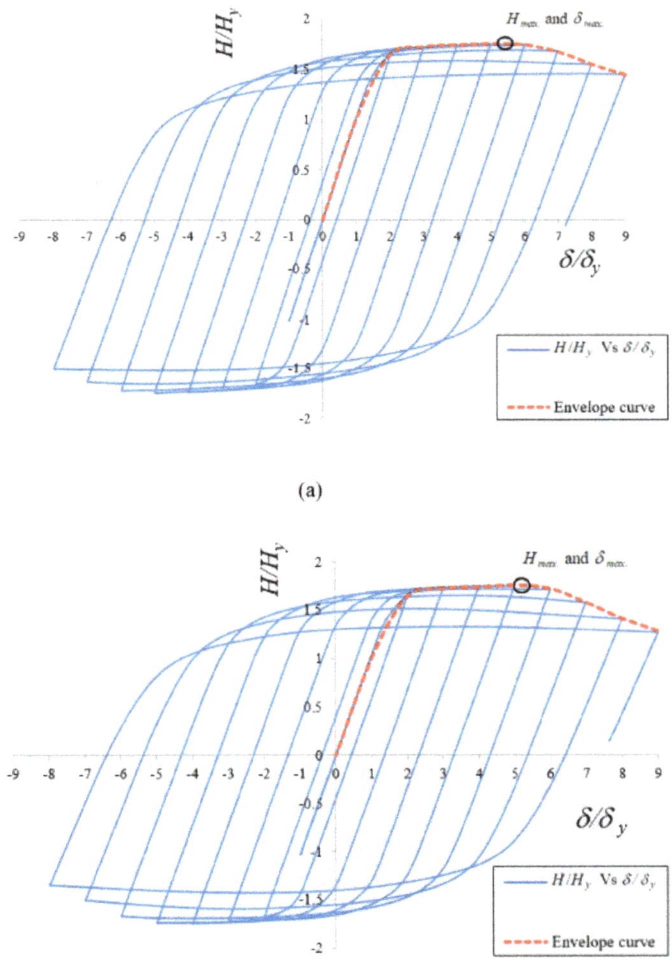

Figure 14. (**a**) FE results for Column 1 (δ_m/δ_y), (**b**) FE results for Column 2 (δ_m/δ_y).

Table 4. Results for design verification.

Column		P/P_y	R_f	λ	t (mm)	H (mm)	H_m/H_y	δ_m/δ_y	$\delta_{0.95}/\delta_y$
1	Proposed Equation	0.15	0.35	0.25	8.220	1956	1.690	5.550	6.567
	FE Analysis						1.745	5.500	7.000
	Accuracy (%)						97.00	100.00	94.00
2	Proposed Equation	0.18	0.3	0.3	9.590	2348	1.600	5.411	6.296
	FE Analysis						1.750	5.200	6.100
	Accuracy (%)						92.00	96.10	97.00

7. Conclusions

This study examined the cyclic behavior of prismatic thin-walled, stiffened steel square box columns, which model bridge piers, using finite element (FE) analysis with the commercial software ABAQUS, focusing on local, global, and local-global interactive buckling phenomena. The analysis employed the 2SM model to capture material nonlinearity under cyclic loading. The 2SM model offers a more accurate representation of the cyclic elastoplastic behavior of structural steel compared to existing material models in ABAQUS. It effectively incorporates cyclic effects within the yield plateau, accounting for their reduction and eventual disappearance, as well as the cyclic strain hardening of the material. Key design parameters investigated included the width-to-thickness ratio (R_f), the slenderness ratio (λ), and the axial load ratio (P/P_y). The study led to the following conclusions:

FE Analysis Accuracy: The FE analysis using ABAQUS and the 2SM model (implemented via a UMAT subroutine) accurately captured column behavior under cyclic loading, closely matching experimental results. The thin-walled columns showed minimal variation when the FE model included a multi-linear kinematic hardening model. Local buckling was observed in thin-walled steel tubes, with geometric attributes influencing cyclic behavior more than material non-linearities, especially for thin-walled sections susceptible to local buckling. Significant differences in strength and ductility were found in compact columns when using different material models, as the material nonlinearity governs the column behavior in the absence of local buckling.

Effect of Axial Loads: Axial loads up to 20% of the yield load (P_y) had a negligible effect on the ultimate strength of the columns. However, higher axial loads (exceeding 30% of P_y) significantly lowered buckling, post-buckling, and ductility of the impacted columns.

Parametric Study Results: The ultimate lateral strength of thin-walled, stiffened steel square box columns was achieved when lateral displacement reached $5\delta_y$. Beyond this point, strength decreased due to the local buckling of the component plates. Increasing both R_f and λ resulted in lower ultimate lateral strength and ductility.

Design Equations: New design equations were introduced to predict ultimate lateral strength, maximum lateral displacement, and post-buckling displacement. These equations offer a practical approach for designing thin-walled, stiffened steel square box columns for bridge piers.

Validation of Design Equations: To validate the proposed design equations, two columns were designed, and their ultimate lateral strength and displacement values were confirmed through FE analysis. The design equations achieved 95% accuracy in predicting ultimate lateral strength and 97% accuracy in predicting ultimate lateral displacements. The discrepancies between the FE predictions and experimental results were 5% for strength and 3% fortool ductility.

Author Contributions: Conceptualization, I.H.P.M.; Methodology, I.H.P.M.; Software, I.H.P.M.; Validation, M.N. and I.H.P.M.; Formal analysis, M.N. and I.H.P.M.; Investigation, I.H.P.M.; Resources, I.H.P.M.; Data curation, M.N. and I.H.P.M.; Writing—original draft, M.N. and I.H.P.M.; Writing—review & editing, I.H.P.M.; Visualization, M.N. and I.H.P.M.; Supervision, I.H.P.M.; Project administration, I.H.P.M.; Funding acquisition, I.H.P.M. All authors have read and agreed to the published version of the manuscript.

Funding: This research received no external funding.

Institutional Review Board Statement: Not applicable.

Informed Consent Statement: Not applicable.

Data Availability Statement: The data presented in this study are available on request from the corresponding author.

Conflicts of Interest: The authors declare no conflict of interest.

References

1. *EN 10219-1*; Cold-Formed Welded Structural Hollow Sections of Non-Alloy and Fine-Grain Steels. Technical Delivery Requirements; European Committee for Standardization (CEN): Brussels, Belgium, 2006.
2. Usami, T. *Interim Guidelines and New Technologies for Seismic Design of Steel Structures*; Committee on New Technology for Steel Structures, Japanese Society of Civil Engineers: Tokyo, Japan, 1996. (In Japanese)
3. Nishikawa, K.; Yamamoto, S.; Natori, T.; Terao, K.; Yasunami, H.; Terada, M. Retrofitting for Seismic Upgrading of Steel Bridge Columns. *Eng. Struct.* **1998**, *20*, 540–551. [CrossRef]
4. Kawashima, K.; Hasegawa, K.; MacRae, G.A.; Ikeuchi, T. Ductility of Steel Bridge Piers from Dynamic Loading Tests. In *Stability of Steel Structures under Cyclic Loading*; Fukumoto, T., Lee, G.C., Eds.; CRC Press: Boca Raton, FL, USA, 1992.
5. MacRae, G.A.; Kawashima, K. Estimation of the Deformation Capacity of Steel Bridge Piers. In *Stability of Steel Structures under Cyclic Loading*; Fukumoto, T., Lee, G.C., Eds.; CRC Press: Boca Raton, FL, USA, 1992.
6. Mamaghani, I.H.P. Cyclic Elastoplastic Behavior of Steel Structures: Theory and Experiments. Ph.D. Dissertation, Nagoya University, Nagoya, Japan, 1996.
7. Meng, X.; Whitfield, S.; Boudjabeur, S.; Gardner, L. Stability of Hot-finished and Cold-formed Steel Hollow Section Columns: A Comparative Study. *Structures* **2024**, *68*, 107192. [CrossRef]
8. Rondal, J.; Maquoi, R. Stub-Column Strength of Thin-walled Square and Rectangular Hollow Sections. *Thin-Walled Struct.* **1985**, *3*, 15–34. [CrossRef]
9. Meng, X.; Gardner, L. Cross-sectional Behaviour of Cold-formed High Strength Steel Circular Hollow Sections. *Thin-Walled Struct.* **2020**, *156*, 106822. [CrossRef]
10. Yang, L.; Yin, F.; Wang, J.; Bilal, A.; Ahmed, A.H.; Lin, M. Local Buckling Resistances of Cold-formed High-Strength Steel SHS and RHS with Varying Corner Radius. *Thin-Walled Struct.* **2022**, *172*, 108909. [CrossRef]
11. Wang, J.; Afshan, S.; Schillo, N.; Theofanous, M.; Feldmann, M.; Gardner, L. Material Properties and Compressive Local Buckling Response of High Strength Steel Square and Rectangular Hollow Sections. *Eng. Struct.* **2017**, *130*, 297–315. [CrossRef]
12. Ma, J.L.; Chan, T.M.; Young, B. Experimental Investigation on Stub-column Behavior of Cold-formed High-Strength Steel Tubular Sections. *J. Struct. Eng.* **2016**, *142*, 04015174. [CrossRef]
13. Wang, J.; Gardner, L. Flexural Buckling of Hot-finished High-strength Steel SHS and RHS Columns. *J. Struct. Eng.* **2017**, *143*, 04017028. [CrossRef]
14. Zhao, O.; Gardner, L.; Young, B. Testing and Numerical Modelling of Austenitic Stainless Steel CHS Beam-Columns. *Eng. Struct.* **2016**, *111*, 263–274. [CrossRef]
15. Goto, Y.; Koyama, R.; Fujll, Y.; Obata, M. Ultimate State of Thin-Walled Stiffened Rectangular Steel Columns under Bi-Directional Seismic Excitations. *Doboku Gakkai Ronbunshuu A* **2009**, *65*, 61–80. [CrossRef]
16. Meng, X.; Gardner, L. Flexural Buckling of Normal and High Strength Steel CHS Columns. *Structures* **2021**, *34*, 4364–4375. [CrossRef]
17. Meng, X.; Gardner, L. Behavior and Design of Normal-and High-strength Steel SHS and RHS Columns. *J. Struct. Eng.* **2020**, *146*, 04020227. [CrossRef]
18. Huang, Y.; Young, B. Structural Performance of Cold-formed Lean Duplex Stainless Steel Columns. *Thin-Walled Struct.* **2014**, *83*, 59–69. [CrossRef]
19. Pournara, A.E.; Karamanos, S.A.; Mecozzi, E.; Lucci, A. Structural Resistance of High-strength Steel CHS Members. *J. Constr. Steel Res.* **2017**, *128*, 152–165. [CrossRef]
20. Zhao, O.; Gardner, L.; Young, B. Behaviour and Design of Stainless Steel SHS and RHS Beam-columns. *Thin-Walled Struct.* **2016**, *106*, 330–345. [CrossRef]
21. Poursadrollah, A.; D'Aniello, M.; Landolfo, R. Experimental and Numerical Tests of Cold-formed Square and Rectangular Hollow Columns. *Eng. Struct.* **2022**, *273*, 115095. [CrossRef]
22. Yuan, H.X.; Wang, Y.Q.; Gardner, L.; Shi, Y.J. Local-overall Interactive Buckling of Welded Stainless-Steel Box Section Compression Members. *Eng. Struct.* **2014**, *67*, 62–76. [CrossRef]
23. Yang, L.; Shi, G.; Zhao, M.; Zhou, W. Research on interactive buckling behavior of welded steel box-section columns. *Thin-Walled Struct.* **2017**, *115*, 34–47. [CrossRef]

24. Schillo, N.; Feldmann, M. Interaction of local and global buckling of box sections made of high strength steel. *Thin-Walled Struct.* **2018**, *128*, 126–140. [CrossRef]
25. Fukumoto, Y. Cyclic Performance Assessment of Stiffened Box Columns with Thickness Tapered Plates. In Proceedings of the 2004 Annual Stability Conference, SSRC, Long Beach, CA, USA, 24–27 March 2004; SSRC 2004 Beedle Award Paper. pp. 1–18.
26. Key, P.W.; Hasan, S.W.; Hancock, G.J. Column Behavior of Cold-formed Hollow Sections. *J. Struct. Eng.* **1988**, *114*, 390–407. [CrossRef]
27. Zhou, F.; Huang, L.; Li, H.T. Cold-Formed Stainless Steel SHS, and RHS Columns Subjected to Local-flexural Interactive Buckling. *J. Constr. Steel Res.* **2022**, *188*, 106999. [CrossRef]
28. Yin, F.; Wang, J.; Yang, L.; Kang, N.; Elliott, F. Testing and Design of Cold-formed High Strength Steel Square Hollow Section Columns. *J. Constr. Steel Res.* **2024**, *213*, 108394. [CrossRef]
29. Zhang, X.Z.; Liu, S.; Zhao, M.S.; Chiew, S.P. Comparative Experimental Study of Hot-formed, Hot-finished, and Cold-formed Rectangular Hollow Sections. *Case Stud. Struct. Eng.* **2016**, *6*, 115–129. [CrossRef]
30. Japanese Road Association (JRA). *Specification for Highway Bridges*; Part V; Japanese Road Association (JRA): Tokyo, Japan, 1996. (In Japanese)
31. Mamaghani, I.H.P. Seismic Design and Ductility Evaluation of Thin-Walled Steel Bridge Piers of Box Sections. *Transp. Res. Rec. J. Transp. Res. Board* **2008**, *2050*, 137–142. [CrossRef]
32. Bruneau, M. Performance of steel bridges during the 1995 Hyogoken–Nanbu (Kobe, Japan) Earthquake—A North American Perspective. *Eng. Struct.* **1998**, *20*, 1063–1078. [CrossRef]
33. Gao, S.B.; Ge, H.B. Numerical Simulation of Hollow and Concrete-filled Steel Columns. *Adv. Steel Constr.* **2007**, *3*, 668–678.
34. Li, H.; Gao, X.; Liu, Y.; Luo, Y. Seismic Performance of New-type Box Steel Bridge Piers with Embedded Energy-Dissipating Shell Plates under Tri-directional Seismic Coupling Action. *Int. J. Steel Struct.* **2017**, *17*, 105–125. [CrossRef]
35. Gao, S.B.; Usami, T.; Ge, H.B. Ductility Evaluation of Steel Bridge Piers with Pipe Sections. *J. Eng. Mech.* **1998**, *124*, 260–267. [CrossRef]
36. Chiew, S.P.; Lee, S.L.; Shanmugam, N.E. Experimental Study of Thin-walled Steel Box Columns. *J. Struct. Eng.* **1987**, *113*, 2208–2220. [CrossRef]
37. Dewolf, J.T.; Pekoz, T.; Winter, G. Local and Overall Buckling of Cold-formed Members. *J. Struct. Eng.* **1974**, *100*, 2017–2036. [CrossRef]
38. Kitada, T.; Nakai, H.; Matsumura, M.; Kagayama, T. Experimental Study on Seismic Retrofitting Method of Stiffened Plates in Existing Steel Bridge Piers under Cyclic Loading. *J. Struct. Eng.* **2000**, *46*, 127–134. (In Japanese)
39. Al-Kaseasbeh, Q.; Mamaghani, I.H.P. Buckling Strength and Ductility Evaluation of Thin-Walled Steel Tubular Columns with Uniform and Graded Thickness under Cyclic Loading. *J. Bridge Eng.* **2019**, *24*, 04018105. [CrossRef]
40. Mamaghani, I.H.P.; Njiru, M. Buckling Strength of Cold-Formed Stiffened Thin-Walled Steel Box Bridge Piers under Cyclic Loading. *Ce/Papers* **2023**, *6*, 1291–1294. [CrossRef]
41. Mamaghani, I.H.P.; Njiru, M. Strength and Ductility Evaluation of Thin-walled Steel Tubular Bridge Piers under Cyclic Loading. In Proceedings of the International Structural Engineering and Construction, Chicago, IL, USA, 14–18 August 2023; Volume 10. [CrossRef]
42. Fukumoto, Y. *Structural Stability Design—Steel and Composite Structures*; Pergamon Press: Oxford, UK, 1997.
43. Fukumoto, Y.; Uenoya, M.; Nakamura, M.; Saya, H. Cyclic Performance of Stiffened Square Box Columns with Thickness Tapered Plates. *Steel Struct.* **2003**, *3*, 107–115.
44. Zheng, Y.; Usami, T.; Ge, H. Ductility of Thin-Walled Steel Box Stub-Columns. *J. Struct. Eng.* **2000**, *126*, 1304–1311. [CrossRef]
45. Tran, K.L.; Douthe, C.; Sab, K.; Dallot, J.; Davaine, L. Buckling of Stiffened Curved Panels under Uniform Axial Compression. *J. Constr. Steel Res.* **2014**, *103*, 140–147. [CrossRef]
46. Usami, T.; Gao, S.; Ge, H. Elastoplastic Analysis of Steel Members and Frames Subjected to Cyclic Loading. *Eng. Struct.* **2000**, *22*, 135–145. [CrossRef]
47. Lyu, F.; Goto, Y.; Kawanishi, N.; Xu, Y. Three-dimensional Numerical Model for Seismic Analysis of Bridge Systems with Multiple Thin-Walled Partially Concrete-Filled Steel Tubular Columns. *J. Struct. Eng.* **2020**, *146*, 04019164. [CrossRef]
48. Ucak, A.; Tsopelas, P. Accurate Modeling of the Cyclic Response of Structural Components Constructed of Steel with Yield Plateau. *Eng. Struct.* **2012**, *35*, 272–280. [CrossRef]
49. Nigel Priestley, M.J.; Kowalsky, M.J.; Calvi, G.M. Displacement-Based Seismic Design of Bridges. In *Bridge Engineering Handbook*, 2nd ed.; Chen, W.F., Duan, L., Eds.; CRC Press: Boca Raton, FL, USA, 2014; pp. 217–252. [CrossRef]
50. Njiru, M.; Mamaghani, I.H.P. Inelastic Behavior of Thin-walled Circular Steel Tubular Columns under Pushover and Cyclic Loading. In Proceedings of the 7th International Conference on Civil, Structural and Transportation Engineering, Niagara Falls, ON, Canada, 5–7 June 2022. [CrossRef]
51. Chatzopoulou, G.; Karamanos, S.A. Numerical Implementation of a Bounding-surface Model for Simulating the Cyclic Inelastic Response of Metal Piping Components. *Finite Elem. Anal. Des.* **2021**, *185*, 103493. [CrossRef]
52. Smith, M. *ABAQUS/Standard User's Manual, Version 6.9*; Dassault Systèmes: Providence, RI, USA, 2009; Available online: https://www.3ds.com/products-services/simulia/products/abaqus/ (accessed on 20 August 2024).
53. Zienkiewicz, O.C. *The Finite Element Method*, 3rd ed.; McGraw-Hill: New York, NY, USA, 1977.
54. Kassimali, A.; Prashant, A. *Structural Analysis*, 4th ed.; SI ed./Prepared by Amit Prashant; Cengage Learning: Stamford, CT, USA, 2011.

55. Wang, W.; Zhang, C.; Li, Z.; Gan, X. A Study on the Application of Two Different Material Constitutive Models Used in the FE Simulation of the Cyclic Plastic Behavior of a Steel Beam-Column T-Stub Connection. *Adv. Civ. Eng.* **2021**, *2021*, 8868015. [CrossRef]
56. Kim, D.K.; Dargush, G.F.; Shin, S.W.; Hu, J.W. A Two-surface Plasticity Model for the Simulation of Uniaxial Ratchetting Response. *J. Mech. Sci. Technol.* **2012**, *26*, 145–152. [CrossRef]
57. Shen, C.; Tanaka, Y.; Mizuno, E.; Usami, T. A two-surface Model for Steels with Yield Plateau. *Struct. Eng.* **1992**, *8*, 179s–188s. [CrossRef]
58. Shen, C.; Mamaghani, I.H.P.; Mizuno, E.; Usami, T. Cyclic Behavior of Structural Steels, II: Theory. *J. Eng. Mech.* **1995**, *121*, 1165–1172. [CrossRef]
59. Banno, S.; Mamaghani, I.H.P.; Usami, T.; Mizuno, E. Cyclic Elastoplastic Large Deflection Analysis of Thin Steel Plates. *J. Eng. Mech.* **1998**, *124*, 363–370. [CrossRef]

applied sciences

MDPI

Article

Efficient Design Optimization of Cable-Stayed Bridges: A Two-Layer Framework with Surrogate-Model-Assisted Prediction of Optimum Cable Forces

Yuan Ma [1], Chaolin Song [1], Zhipeng Wang [1,2], Zuqian Jiang [1], Bin Sun [1,*] and Rucheng Xiao [1]

1 Department of Bridge Engineering, Tongji University, Shanghai 200092, China; 2011156@tongji.edu.cn (Y.M.); songcl@tongji.edu.cn (C.S.); w_zhipeng@163.com (Z.W.); zuqianjiang@tongji.edu.cn (Z.J.); xiaorc@tongji.edu.cn (R.X.)

2 CCCC Highway Consultants Co., Ltd., Beijing 100010, China

* Correspondence: sunbin@tongji.edu.cn

Abstract: Cable-stayed bridges have commonly been built for crossing large-span obstacles, such as rivers, valleys, and existing structures. Obtaining an optimum design for a cable-stayed bridge is challenging, due to the large number of design variables and design constraints that are typically nonlinear and usually conflict with each other. Therefore, it is a reasonable alternative to turn the large and complex optimization problem into two sub-problems, i.e., optimizing the internal force distribution by adjusting the cable prestressing forces, and optimizing the other sizing or geometrical parameters. However, conventional methods are lacking in efficiency when dealing with the problem of optimization of cable forces in the first sub-problem, under the circumstance that iteration between the two sub-problems is required. To address this, this paper presents a surrogate-model-assisted method to construct a cable forces predictor ahead of the structural optimization process, so that cable forces can be effectively predicted rather than optimized in each iterative round. Additionally, B-spline interpolation curve is adopted for variable condensation when sampling for the surrogate model. Finally, the structure optimization in the second sub-problem is performed by leveraging an optimization program based on particle swarm optimization method. The performance of the proposed framework is tested with a practical engineering application. Results show that the proposed method showcases good efficiency and accuracy. The theoretical raw material consumption of the towers and the cables is 32% lower than the original design.

Keywords: cable-stayed bridges; optimum design; surrogate model; cable forces predictor; particle swarm optimization

1. Introduction

Cable-stayed bridges have been widely constructed to span roads and rivers. As a highly redundant structure, cable-stayed bridges have advantages in terms of the stiffness, wind-load resistance, maintenance, and span-crossing ability. As a result, cable-stayed bridges have exceeded a span of 1 km in just 60 years since the first cable-stayed bridge was built. The Chang-Tai Yangtze River Bridge in China, with a total length of 5.3 km and a main span of 1176 m, is the world's longest cable-stayed bridge for both highway and railway. The construction of such a huge engineering structure is extremely challenging and costly. Therefore, it is becoming increasingly critical to carry out detailed structural optimization and comparison at the design stage.

For a practical engineering project, the implementation of bridge design optimization can face a couple of challenges. First, the involved design parameters for optimization of a cable-stayed bridge are typically high-dimensional. As a highly redundant structure, there are numerous design parameters to be optimized in the bridge design process. Variables involved can be classified into mechanical, sizing, geometrical, and topological [1]. The optimization problem becomes highly dimensional due to the increasing number of variables, which hinders efficiency and consumes more time. The interdependency among design variables exacerbates nonconvexity and nonlinearity, making the problem difficult and challenging. In addition, various constraints need to be considered in the optimization design of cable-stayed bridges. Due to the complexity of real-world environments and bridge-operating conditions, engineers should consider the varieties of site, structure, material, member, section type, load cases, and the complex verification items of specification. These considerations, whether linear or nonlinear, make the feasible domain very limited within the design space. Moreover, the mechanical behavior of cable-stayed bridges must be properly simulated. Cable-stayed bridges transmit loads with cables. They are relatively flexible structures compared to girder and arch bridges. Therefore, the geometric nonlinearity caused by sag effect of cables, large displacement effect, and p-Δ effect, should be considered. Though current commercial software can handle these non-linearity effects by finite element analysis, the simulation can be time-consuming and undermine the feasibility of optimization.

To address these challenges, a lot of research has been conducted to advance the computational methods or strategies for the design optimization of cable-stayed bridges. Early studies conducted by Feder [2] introduced an optimality-criteria-based method to determine the prestressing forces of cables in steel bridges. Similar early studies involved plenty of assumptions and addressed the problem using simplified mathematical formulas. Sung et al. [3] minimized the total strain energy expressed as a quadratic function of the post-tensioning cable forces with an influence matrix. Baldomir et al. [4] optimized cable areas for a long-span steel bridge with the finite differences sensitivity analysis method and solved the problem through a gradient-based sequential quadratic programming algorithm. A three-stage algorithm was presented by Ha et al. [5] to optimize the cable prestressing tensions with a nonlinear inelastic analysis. Besides the above work on optimizing single type of variables, research has also been carried out to pursue a more comprehensive "optimum design" of the bridge. The structural design problem is formulated with various variables including not only mechanical but also sizing, geometrical, and topological variables. Lute et al. [6] proposed an optimization method for cable-stayed bridges that utilized a genetic algorithm to minimize costs while considering geometrical parameters and cross-sectional dimensions as design variables. A support vector machine was utilized for constraint verification, and the presented method was proven to be accurate and computationally efficient for prediction purposes. Gao et al. [7] obtained the optimum design of prestressed concrete bridges. Design variables included the number of prestressing tendons, cable forces, cable areas, and girders' and towers' sectional dimensions. Cid et al. [8] examined multi-span cable-stayed bridges while considering geometric nonlinearity effects. The variables included anchorage positions, cable forces, and cable section areas. The SQP algorithm was utilized to minimize the total cost of steel, and sensitivity analysis was conducted using the finite difference method.

However, for the implementation of optimizing a practical cable-stayed bridge, handling all types of variables simultaneously is typically not the most effective strategy, because this formula can significantly lead to the increase in the dimensionality, nonconvexity, and nonlinearity of the problem. Moreover, it should be noted that optimizing the distribution of internal forces is not intrinsically contradictory to the optimization of the cost. Taking advantage of this feature, the optimization can be separated into a hierarchical layout, i.e., sequentially optimizing the mechanics-related variables and the other ones. Following this idea, [9] is one of the earliest works that introduced surrogate functions that reveal the potential connection between mechanical variables and other sizing and geometrical variables, therefore decoupling the cable forces optimization from the structural design optimization. To develop the functions, polynomial regression with the ordinary least square method was adopted. The necessary data for the regression were collected from a large parametric study conducted by repeating the finite element technique, while varying three parameters. The surrogate functions, expressed as quadratic polynomials, explicitly related cable forces to three variables concerning the span length, the total length, and the upper structure height. With these functions, the cable forces can be easily determined for different variable values to achieve the optimum post-tensioning distribution, which can minimize the deflection in both the deck and pylon. In later research [10], these functions are used to facilitate the optimum design of a cable-stayed bridge, considering variables such as the cables' section areas, prestressing forces, and cross-sectional dimensions of the girder.

Although the post-tensioning functions were assessed as accurate in the previous investigation [10], there are still drawbacks when it comes to optimum design problems with more complex design conditions. Firstly, the variables included in the functions are limited in terms of the number and the types, which may lead to inaccurate predictions for the bridges if some of the other design conditions are changed. Secondly, the variables of the surrogate functions are supposed to be consistent with those of the optimum design problem. Therefore, there is a demand for reconstructing the surrogate function if some new variables that may have a visible influence on the post-tensioning distribution are introduced. Thirdly, as the number of design variables increases, the nonlinearity of the problem significantly grows. The polynomial regression will be insufficient to give accurate predictions, due to its poor anti-interference and local fitting ability, compared with other regression methods.

To this end, machine learning techniques have recently garnered significant attention. Researchers have explored various tools, including Random Forest [11], Support Vector Machines [12], and Physics-informed or Data-driven Neural Networks [13], to predict structural performance or responses. In a broad sense, as these tools are strategically designed as a cheaper-to-evaluate substitute for the original sophisticated computational model, they are also typically called surrogate models or meta-models [14]. The conceptual illustration of the construction and the utilization of a surrogate model is shown in Figure 1. Multidisciplinary applications have been conducted based on surrogate models to replace the original time-consuming processes or high-cost experiments. These methods have been successfully applied in a variety of research fields, such as the hydro-environment [15], rock and soil mechanics [16], and bridge engineering [17].

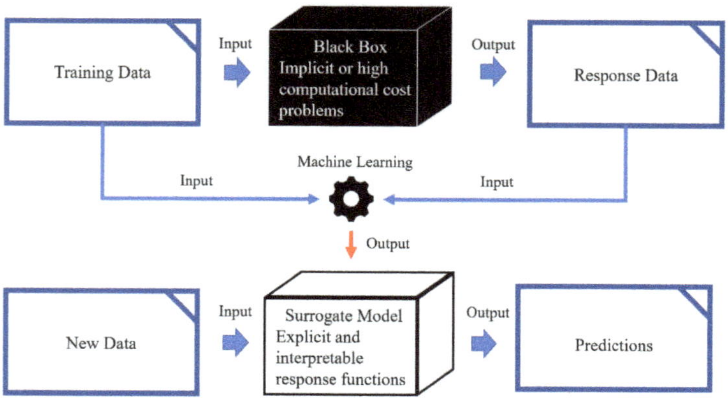

Figure 1. Construction and utilization of a surrogate model.

Motivated by the above advancements, the surrogate model from machine learning is used in this research to replace the surrogate functions and is expected to exhibit greater adaptability to a large number and diverse types of variables. Four regression models are studied in this work: Polynomial Regression (PR), Gaussian Process Regression (GPR), Regression Tree (RT), and Support Vector Regression (SVR). To construct the surrogate model, samples are collected by means of a full factorial experiment, concerning the modest number of target sample count and acceptable computational cost. To test the surrogate model, the predicted results and optimized results of a design point distinguished from the samples are compared. Thereafter, the surrogate model is combined with heuristic algorithms to demonstrate its efficiency in the optimum design. By introducing such a predictor, the force variables can be decoupled from other sizing and geometrical variables in the optimization routine, thus enabling a substantial reduction in the problem complexity and boosting the efficiency of optimization. Additionally, a practical application of a cable-stayed bridge project with a main span of 818 m has been used to validate the performance of the proposed framework.

The rest of the paper is organized as follows: Section 2 displays our improved formula of the two sub-problems. Section 3 presents the proposed methods and their integration in the overall procedure. Section 4 shows the implementation of the methods on a (358 + 818 + 358)-meter-long cable-stayed bridge example. Section 5 displays the optimization results and complementary checks. Conclusions are drawn in Section 6.

2. Problem Formula

In a general cable-stayed bridge design, engineers need to carry out comprehensive design and optimization from the structure configuration to the detail members. In terms of the configuration, crucial parameters including the length of the main and the side span, the width of the deck, the height of the towers, and the anchorage position of the cables should be well determined. In terms of the members, cross-sectional dimensions of the main members including the towers, the cables, and the deck are supposed to be well designed. The typical optimum design formula with all design variables optimized altogether can be stated as

$$\begin{aligned} &\min_{\mathbf{X},\mathbf{Q}} \Phi = \sum_{j=1}^{m} C_j V_j \\ &s.t. x_{il} \leqslant x_i \leqslant x_{iu} \text{ for all } i \in N_X \\ &q_{kl} \leqslant q_k \leqslant q_{ku} \text{ for all } k \in N_Q \\ &\sigma_{jl} \leqslant \sigma_j^{(p)} \leqslant \sigma_{ju} \text{ for all } j \in M, \text{ all } p \in L \\ &f_j^{(p)} \leqslant f_{ju} \text{ for all } j \in M, \text{ all } p \in L \\ &K_j^{(p)} \leqslant K_{ju} \text{ for all } j \in M, \text{ all } p \in L \end{aligned} \tag{1}$$

where Φ represents the total cost of members, $\mathbf{X} = \{x_1, x_2, x_3, \ldots, x_{n_X}\}$ stands for sizing (e.g., cross-sectional dimensions) and geometrical (e.g., side-span ratio, height-span ratio) design variables, and $\mathbf{Q} = \left\{q_1, q_2, q_3, \ldots, q_{n_Q}\right\}$ stands for mechanical (e.g., cable prestressing forces, tendon prestressing forces in reinforced concrete beams) design variables. C_k and V_k represent the cost coefficient and the volume of the j-th member, respectively. $M = \{1, 2, 3, \ldots, m\}$ is the member number set. The next two constraints represent the lower and upper boundaries of the i-th variable in $\mathbf{X}$ and the k-th variable in $\mathbf{Q}$, where $N_X = \{1, 2, 3, \ldots, n_X\}$ and $N_Q = \{1, 2, 3, \ldots, n_Q\}$. The following three constraints represent, respectively, the strength, stiffness, and stability constraints of the structure. The first constraint of these represents the upper bound of the j-th member's stress response, where p is the load case identifier and $L = \{1, 2, 3, \ldots, l\}$ is the load case number set. The second represents the boundary of the deflection response, and the third represents the boundary of the stability coefficient.

It is worth noting that wind or seismic resistance is also critical, especially in the design of long-span bridges; however, due to the limit of the paper length, such dynamic performance is not discussed in this paper. Therefore, the girder sizes are excluded from $\mathbf{X}$ and not optimized in the later example. Meanwhile, considering that cable prestressing forces are more dominant than tendon prestressing forces in preliminary design, $\mathbf{Q}$ consists only of cable forces in the later example.

In this optimization, **X** and **Q** are variables optimized together, where n_q is generally much larger than n_X due to the high density of cables. However, an optimization process dealing with **X** and **Q** simultaneously is highly complex and computationally costly, due to their nonlinear and conflicting (i.e., coupling) feature [1]. Thus, in this work, a reasonable two-layer framework is adopted to decouple them into two sub-problems in one optimization round.

Sub-problem one (Sp1) is an internal force distribution optimization problem where **Q** is optimized with fixed $\hat{\mathbf{X}}$, as most of the "cable forces optimization" problems [18,19]. The most common and practical objective function of Sp1 is the weighted sum of members' strain energy. The formula of Sp1 can be stated as

$$\begin{array}{ll} \min\limits_{\mathbf{Q}} U = & \min\limits_{\mathbf{Q}} \int_S \frac{M^2(S,\mathbf{Q})}{2EI} dS \\ q_{kl} \leqslant q_k & \leqslant q_{ku} \text{ for all } k \in N_Q \\ f^{(deadload)} & \leqslant f_u \quad \text{for all } p \in L \end{array} \tag{2}$$

where U represents the total bending strain energy of members, usually including both the main girder and the towers, E is Young's modulus, and I is the bending moment of inertia. The first constraint is lower and upper bounds set to ensure that the cable force magnitude does not turn negative or exceed its design strength. In the second constraint, $f^{(deadload)}$ is the deflection under the dead load case. It is set to ensure that the configuration of the bridge meets the design upon its completion.

Sub-problem two (Sp2) is a sizing and geometry optimization problem where **X** is optimized with $\hat{\mathbf{Q}}$ determined in Sp1. The formula of Sp2 can be stated as

$$\begin{aligned} &\min_{\mathbf{X}} \Phi = \sum_{j=1}^{m} C_j V_j \\ &s.t. x_{il} \leqslant x_i \leqslant x_{iu} \text{ for all } i \in N_X \\ &\sigma_{jl} \leqslant \sigma_j^{(p)} \leqslant \sigma_{ju} \text{ for all } j \in M, \text{ all } p \in L \\ &f_j^{(p)} \leqslant f_{ju} \text{ for all } j \in M, \text{ all } p \in L \\ &K_j^{(p)} \leqslant K_{ju} \text{ for all } j \in M, \text{ all } p \in L \end{aligned} \tag{3}$$

Conceptual differences are shown in Figure 2.

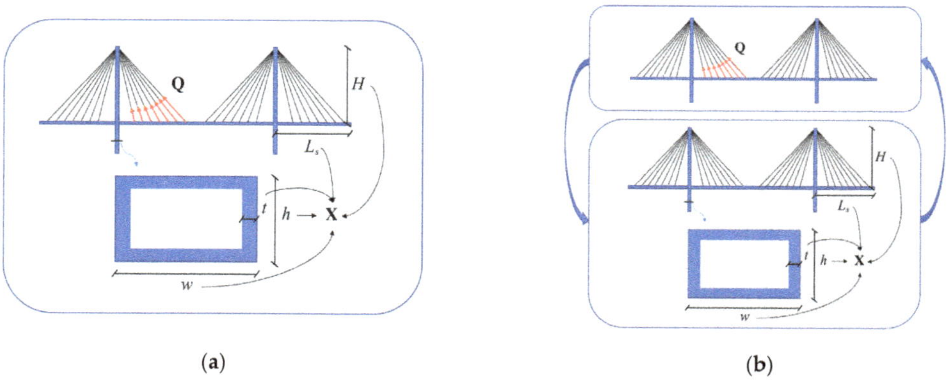

Figure 2. Illustration of conceptual differences. (**a**) is the illustration of Formula (1), presenting the united optimization problem, with **X** and **Q** optimized together; (**b**) is illustration of Formulas (2) and (3), presenting the decoupled optimization problem. The box above means the sub-problem 1 optimizing **Q** with fixed $\hat{\mathbf{X}}$. The box below means the sub-problem 2 optimizing **X** with fixed $\hat{\mathbf{Q}}$. The two sub-problems are processed iteratively.

Formula (1) has a drawback in dealing with multiple interdependent variables together, whereas the routine methods dealing with Formula (2) are overly time-consuming for the proposed iterative approach. Thus, this paper focuses on how to solve Sp1 in Formula (2) with greater efficiency and to make the proposed methods more adaptable to different and complex circumstances. Rather than directly optimizing the cable forces in each iteration round, this research seeks to solve Sp1 with a surrogate model from machine learning, by predetermining **Q** with a cable forces predictor trained in advance of the optimization.

3. The Proposed Method

3.1. Cable Forces Optimization

In the beginning of this section, the method for determining the optimum cable forces, i.e., Formula (2), is discussed, which builds a foundation for establishing the training database of the force predictor. It should be noted that a rational distribution of cable forces in cable-stayed bridges is generally smoothly distributed without sudden changes. Therefore, B-spline curve presented by French engineer Pierre Bézier, is handy for fitting the cable force distribution in cable-stayed bridges [20–22]. Compared to the precise optimization of each original cable force, introducing B-spline greatly condenses design variables into a few control points, while causing only a limited impact on the overall structural behavior.

Suppose a p-th degree B-spline curve is introduced, the original dense cable prestressing forces $\mathbf{Q} = \{q_1, q_2, q_3, \ldots\ldots q_{n_q}\}$ can be replaced by its control points $\mathbf{P} = \{p_1, p_2, p_3, \ldots\ldots p_{n_p}\}$, where n_p is far less than n_q. The relation between **Q** and **P** can be described as

$$q_k = \sum_{i=0}^{n_p} N_{i,p}(u_k) p_i \text{ for all } k \in \{1, 2, 3, \ldots\ldots, n_q\} \tag{4}$$

where u_k is the relative position of the k-th cable ($0 \leqslant u_k \leqslant 1$), and $N_{i,p}$ is the shape function. More detail can be found in [22].

In this work, assisted by the B-spline technique, the following design variables are considered: four fitting points (d_1–d_4) in the side span, four fitting points (d_5–d_8) in the main span, the force of the No. 1 cable (d_9), the range of the counterweight (d_{10}), and the load magnitude (d_{11}), as shown in Figure 3. The counterweight and its distribution range are set as variables as a complement to deal with the unbalanced self-weight between the main span and the side.

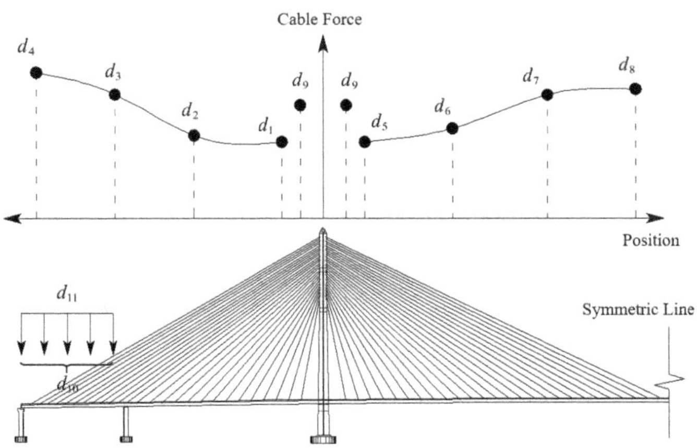

Figure 3. Cable forces optimization with B-spline interpolation curve.

Since Formula (2) is a typical nonlinear optimization problem and the number of design variables is condensed with B-spline curve, it can be efficiently solved with several optimization methods, such as the differential evolution and surrogate-model-assisted differential evolution [17]. The sequential quadratic programming (SQP) solver [23] is adopted in this research due to its high efficiency (note that the initial point should be carefully selected to avoid local optimum solutions).

During the sampling to construct the predictor, Formula (2) is solved in collaboration with MATLAB R2018b and ANSYS 2022 R1 APDL. For each sample, modeling parameters are derived in the beginning. An initial FEM is built with these modeling parameters in ANSYS. When the cable forces and counterweight are updated during the cable forces optimization in MATLAB, the initial strains of the cables are modified for the FEM and the counterweight is imposed in ANSYS. Then, the static analysis is launched to obtain the bending energy. If convergence is checked, the result of the sample is output to a text file. The sample process is completed when it reaches the target sample count.

3.2. Surrogate Model Assisted Predictor

Because of the high computational cost to solve the cable forces optimization problem, it is appropriate to introduce the surrogate model method to predict the cable forces of the desired completion state. As shown in Figure 4, the following steps are developed in this work to train and select a proper surrogate model.

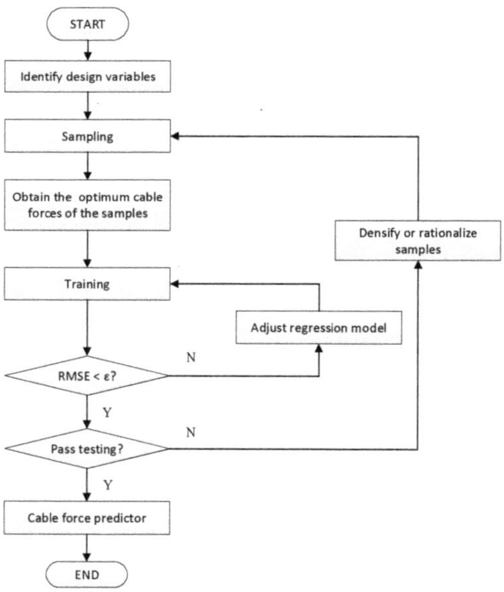

Figure 4. Steps to construct a cable-force predictor with surrogate model method.

1. **Sampling**: Generate training data (or samples) from the design of experiment (DOE) [24]. In this work, the samples are collected by means of a classic full factorial experiment. As for the cable-force predictor, the input data are the identified design variables, and the output data are the optimum cable forces taking the form of controlling points mentioned in Section 3.1. Considering that the cable forces optimization problem is well solved with the minimum bending energy method and the proper algorithm, there should be no anomalous sample.
2. **Training**: Configure the output function based on the machine learning algorithm. The property of the chosen surrogate model can be evaluated with root mean square error (RMSE) and training time. The surrogate model with the least RMSE and an

acceptable training time is the best solution. In this work, the following models are considered and compared: Polynomial Regression (PR) [25], Gaussian Process Regression (GPR) [26], Regression Tree (RT) [27], and Support Vector Regression (SVR) [28].

3. **Testing**: Create new testing data, conduct the experiment and obtain the actual responses. Expose the new data to the output function and get the predictions. Compare the actual responses and the predictions, then evaluate the accuracy of the surrogate model. If the test fails, it means that the discretization of the variables is either not dense enough or not rational; so, the design of the experiment should be adjusted.

With the predictor constructed, the optimized cable forces in each iteration round can be predetermined in subsequent Sp2. The introduction of the surrogate model significantly speeds up the iteration process, thus enabling efficient design optimization later.

3.3. Modification Strategies

To simplify some parts of the overall optimization procedure, two modification strategies are further adopted to improve the accuracy of simulations without significantly reducing efficiency.

3.3.1. Elastic Modulus of the Cables

The first modification is made on the elastic modulus of the cables. Among the three typical expressions of geometric nonlinearity, namely, sag effect, p-Δ effect, and large geometric deformation [29], sag effect has the most obvious impact on the optimized cable forces under dead load. Instead of performing the time-consuming nonlinear analysis, the Ernst formula modifying the elastic modulus of cable is adopted.

$$E_{eq} = \frac{E}{1 + \frac{EA}{12T^3}(qH)^2} \tag{5}$$

where E_{eq} is the effective elastic modulus of the cable, E is the material elastic modulus, A is the section area, T is the cable prestressing force, q is the cable weight per meter, and H is the horizontal projection length of the cable.

3.3.2. Section Area of the Cables

The second modification is made on the section area of the cables. Instead of identifying areas for each cable as a variable, 3 to 5 adjacent cables are grouped and share the same section area. However, there are still too many area variables that hinder the optimization efficiency. Therefore, an iterative method is adopted in this paper to estimate each cable area with a certain designated safety factor and calculate the cable force changes caused by the modification.

The steps are as follows:

1. Calculate objective cable stress $\sigma_{obj} = \sigma_s/\eta$, where σ_s is the design strength of cable material and η is the designated safety factor ($\eta = 3$ in this paper).
2. Designate initial section area A_{i0} for each cable. Initialize $\Delta A_i^{(k)} = 0$.
3. Get the cable force determined with the cable forces predictor q_i.
4. Calculate the cable stress $\sigma_{id}^{(k)}$ under dead load according to $\sigma_{id}^{(k)} = q_i/\left(A_{i0} + \Delta A_i^{(k)}\right)$.
5. Impose load cases, launch analysis and combine the results. Get the cable stress response $\sigma_{il}^{(k)}$ under live load.
6. Calculate $\Delta A_i^{(k+1)}$ with: $\left(A_{i0} + \Delta A_i^{(k+1)}\right) \cdot \sigma_{obj} = A_{i0} \cdot \left(\sigma_{id}^{(k)} + \sigma_{il}^{(k)}\right) + \Delta A_i^{(k+1)} \cdot l_i \cdot \gamma$, where l_i is the length of the cable and γ is the volumetric weight of the cable material.
7. Designate $A_{i0} + \Delta A_i^{(k+1)}$ for each cable area.

8. Check convergence of $\Delta A_i^{(k+1)}$. If convergence is achieved, perform the last load case analysis; otherwise, return to step 4 and update $\Delta A_i^{(k)}$ with $\Delta A_i^{(k+1)}$.

The cable areas and stresses are supposed to converge after only a few iterations (two times in our later example). This method benefits the optimization by cutting off the design variables of cable section areas, thus reducing the complexity of the optimization problem.

It is worth pointing out that the presented method causes a slight disturbance to the prediction results due to the redistribution effect of stiffness change. This disturbance is closely connected with the initial section area A_{i0}. The closer A_{i0} is to its convergent value $A_{i0} + \Delta A_i^{(k)}$, the less the disturbance and the more desirable the corresponding completion state will be.

3.4. The Proposed Optimization Framework

The optimum design of cable-stayed bridges concerning varieties of variables is typically a nonconvex problem. Gradient algorithms can often be trapped in local optima. Therefore, it is advisable to use heuristic algorithms with global search strategies to obtain the global optimum. Among the commonly used heuristic algorithms, Particle Swarm Algorithm (PSO) stands out due to its fast convergence, few parameters, and easy-to-implement formula. It is effective for high-dimensional optimization problems and converges quickly to the optimum solution. Thus, PSO is chosen as the optimization method to solve Formula (3) in Sp2. Based on the above considerations, the PSO-integrated overall procedure, presented in Figure 5, is as follows.

1. **Problem Definition.** Basic information of the optimization problem is identified, including variables, objective, and constraints. In this work, the problem is defined as follows:
    - Objective function—the total theoretical material cost of the towers and the cables;
    - Design variables—several parameters determining the volume or the weight of the towers and the cables;
    - Constraints—variable boundaries to achieve practical design and strength, stiffness and stability verifications;
    - Load cases—dead load, vehicle load, wind load, as well as their combinations according to Chinese Design Code.
2. **Predictor Construction.** The cable forces predictor is constructed with the surrogate model method. Detailed construction steps can be found in Section 3.2.
3. **Design Optimization.** An optimization program is made based on PSO algorithm. Readers can refer to Appendix A for the basic theory of PSO and Appendix B for the detailed steps of the program. The program consists of three modules:
    - Initialization module—particles are generated randomly within the search area. Those that pass the constraint verification remain in the swarm. The Pbest, the Gbest, and their fitness are initialized.
    - Iteration module—velocity and position of each particle are updated in each iteration round. Given that the objective function (cost) can be calculated explicitly and directly by design variables, constraint verification is only performed for particles whose cost is lower than the previously lowest. The improvement leads to a reduction in verification times. If the updated particle passes the constraint verification, the Pbest, the Gbest, and their fitness are updated.
    - Structural analysis module—constraints relating to design code verifications are verified by calling the structural analysis module. In the analysis module, a finite element model is established, the desired completion state is obtained utilizing the cable forces predictor, the responses under various load combinations are analyzed, and finally, the constraints are verified according to the code.

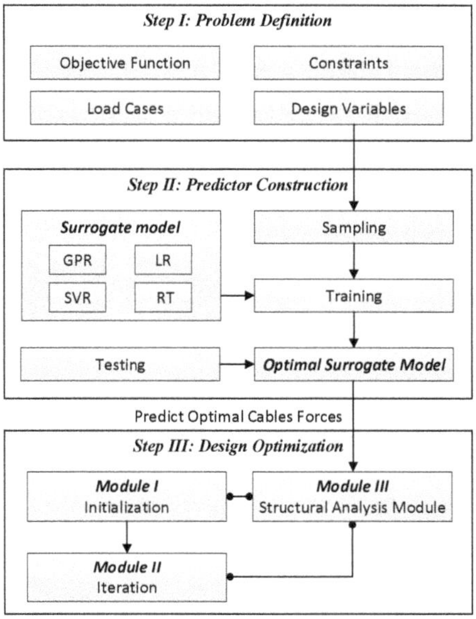

Figure 5. The improved overall procedure in this paper.

4. Implementation

4.1. Description

A (358 + 818 + 358)-meter-long cable-stayed bridge was chosen as the example to implement our formula and methods, as shown in Figure 6. It was a five-span cable-stayed bridge with two towers. The geometry of the vehicle lanes was planarly straight and vertically circular (with a radius of 29,000 m). The towers were both in a typical H-shape. As for the main girder, it was 38.9 m in width and 3.6 m in height.

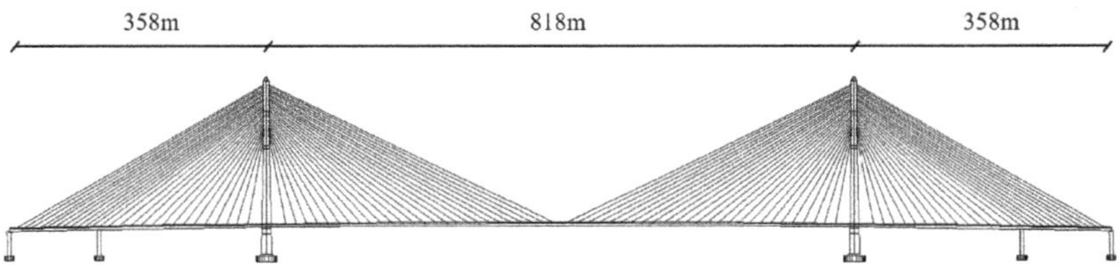

Figure 6. The geometry of the cable-stayed bridge example.

4.2. Objective Function

The material cost constitutes a large proportion of the total cost of a construction project. The theoretical material cost of bridge towers and cables was taken as the objective function in this paper. The cost of the beam was excluded, because its cross-section should be designed with priority to satisfy the wind resistance performance. The objective function is stated as

$$\Phi = \Phi_t + \Phi_c = \sum\nolimits_{i=1}^{N_t} C_t L_{ti} A_{ti} + \sum\nolimits_{j=1}^{N_c} C_c L_{cj} A_{cj} \gamma_c \tag{6}$$

where C is the unit steel weight or concrete volume cost, L is the length of the member, A is the section area of the member, γ is the volumetric weight of material, and N is the number of elements.

It is worth noting that the theoretical cost only considers the raw material consumption, and ignores the other costs including rebars, detailing, labor, and machinery.

4.3. Design Variables

Parameters directly determining the volume or the weight were supposed to be chosen as design variables, because the optimization objective was to reduce the total cost of towers and cables. In the implementation, six variables were chosen for the example in Table 1. These variables covered different variable formats and types and were a suitable example for the optimum design problem.

Table 1. Design variables.

Illustration	Variable	Format	Type	Symbol
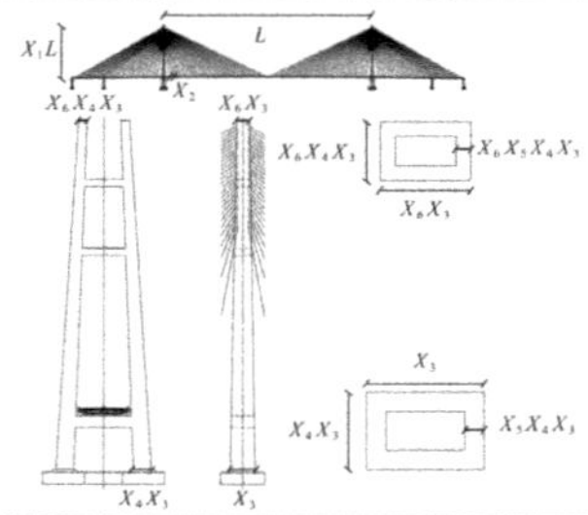	Ratio of tower height to span	Continuous	Geometrical	X_1
	Cable anchorage distance on girder	Discrete	Topology	X_2
	Longitudinal length at tower bottom	Continuous	Sizing	X_3
	Ratio of cross-sectional traverse width to longitudinal length	Continuous	Sizing	X_4
	Ratio of cross-sectional thickness to traverse width	Continuous	Sizing	X_5
	Cross-sectional size reduction ratio of tower top to tower bottom	Continuous	Sizing	X_6

After identifying the design variables, other dependent parameters essential for forming the parametric model were obtained according to the derivations in Table 2.

Table 2. Determination of dependent design parameters.

Illustration

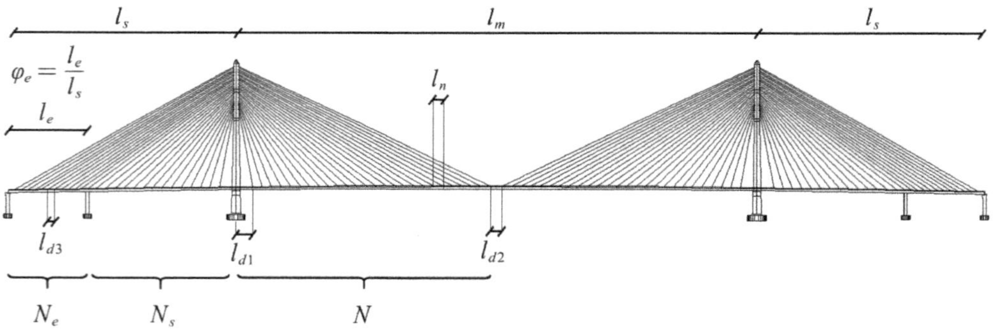

Parameter	Type	Symbol	Derivation
Length of the side span	Set	l_s	$= 358$ m
Length of the main span	Set	l_m	$= 818$ m
Distance of standard anchorage	Variable	l_n	$= X_2$ in Table 1
Ratio of densification length	Set	φ_e	$= \frac{l_e}{l_s} = 0.65$
Half count of mid-span cables	Dependent	N	$= \text{round}\left(\frac{l_m}{2 \cdot l_n}\right) - 1$
Count of symmetric cables	Dependent	N_s	$= \text{round}\left(\frac{l_s \cdot (1-\varphi_e)}{l_n}\right)$
Count of densified cables	Dependent	N_e	$= N - N_s$
Length of none-cable area near the towers	Dependent	l_{d1}	$= \left(\frac{l_m - l_{d2}}{2l_n} - \text{round}\left(\frac{l_m - l_{d2}}{2l_n}\right) + 1\right) \cdot l_n$
Length of none-cable area in the middle	Dependent	l_{d2}	$= l_n$
Distance of densified anchorage	Dependent	l_{d3}	$= \frac{l_s \cdot \varphi_e}{N_e}$

4.4. Load Cases

To cover all the required verifications in the Chinese Design Code, corresponding load and their combination cases were included in the example.

1. Dead Load (DL). The dead load consisted of the self-weight, the secondary loads, the cable tension forces, and the counterweight in the side span.
2. Vehicle Load (VL). The vehicle loads consisted of uniformly distributed loads (q, 40.2 kN/m) and concentrated loads (P, 1377.1 kN). As simplified static loads, four static load cases with different load layouts were adopted, as shown in Figure 7.

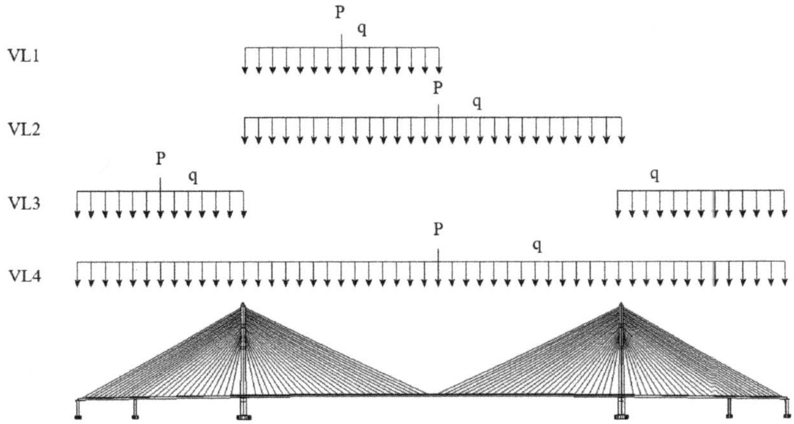

Figure 7. Illustration of vehicle loads. P = 1377.1 kN, and q = 40.2 kN/m.

3. Wind Load (WL). The wind speed for the bridge site was 30.1 m/s. Two wind loads with different return periods were imposed as the equivalent static gust wind loads on the main girder, the towers, and the cables, according to the Chinese Design Code. The return period of the design wind speed of WL1 was 10 years, and that of WL2 was 100 years.
4. Load Combination (LCB). A total of 11 different combinations of the dead load, the vehicle load, and the wind load were considered according to the Chinese Design Code.

4.5. Constraints

Constraints regarding the value ranges of the design variables could be selected based on engineering experience. They were necessary to make sure the results did not deviate too much from practical engineering sizes. Note that the ranges here also defined the boundaries of the samples when constructing the predictor. The consistency ensures the accuracy of the prediction results for cable forces. Meanwhile, constraints regarding strength, stiffness, and stability were defined in accordance with the Chinese Design Code. These constraints chosen to be applied here are the most important items in the preliminary design of cable-stayed bridges. They are the most typical representatives of the strength, stiffness, and stability of the bridge. The constraints are shown in Table 3.

Table 3. Constraints and boundaries.

Name	Type	Load Cases	Boundary
X_1	Variable	All	0.20~0.26
X_2	Variable	All	12 m~18 m
X_3	Variable	All	13 m~14 m
X_4	Variable	All	0.65~0.79
X_5	Variable	All	0.15~0.29
X_6	Variable	All	0.5~0.9

Table 3. *Cont.*

Name	Type	Load Cases	Boundary
Concrete Design Strength of the towers	Strength	LCB1~LCB5	−22.4 MPa~1.83 MPa
Steel Design Strength of the main girder	Strength	LCB1~LCB5	−270 MPa~270 MPa
Safety Stress Factor of the cables	Strength	LCB1~LCB5	≥2.5
Vertical Deflection of the main girder	Stiffness	VL1~VL4	≤2.045 m
Elastic Buckling Stability Factor	Stability	LCB6~LCB11	≥4

4.6. Cable Forces Predictor

4.6.1. Sampling

A modest number of 972 samples were generated from uniformly discretizing the design variables listed in Table 1, as shown in Table 4.

Table 4. A total of 972 samples generated from design variables.

Variable	Boundary	Discretization	Detail
X_1	0.20~0.26	4	0.20, 0.22, 0.24, 0.26
X_2	12 m~18 m	3	12 m, 15 m, 18 m
X_3	13 m~14 m	3	13 m, 13.5 m, 14 m
X_4	0.65~0.79	3	0.65, 0.72, 0.79
X_5	0.15~0.29	3	0.15, 0.22, 0.29
X_6	0.5~0.9	3	0.5, 0.7, 0.9

Constraints of the cable forces optimization problem mentioned in Section 3.1 included the following: 1000 kN to 10,000 kN cable force boundary for all the cable forces derived from variables d_1 to d_9; 0 to half side span length for variable d_{10}; 0 to the load magnitude when the main girder was filled with steel grit concrete for variable d_{11}. After defining the optimization problem, the fmincon function was called in MATLAB to solve it.

4.6.2. Training

The data (i.e., 972 samples including their optimized cable force and counterweight results) were imported into MATLAB and trained with the Statistics and Machine Learning Toolbox. The RMSE and training time of variable d_1 are shown in Table 5. Results show that Gaussian Process Regression should be the most suitable training model for our problem, because the RMSE is the least among the four concerned models. Though the training time of GPR is the longest, it was still acceptable in terms of absolute timespan. Polynomial Regression was inadequate in terms of accuracy to deal with optimum design problems, which further highlights the motivation of this work. Regression Tree was both relatively accurate in prediction results and efficient in training time. The fast training advantage of RT is rooted in its efficient binary tree data hierarchy, but it did not show up much because of the modest number of training data. Support Vector Regression was also inaccurate in predicting the results, demonstrated by its relatively large RMSE.

Table 5. Training model comparison.

Training Model	RMSE	Training Time (s)
PR	1964.4	5.8
RT	253.6	**2.9**
SVR	761.7	3.4
GPR	**168.7**	11.6

4.6.3. Testing

Selecting a testing design distinguished from our samples (X_1 = 0.23, X_2 = 15 m, X_3 = 13.2 m, X_4 = 0.69, X_5 = 0.25, X_6 = 0.85), a test was performed by comparing its predicted results to individually optimized results. Results in Table 6 and Figure 8 show that the predictor was accurate enough to determine the optimized cable forces and counterweight of our example. The predicted bending moment was close to the optimized one over the entire length. The maximum error of the moment did not exceed 5%, and the relatively large errors were located at the positions where the moment was small in terms of absolute value.

Table 6. Prediction error of design variables for the testing design.

Design Variables	Predicted Result	Individually Optimized Result	Error (%)
d_1	2,690,799.02 N	2,693,537.33 N	−0.102
d_2	4,233,299.06 N	4,235,208.77 N	−0.045
d_3	6,032,463.85 N	6,032,898.59 N	−0.007
d_4	7,723,535.68 N	7,725,499.78 N	−0.025
d_5	2,033,779.01 N	2,033,014.80 N	0.038
d_6	3,360,536.53 N	3,359,055.61 N	0.044
d_7	5,259,640.37 N	5,259,081.00 N	0.011
d_8	6,625,880.87 N	6,624,414.47 N	0.022
d_9	3,557,614.41 N	3,557,905.04 N	−0.008
d_{10}	102.91 m	102.91 m	−0.036
d_{11}	162,867.39 N/m	162,925.93 N/m	0.005

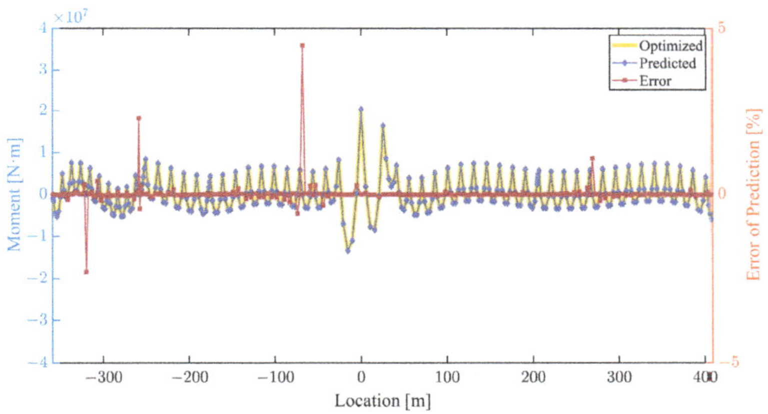

Figure 8. Predicted moments of the testing design utilizing cable forces predictor and the error compared with the optimized moments.

4.7. Optimization Settings

Problem-defining parameter settings for PSO (as shown in Table 7) were referenced from Cao [30], who dealt with the cable-supported bridge optimization problem with PSO as well.

Table 7. Problem-defining parameters settings for PSO algorithm.

Parameter	Symbol	Value
Population	N_p	30
Iteration Count	$T_{\max}$	500
Learning Factor	c	$c_1 = 2.0$, $c_2 = 2.0$
Inertia Weight	ω	$\omega_{\max} = 0.9$, $\omega_{\min} = 0.4$

The results recorded the Gbest (refer to Appendix A) during the 500 iteration rounds. The values of variables were stable after around 320 iterations. Meanwhile, there were a total of 20 price decreases, as shown in Figure 9. After seven times of decrease, the theoretical material cost nearly stopped decreasing. The results indicate the fast convergence speed of PSO algorithm in optimum design problems.

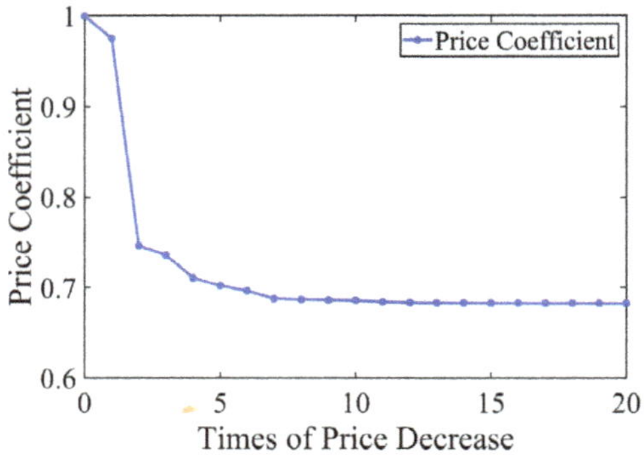

Figure 9. Convergence process of PSO.

5. Results

The optimized results of the design variables are shown in Table 8. Compared to the initial values, there was a significant decrease (32%) in the theoretical material cost of the optimum design, demonstrating the efficiency of our improved formula.

Table 8. The optimum design.

Design Variable	Boundary	Initial	Optimized
X_1	0.20~0.26	0.253	0.244
X_2	12 m, 15 m, 18 m	12 m	18 m
X_3	13 m~14 m	13.5 m	13.012 m
X_4	0.65~0.79	0.746	0.670
X_5	0.15~0.29	0.256	0.150
X_6	0.5~0.9	0.804	0.511
Objective Function Φ (Relative)	-	1	0.68

The internal force distribution of the bridge in the obtained optimum solution was checked to validate the rationality of the design. As shown in Figure 10, the predicted moment of the main girder was in a jagged shape without any sudden changes. Therefore, the completion stage of the optimum design was in the desired state.

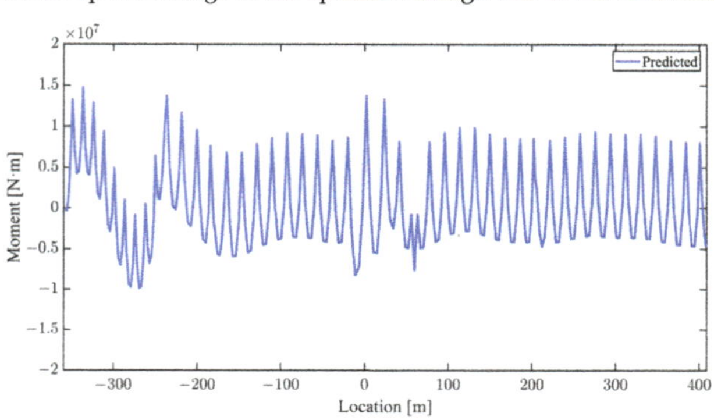

Figure 10. Check of the main girder's bending moment in the optimum design.

Checks were performed to identify if strength, stiffness, and stability constraints were met in the optimum design. As shown in Table 9, all constraints were met. Meanwhile, the stress of the towers served as the controlling structural factor in our optimization.

Table 9. Check of strength, stiffness, and stability constraints.

Constraints	Boundary	Load Cases	Optimum Design
Concrete Design Strength of the towers	−22.4 MPa~1.83 MPa	LCB1~LCB5	−18.6 MPa~1.83 MPa
Steel Design Strength of the main girder	−270 MPa~270 MPa	LCB1~LCB5	−144.51 MPa~52.1 MPa
Safety Stress Factor of the cables	≥2.5	LCB1~LCB5	2.7
Vertical Deflection of the main girder	≤2.045 m	VL	0.953 m
Elastic Buckling Stability Factor	≥4	LCB6~LCB11	7.22

Detailed checks of strength, stiffness, and stability constraints under each load combination are listed in Table 10. The results show that the stress of towers under LCB3 was close to the boundary (1.83 MPa), which means LCB3 (LCB3 = 1.1 × (1.2 × DL + 1.4 × VL3 + 0.75 × 1.1 × WL1)) served as the controlling load factor of our optimization. This was because the LCB3 was composed of all kinds of load types and the VL3 had a side-span layout, which was unfavorable for the towers.

Table 10. Check of constraints under each load combination.

Type	Verification Item	Load Cases					
Strength		LCB1	LCB2	LCB3	LCB4	LCB5	
	Tensile stress of the main girder (MPa)	32.26	49.2	52.1	36.45	17.11	
	Compressive stress of the main girder (MPa)	−126.73	−129.43	−122.97	−144.51	−118.39	
	Tensile stress of the towers (MPa)	0.66	0	1.83	0.42	0.69	
	Compressive stress of the towers (MPa)	−18.2	−18.13	−18.65	−18.66	−17.76	
	Maximum stress of the cables (MPa)	571.33	619.74	572.6	566.61	526.1	
Stiffness		VL1	VL2	VL3	VL4		
	Maximum deflection of the main girder (m)	0.611	0.952	0.429	0.551		
Stability		LCB6	LCB7	LCB8	LCB9	LCB10	LCB11
	Safe factor	8.13	7.63	7.44	7.79	7.22	8.13

The stresses and deflections were obtained as the maximum over the whole girder or towers. The safe factor of stability was obtained as the elastic buckling stability factor that corresponds to the first modality. Elements of were generated by meshing the line between geometry key points with a discretization number of five for the girder and towers and one for the cables. (In Ansys 2022 R1) The element type of the girder and the towers was Beam4. The element type of the cables was Link10.

Finally, the complexity of the optimization was evaluated and compared for the improved formula. In this example, the particle swarm optimization algorithm consisted of 30 particles and 500 iterations. If the routine method was adopted, the number of cable forces optimization would be 15,000 times. By means of a surrogate-model-assisted cable forces predictor, only 972 times were needed for the samples when constructing the predictor before structural optimization. During the structural optimization process, the optimum cable forces and counterweight of the particles were determined with the predictor, eliminating the need for cable forces optimization.

6. Conclusions

In the preliminary design of cable-stayed bridges, it is strategically significant to properly determine the design parameters. However, it is challenging to obtain the optimum design of a cable-stayed bridge because of numerous variables, multiple load cases, and diverse constraints. Integrated methods are adopted in this paper to enable efficient design optimization of cable-stayed bridges:

- to simplify the complexity of the problem, an improved two-layer framework is presented for cable-stayed bridge optimum design problems. The formula consists of two mutually iterative sub-problems: optimizing the internal force distribution by adjusting the cable prestressing forces and optimizing the sizing and geometrical parameters. The sub-problems exhibit fewer variable coupling features, making them easier to solve.
- to decrease the dimension of the design variables, B-spline interpolation curve is adopted to condense the variables, instead of setting all cable forces as variables. B-spline curve stands out when confronting large-span cable-stayed bridges with dense cables, because it can fit cable force distribution in cable-stayed bridges with only a few controlling points or fitting points.
- to improve optimization efficiency, a surrogate model-assisted predictor for optimum cable forces is constructed. The predictor addresses the time-consuming problem of determining the optimum cable prestressing forces in each of the iteration rounds in the optimization problem. This predictor is expected to be the highlight of this paper.
- to deal with the nonconvex optimization problem, an optimization program consisting of the initialization module, iteration module, and structural analysis module is made based on PSO algorithm. PSO is well-known for its global searching ability. The global searching ability is enhanced by the following measures: a moderate population number of 30 which is five times to the number of the design variables; well-set defining parameters of the algorithm; and the dual strategy (refer to Appendix B) adopted when initializing particles.

Finally, the optimum design of a (358 + 818 + 358)-meter-long cable-stayed bridge was chosen as the implementation of the proposed methods. First and foremost, the theoretical material cost was set as the objective function. Six design parameters closely connected with cost of the towers and the cables were identified as design variables. Constraints were set consistently with the Chinese Design Code. Then, 972 samples were generated by uniformly discretizing the design variables. After training and comparison, Gaussian Process Regression was demonstrated as the best surrogate model for prediction. The cable forces predictor was successfully constructed when GPR passed the testing. Afterward, the optimization program was launched to obtain the optimum design of the variables. The cable forces predictor was utilized to determine the optimum cable tensions and counterweight in the program. The predictor eliminated the need to solve the cable forces optimization problem in each of the iteration rounds, resulting in improved efficiency. The results show that the theoretical material cost of the optimum design is 32% lower than the original design. The feasibility and reliability of the structural optimization process were verified by several checks.

Despite our proposed method demonstrated accuracy and efficiency in our example, there are still some possible improvements when confronting larger scale problems. For instance, the surrogate model is naturally a regression trained with given samples, which indicates that if some new design variables are introduced, or if a wider searching range is to be explored, the model needs to be updated for the sake of accuracy. Therefore, it is beneficial to construct a larger version of the cable forces predictor covering a broader range of potential design variables for cable forces optimization problems. In addition, the sampling strategy and different surrogate models can also be more comprehensively compared. These directions can be potentially investigated in future work.

Author Contributions: Conceptualization, Y.M. and C.S.; methodology, Y.M. and C.S.; software, Z.W.; validation, Z.W.; formal analysis, Y.M.; investigation, C.S.; resources, R.X.; data curation, Z.W.; writing—original draft preparation, Y.M.; writing—review and editing, Y.M. and C.S.; visualization, Z.J.; supervision, R.X.; project administration, B.S.; funding acquisition, B.S. All authors have read and agreed to the published version of the manuscript.

Funding: This research was funded by National Key R&D Program of China, grant number 2021YFB1600300, National Natural Science Foundation of China, grant number 52378185, and the Science and Technology Committee of Shanghai, China, grant number 21DZ1202900. The opinions and statements do not necessarily represent those of the sponsors.

Institutional Review Board Statement: Not applicable.

Informed Consent Statement: Not applicable.

Data Availability Statement: The data presented in this study are available on request from the corresponding author. The data are not publicly available due to privacy.

Conflicts of Interest: Author Zhipeng Wang was employed by the company CCCC Highway Consultants Co., Ltd. The remaining authors declare that the research was conducted in the absence of any commercial or financial relationships that could be construed as a potential conflict of interest.

Appendix A. Basic Theory of Particle Swarm Method

The idea of Particle Swarm Method (PSO) [31] originated from the study of bird flock foraging behavior. This enables the group to collectively share information and find the optimum destination.

The most important iterative parameters are as follows:

- Position (x_i): The values of the design variables for each particle.
- Velocity (v_i): Moving distance and direction for each particle.
- Fitness (f_i): Magnitude of the objective function.
- Pfitness (f_{pi}): The fitness where the historical best was found for each particle.
- Gfitness (f_g): The fitness where the historical best was found within the whole swarm.
- Pbest ($p_{i,pbest}$): The position where the Pfitness is for each particle.
- Gbest (p_{gbest}): The position where the Gfitness is within the whole swarm.

The problem defining parameters are as follows:

- Population (N_p): Count of the particles. A smaller population size may result in falling into the local optimum, while a larger population size can improve convergence and find the global optimum solution faster.
- Iteration Count ($T_{\max}$): Maximum number of generations, serving as the ending criterion.
- Inertia Weight (ω): The influence of the previous generation's velocity on the current generation's velocity, which was introduced by Eberhart [32]. A larger value of ω enhances the particle's ability to explore new regions and conduct global optimization searches but weakens its ability to conduct local optimization searches.
- Learning Factor (c): c_1 is the particle learning factor. c_1 represents the weight of the particle's next action based on its own experience, indicating its attraction to its PBest. c_2 is the swarm learning factor. c_2 represents that based on the others' experience, indicating its attraction to the Gbest.

The v_i and x_i are updated in the new generation for each particle as

$$v_i^{(k+1)} = \omega v_i^{(k)} + c_1 r_1 \left(p_{i,pbest}^{(k)} - x_i^{(k)} \right) + c_2 r_2 \left(p_{gbest}^{(k)} - x_i^{(k)} \right) \tag{A1}$$

$$x_i^{(k+1)} = x_i^{(k)} + v_i^{(k+1)} \tag{A2}$$

where r_1 and r_2 are random numbers from 0 to 1 to increase the randomness of the search. The update of particle's velocity and position is illustrated in Figure A1.

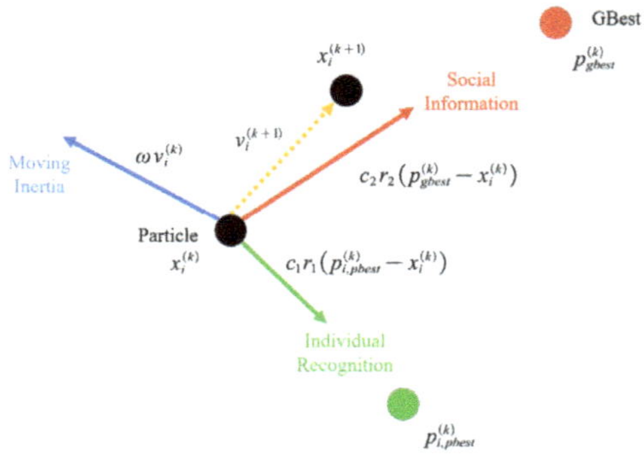

Figure A1. Update of particle's velocity and position. The update formulas suggest the searching direction of the particle is determined by the social information, the individual recognition, and its moving inertia.

Appendix B. PSO-Based Optimization Program

A PSO-based optimization program consisting of three modules is made to realize optimization design. Detailed steps in each module are shown in Figure A2.

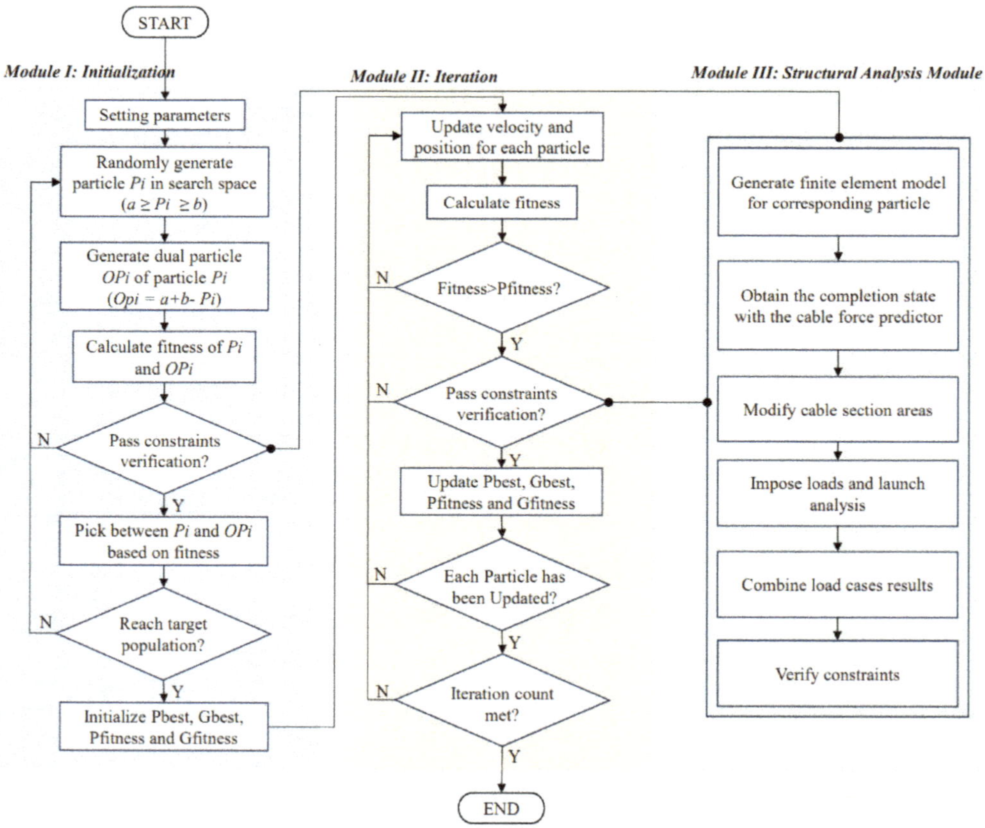

Figure A2. The optimization program utilizing particle swarm method and cable forces predictor. It consists of three modules: the initialization module, the iteration module, and the structural analysis module.

Appendix B.1. Initialization Module

Particle initialization is a crucial step that affects the speed and direction of convergence in the optimization process. Selecting an appropriate initialization strategy can reduce the optimization's convergence time and prevent it from getting trapped in a local optimum. To achieve this, a dual generation strategy is adopted. For any value $x_i \in \mathbb{R}, x_i \in [a_i, b_i]$, its dual value is $x_i^O = a_i + b_i - x_i$.

Detailed steps of the initialization module are as follows:

1. Randomly generate particle Pi in the search area.
2. Generate the dual particle OPi. Each value within the variable vector of OPi is the dual value of Pi.
3. Calculate the fitness of Pi and OPi.
4. Verify constraints for Pi and OPi. If both particles pass the verification, the particle with better fitness is selected in the swarm. If only one of them passes the verification, the passing particle is selected. If neither of them passes the verification, return to step 1 and regenerate particle Pi.
5. Repeat step 1~step 4 until the target swarm population is reached.
6. Initialize Pbest, Gbest, Pfitness, and Gfitness.

Appendix B.2. Iteration Module

During the iterative process of the particle swarm optimization algorithm, it is essential to confine all particles within the feasible region. This means that the constraints of each particle must be checked. However, verifying the constraints with the finite element analysis is time-consuming. To improve this, constraint verification is only performed for particles whose cost is lower than the previously lowest, given that the objective function (theoretical material cost) can be explicitly and directly calculated by design variables. The improvement leads to a significant reduction in verification times.

Detailed steps of the iteration module are:

1. Initialize random velocity for each particle in the swarm.
2. Initialize Pfitness, Gfitness, Pbest, and Gbest.
3. Update velocity and position for each particle with Formula (A1) and (A2).
4. Calculate fitness for each particle. If the fitness is worse than Pfitness, undo the update of position and wait for the next evolution. If the fitness is better than Pfitness and verification is checked, then update Pbest. If the fitness is also better than Gfitness with verification checked, then update Gbest.
5. Repeat step 3 and step 4 until each particle in the swarm has been updated.

Appendix B.3. Structural Analysis Module

During initialization and iteration, constraints relating to code verifications are verified by calling the structural analysis module. The modeling and analysis are performed with ANSYS 2022 R1 APDL.

Detailed steps of the structural analysis module are as follows:

1. Generate corresponding finite element model in ANSYS for particle.
2. Obtain the optimum cable forces with the predictor constructed in Section 3.2. Change the initial strain of cable elements.
3. Modify elastic modulus and section areas of cables, with the strategy mentioned in Section 3.3.
4. Impose loads, launch static analysis, and extract the results.
5. Combine the results according to combination cases.
6. Verify the constraints according to the code.

References

1. Martins, A.M.B.; Simões, L.M.C.; Negrão, J.H.J.O. Optimization of Cable-Stayed Bridges: A Literature Survey. *Adv. Eng. Softw.* **2020**, *149*, 102829. [CrossRef]
2. Feder, D. Optimization of the Prestressing in the Cables of a Cable-Stayed Bridge. In Proceedings of the 10th Congress of IABSE, Tokyo, Japan, 6–11 September 1976. [CrossRef]
3. Sung, Y.-C.; Chang, D.-W.; Teo, E.-H. Optimum Post-Tensioning Cable Forces of Mau-Lo Hsi Cable-Stayed Bridge. *Eng. Struct.* **2006**, *28*, 1407–1417. [CrossRef]
4. Baldomir, A.; Hernandez, S.; Nieto, F.; Jurado, J.A. Cable Optimization of a Long Span Cable Stayed Bridge in La Coruña (Spain). *Adv. Eng. Softw.* **2010**, *41*, 931–938. [CrossRef]
5. Ha, M.-H.; Vu, Q.-A.; Truong, V.-H. Optimum Design of Stay Cables of Steel Cable-Stayed Bridges Using Nonlinear Inelastic Analysis and Genetic Algorithm. *Structures* **2018**, *16*, 288–302. [CrossRef]
6. Lute, V.; Upadhyay, A.; Singh, K.K. Computationally Efficient Analysis of Cable-Stayed Bridge for GA-Based Optimization. *Eng. Appl. Artif. Intell.* **2009**, *22*, 750–758. [CrossRef]
7. Gao, Q.; Yang, M.-G.; Qiao, J.-D. A Multi-Parameter Optimization Technique for Prestressed Concrete Cable-Stayed Bridges Considering Prestress in Girder. *Struct. Eng. Mech.* **2017**, *64*, 567–577. [CrossRef]
8. Cid, C.; Baldomir, A.; Hernández, S. Optimum Crossing Cable System in Multi-Span Cable-Stayed Bridges. *Eng. Struct.* **2018**, *160*, 342–355. [CrossRef]
9. Hassan, M.M.; Nassef, A.O.; Damatty, A.A.E. Surrogate Function of Post-Tensioning Cable Forces for Cable-Stayed Bridges. *Adv. Struct. Eng.* **2013**, *16*, 559–578. [CrossRef]
10. Hassan, M.M.; Nassef, A.O.; El Damatty, A.A. Optimal Design of Semi-Fan Cable-Stayed Bridges. *Can. J. Civ. Eng.* **2013**, *40*, 285–297. [CrossRef]
11. Li, Y.; Zou, C.; Berecibar, M.; Nanini-Maury, E.; Chan, J.C.-W.; Van Den Bossche, P.; Van Mierlo, J.; Omar, N. Random Forest Regression for Online Capacity Estimation of Lithium-Ion Batteries. *Appl. Energy* **2018**, *232*, 197–210. [CrossRef]
12. Song, C.; Shafieezadeh, A.; Xiao, R. High-Dimensional Reliability Analysis with Error-Guided Active-Learning Probabilistic Support Vector Machine: Application to Wind-Reliability Analysis of Transmission Towers. *J. Struct. Eng.* **2022**, *148*, 04022036. [CrossRef]
13. Zhang, C.; Shafieezadeh, A. Nested Physics-Informed Neural Network for Analysis of Transient Flows in Natural Gas Pipelines. *Eng. Appl. Artif. Intell.* **2023**, *122*, 106073. [CrossRef]
14. Han, Z.-H.; Görtz, S. Hierarchical Kriging Model for Variable-Fidelity Surrogate Modeling. *AIAA J.* **2012**, *50*, 1885–1896. [CrossRef]
15. Meert, P.; Pereira, F.; Willems, P. Surrogate Modeling-Based Calibration of Hydrodynamic River Model Parameters. *J. Hydro-Environ. Res.* **2018**, *19*, 56–67. [CrossRef]
16. Furtney, J.K.; Thielsen, C.; Fu, W.; Le Goc, R. Surrogate Models in Rock and Soil Mechanics: Integrating Numerical Modeling and Machine Learning. *Rock Mech. Rock Eng.* **2022**, *55*, 2845–2859. [CrossRef]
17. Song, C.; Xiao, R.; Sun, B.; Wang, Z.; Zhang, C. Cable Force Optimization of Cable-Stayed Bridges: A Surrogate Model-Assisted Differential Evolution Method Combined with B-Spline Interpolation Curves. *Eng. Struct.* **2023**, *283*, 115856. [CrossRef]
18. Asgari, B.; Osman, S.A.; Adnan, A. A New Multiconstraint Method for Determining the Optimal Cable Stresses in Cable-Stayed Bridges. *Sci. World J.* **2014**, *2014*, 503016. [CrossRef]
19. Sun, S.J.; Gao, J.; Huang, P.M. Forward-Calculating Optimization Method for Determining the Rational Construction State of Cable-Stayed Bridges. *Adv. Mater. Res.* **2013**, *671–674*, 980–984. [CrossRef]
20. Guo, J.; Yuan, W.; Dang, X.; Alam, M.S. Cable Force Optimization of a Curved Cable-Stayed Bridge with Combined Simulated Annealing Method and Cubic B-Spline Interpolation Curves. *Eng. Struct.* **2019**, *201*, 109813. [CrossRef]
21. Hassan, M.M. Optimization of Stay Cables in Cable-Stayed Bridges Using Finite Element, Genetic Algorithm, and B-Spline Combined Technique. *Eng. Struct.* **2013**, *49*, 643–654. [CrossRef]
22. Song, C.; Xiao, R.; Sun, B. Optimization of Cable Pre-Tension Forces in Long-Span Cable-Stayed Bridges Considering the Counterweight. *Eng. Struct.* **2018**, *172*, 919–928. [CrossRef]
23. Boggs, P.T.; Tolle, J.W. Sequential Quadratic Programming. *Acta Numer.* **1995**, *4*, 1–51. [CrossRef]
24. Weissman, S.A.; Anderson, N.G. Design of Experiments (DoE) and Process Optimization. A Review of Recent Publications. *Org. Process Res. Dev.* **2015**, *19*, 1605–1633. [CrossRef]
25. Montgomery, D.C.; Peck, E.A. *Introduction to Linear Regression Analysis*, 2nd ed.; Wiley Series in Probability and Mathematical Statistics Applied Probability and Statistics; Wiley: New York, NY, USA, 1992; ISBN 978-0-471-53387-0.
26. Jones, D.R.; Schonlau, M.; Welch, W.J. Efficient Global Optimization of Expensive Black-Box Functions. *J. Glob. Optim.* **1998**, *13*, 455–492. [CrossRef]
27. Gordon, A.D.; Breiman, L.; Friedman, J.H.; Olshen, R.A.; Stone, C.J. Classification and Regression Trees. *Biometrics* **1984**, *40*, 874. [CrossRef]
28. Schölkopf, B.; Smola, A.J.; Williamson, R.C.; Bartlett, P.L. New Support Vector Algorithms. *Neural Comput.* **2000**, *12*, 1207–1245. [CrossRef]
29. Nazmy, A.S.; Abdel-Ghaffar, A.M. Three-Dimensional Nonlinear Static Analysis of Cable-Stayed Bridges. *Comput. Struct.* **1990**, *34*, 257–271. [CrossRef]

30. Cao, H.; Qian, X.; Chen, Z.; Zhu, H. Layout and Size Optimization of Suspension Bridges Based on Coupled Modelling Approach and Enhanced Particle Swarm Optimization. *Eng. Struct.* **2017**, *146*, 170–183. [CrossRef]
31. Eberhart, R.; Kennedy, J. A New Optimizer Using Particle Swarm Theory. In Proceedings of the MHS'95: Proceedings of the Sixth International Symposium on Micro Machine and Human Science, Nagoya, Japan, 4–6 October 1995; pp. 39–43.
32. Eberhart, R.C.; Shi, Y. Comparison between Genetic Algorithms and Particle Swarm Optimization. In *Evolutionary Programming VII*; Porto, V.W., Saravanan, N., Waagen, D., Eiben, A.E., Eds.; Lecture Notes in Computer Science; Springer: Berlin/Heidelberg, Germany, 1998; Volume 1447, pp. 611–616. ISBN 978-3-540-64891-8.

applied sciences

MDPI

Review

3D-Printed Concrete Bridges: Material, Design, Construction, and Reinforcement

Zahra Sadat Miri, Hassan Baaj * and Maria Anna Polak

Department of Civil and Environmental Engineering, University of Waterloo, Waterloo, ON N2L 3G1, Canada; zsmiri@uwaterloo.ca (Z.S.M.); polak@uwaterloo.ca (M.A.P.)
* Correspondence: hbaaj@uwaterloo.ca

Abstract: 3D Concrete Printing (3DCP) technology is rapidly gaining popularity in the construction industry, particularly for transportation infrastructure such as bridges. Unlike traditional construction methods, this innovative approach eliminates the need for formwork and enhances both economic efficiency and sustainability by lowering resource consumption and waste generation associated with formwork. This paper examines current research on 3D-printed concrete bridges, highlighting key areas such as concrete mixtures, design processes, construction techniques, and reinforcement strategies. It delves into computational methods like topology optimization and iterative "design by testing" approaches, which are crucial for developing structurally efficient and architecturally innovative bridges. Additionally, it reviews specific admixtures or additives within the concrete mix, assessing how they improve essential properties of printable concrete, including extrudability, buildability, and interlayer bonding. Moreover, it shows that the primary construction approach for 3DCP bridges involves prefabrication and on-site assembly, with robotic arm printers leading to scalability and precision. Reinforcement continues to be challenging, with the most commonly used strategies being post-tensioning, hybrid techniques, and fiber reinforcement. This paper offers insights into the advancements and challenges in 3D-printed concrete bridge construction, providing valuable guidance for future research and development in this field.

Keywords: 3D concrete printing (3DCP); 3D-printed bridges; 3DCP bridge design; 3DCP bridge construction; 3DCP bridge reinforcement

1. Introduction

Additive manufacturing (AM), also known as three-dimensional (3D) printing, is a group of emerging techniques for manufacturing 3D structures directly from a digital model in successive layers [1]. The adoption of AM techniques in the construction industry has the potential to revolutionize this industry, transforming its image from a labor-intensive field often associated with safety challenges and sustainability concerns into a more efficient, precise, and environmentally friendly practice [2]. As the construction industry shifts towards AM, cementitious material extrusion has emerged as one of the most widely adopted techniques [3]. Its popularity stems from the affordability and accessibility of open-source extrusion-based printers and the widespread use of Portland cement, which offers reliable mechanical properties at a low cost. Cementitious materials provide a cost-effective alternative to metallic materials and greater functionality than polymers for structural applications [3]. Among these materials, concrete stands out as particularly well-suited for extrusion-based systems. In 3D Concrete Printing (3DCP), a structure is fabricated layer by layer with a nozzle that deposits material along a defined path. Compared to conventional concrete construction methods, 3DCP offers several significant advantages [4]: the reduction of construction costs through the elimination of expensive formwork; improved safety by reducing the need for dangerous jobs; faster construction times due to the constant operational rate of 3D printing; minimized errors through highly precise material deposition; enhanced sustainability by reducing formwork waste; and greater architectural freedom, enabling the creation of more sophisticated and innovative designs.

Bridges make up a significant portion of operational 3D-printed concrete structures globally. Currently, all existing 3D-printed bridges are designed specifically for pedestrian and bicycle traffic. However, researchers are optimistic that future advancements will enable the development and exploration of 3D-printed bridges that can support motor vehicle traffic as well [5]. The first 3D-printed concrete bridge was built in 2017 in Gemert, Netherlands [6,7]. This 8-m-long bicycle bridge results from a collaboration between BAM Infra and the Technological University of Eindhoven (TU/e). It spans 3.5 m in width and has a thickness of 0.9 m, constructed from 800 layers of 3D-printed concrete. In January 2019, a 26.3-m long 3D-printed concrete pedestrian bridge was completed at Wisdom Bay Park in Shanghai [8]. This bridge stands as the second longest of its kind in the world. Designed by Professor Xu Weiguo from Tsinghua University, it features handrails that resemble flowing ribbons along its arch. With a width of 3.6 m, the bridge was constructed using two robotic arms in a process that took over 450 h to complete. That same year, the world's longest 3D-printed concrete bicycle bridge was unveiled in Nijmegen, Netherlands, measuring an impressive 29 m in length. This project was a collaboration between the TU/e and the 3D printing center at Saint-Gobain Weber Beamix, incorporating multiple BAM robotic arms. The design showcased rounded, natural shapes, seamlessly integrating with its outdoor environment [9]. In July 2021, the "Striatus" bridge was constructed in Venice, Italy (see Figure 1a). This project, a collaboration between the Block Research Group and the Computation and Design Group at Zaha Hadid Architects, resulted in a 12-by-16-m arched footbridge located in a park. Notably, the bridge was built entirely without reinforcement, utilizing printed concrete blocks arranged to form an arch, which echoed the appearance of traditional masonry bridges [10]. Two years later, in 2023, the "Phoenix" Bridge, shown in Figure 1b, was constructed as an evolved iteration of the "Striatus" bridge. This 3D-printed concrete masonry structure was constructed using 10 tons of recycled materials, including aggregates derived from the original Striatus blocks. The project was a collaborative effort among Holcim, the Block Research Group at ETH Zurich, the Computation and Design Group at Zaha Hadid Architects, and incremental3D [11,12].

(a) (b)

Figure 1. Examples of 3D-printed concrete bridges: (**a**) Striatus bridge in Venice, Italy [13], and (**b**) Phoenix bridge in Lyon, France [11,14].

This paper provides a literature review on research related to 3D-printed concrete bridges, offering valuable insights for designers and researchers aiming to advance this emerging field. The review focuses on key aspects, including material properties, design processes, construction techniques, and reinforcement strategies. Specifically, the paper addresses the following critical questions:

- Design process: What are the design processes for 3D-printed concrete bridges?
- Material: What additives/admixtures are utilized in the mixture, and how do they enhance the performance and printability of 3D-printed concrete for bridges?
- Construction methods: What construction techniques are implemented, including prefabricated and in situ approaches, and what 3D printing systems, such as gantry-based or robotic arm setups, are used in bridge construction?
- Reinforcement: What reinforcement strategies, such as embedded bars or fiber-reinforced concrete, are applied in 3D-printed concrete bridges, and how is reinforcement integrated?

By addressing these questions, the paper aims to provide a foundational understanding of the current research in 3D-printed concrete bridges and guide future advancements in this innovative field.

2. Design Process of 3D-Printed Concrete Bridges

The conceptual design of a bridge relies heavily on the designer's intuition and their ability to understand how various structural components interact to transfer weight and loads to the ground efficiently and safely [15]. Specifically, in the context of 3D-printed concrete bridges, the design process demands additional considerations due to the unique fabrication methods that impact material behavior, geometric possibilities, and structural performance. This section explores the design approaches discussed in the literature, emphasizing how designers leverage the capabilities of additive manufacturing to develop innovative and efficient bridge designs.

A common computational approach in the design of 3D-printed concrete bridges is Topology Optimization (TO). This technique aims to optimize material distribution within a specified design domain to achieve objectives such as maximizing structural performance or minimizing material usage while adhering to predefined constraints [16]. Vantyghem et al. [16] applied the density-based TO approach to the design and manufacture of a simply supported girder subjected to a uniform load. The density-based structural topology considers a collection of density values ranging between 0 (void) and 1 (material) at discrete points within the design domain. To this end, the beam was discretized into a grid of square finite elements, and the initial domain was initialized with a uniform density value of 0.5, while the top surface density was fixed at 1 to maintain the integrity of the loading area. The setup for the topology optimization is shown in Figure 2a. The post-tensioning tendon was incorporated into the design as an initial geometric constraint. The optimization aimed to minimize displacements at the top surface by iteratively adjusting density values to determine the most efficient material distribution while satisfying constraints on volume fraction and support locations. After optimization, the conceptual 2D design was translated into a practical 3D-printable structure through an extensive post-processing phase. Key modifications included shaping the lower chord circularly to ensure even encasement of the post-tensioning cable, widening the upper chord to create a pedestrian deck surface while limiting its width to reduce transverse tensile forces, and incorporating end blocks to anchor post-tensioning forces. Lastly, a 3D Finite Element (FE) analysis was conducted to validate the design, providing insights into stress distribution and the required reinforcement layout under various loading conditions. This foundational use of TO establishes a basis for efficient material usage and structural performance, linking computational methods to practical applications. The final design is shown in Figure 2a, with black parts representing concrete, white representing void, and cyan representing the tendon.

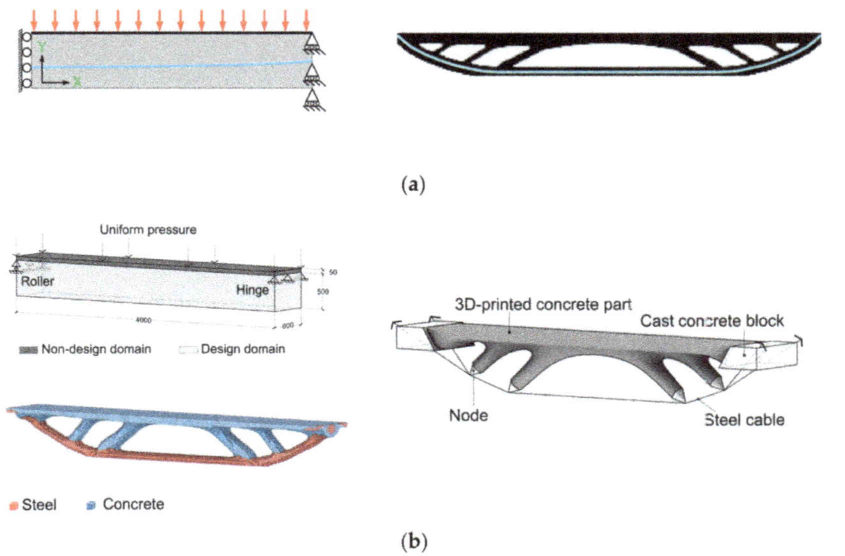

Figure 2. Examples of Topology optimization for design of 3D-printed concrete bridges: (**a**) Simply-supported girder design [14] (**b**) FloatArch bridge design [16].

Building upon this work, Ooms et al. [17] used a similar girder design that integrated a wider top surface for improved bridge applications. The design featured two longitudinal, topology-optimized, post-tensioned concrete girders supporting a continuous bridge deck that was supported at four corners.

The FloatArch bridge was also designed using topology optimization, specifically employing the Bi-directional Evolutionary Structural Optimization (BESO) method to distribute concrete and steel within a defined design space efficiently [18]. This approach leverages concrete's compressive strength and steel's tensile capacity by iteratively refining material placement within a defined design space, concentrating concrete in compression zones and steel in tension zones. In their final design, concrete components had hollow sections to reduce self-weight while maintaining structural efficiency, and steel cables were simplified and anchored to the concrete via steel nodes (see Figure 2b). This example transitions TO principles to bridges with mixed-material systems, showcasing the versatility of computational design.

The Castilla-La Mancha Park footbridge in Alcobendas, Madrid, is another example of how topology optimization can be applied to form-finding in the design of 3D-printed bridges [19]. This project employed the D-shape, particle-bed printing method that differs from conventional extrusion-based 3D concrete printing. Parametric design played a central role in optimizing material distribution, minimizing waste through recycling raw materials during production, and enhancing structural performance by placing material only where necessary. Additionally, generative algorithms were used to maintain porosity while enabling the creation of complex geometries, demonstrating the synergy between advanced computational tools and innovative manufacturing techniques [20].

The "Striatus Bridge" and its evolution, the "Phoenix Bridge", exemplify innovative, iterative design processes that integrate form-finding, performance-based optimization, and rigorous structural analysis to address both aesthetic and structural challenges in 3D-printed concrete bridges [13,21]. Both bridges share a common goal of minimizing tensile stress in unreinforced concrete by utilizing arch forms, which function as transverse span structures designed to transfer loads primarily through compression (see Figure 3a) [22]. The design of the Striatus Bridge began with a skeletal graph connecting its five supports, which served as the foundation for generating a 2D mesh. This mesh was refined using Thrust Network Analysis (TNA) to establish a 3D thrust network of compressive forces in equilibrium with predefined loads, as shown in Figure 3b. The resulting form-found mesh was discretized into individual voussoir blocks, with joints carefully positioned to align with the compressive force flow. Stability was verified through Discrete Element Modeling (DEM), which evaluated the dry-assembled structure under various conditions. Iterative adjustments to block interfaces and thicknesses ensured that the final design met both aesthetic and structural requirements [13]. Integrating form-finding and structural analysis demonstrates the iterative refinement necessary for compression-dominant designs.

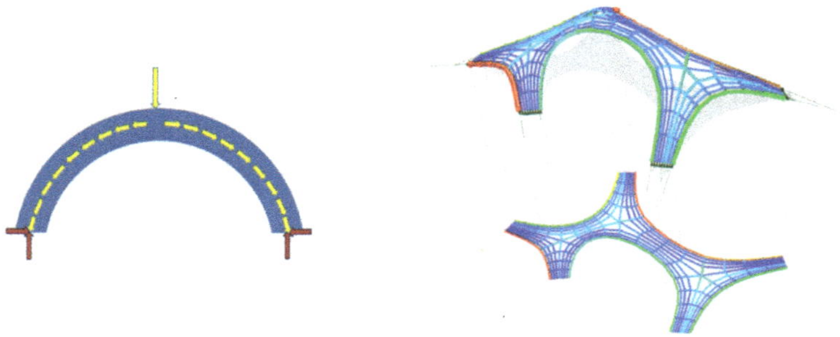

(a) (b)

Figure 3. Examples of form-finding and performance-based design of 3D-printed concrete bridges: (**a**) Force transmission path applied to arch structures [21] and (**b**) TNA form finding of the deck's mesh in Striatus Bridge [13].

Building on the Striatus design, the Phoenix Bridge employed a similar iterative approach while incorporating traditional masonry-inspired techniques [21]. The design process emphasized optimizing the arch axis to minimize bending moments, ensuring that the cross-section primarily experienced compressive stress. Finite Element Analysis (FEA) was used in evaluating stress distribution, allowing refinements to the arch geometry and block dimensions to balance structural performance with aesthetic considerations.

The "Design by Testing" methodology has also been a key approach in several projects. This methodology can complement TO by ensuring experimental validation of design choices and linking computational and physical testing. Salet et al. [23] utilized the "Design by Testing" approach to create a 3D-printed bicycle bridge. This approach involved comprehensive testing of materials and structures at every phase, which included characterizing materials (such as assessing compression, tension, modulus of elasticity, creep, and shrinkage), conducting destructive tests on a 1:2 scaled model, and carrying out assembly trials. The design phase began with understanding the constraints their 3D concrete printer posed. Although the entire bridge could theoretically fit within the printer's build space, the team opted for a modular design to ease transportation and assembly. This modularity also enabled optimization of the bridge's cross-section, exploiting the printer's features. The insights gathered during testing led to notable enhancements, such as a 4% reduction in both print path length and material consumption.

Ahmed et al. [24] implemented a comparable "Design by Testing" strategy to construct the world's longest bicycle bridge. Their approach merged experimental testing with parametric design to enhance printability, material performance, and structural demands.

Tests on material properties verified the structural integrity and evaluated long-term behavior, establishing a foundation for design endorsement. Parametric design enabled the creation of digital models that allowed iterative optimization, resulting in a nature-inspired double-curved deck spanning tapered columns, which appeared to sprout organically from the deck. Zhan et al. [25] also adopted this method to design a 3D-printed prestressed concrete bridge. Their process began with material research and testing to guarantee both printability and structural soundness. Critical properties were assessed, including a maximum overhang angle for inclined printing. Computational tools were later utilized for digital design and optimization, encompassing toolpath refinement to achieve continuous extrusion and focusing material in high-stress regions. Section optimization was also executed to ensure a compressive state during prestressing.

Another innovative approach called "Minimass" technique was developed by Net Zero Projects (NZPs) [26]. This approach focuses on creating bending structures, such as beams, with optimized geometries using 3DCP. The design principles prioritize axial tension and compression, forming a "truss-like" structure rather than relying solely on material bending strength. A concrete top chord resists compression, while a steel cable bottom chord (standard post-tension tendons) resists tension. These elements are separated by concrete webs, with the arrangement of webs and cable geometry tailored to the applied loads. In this application, the concrete top chord was reinforced with mild steel, enabling it to function as a beam column rather than just an axial compression member. The final design included primary beams spanning the bridge's length, with thinner slabs placed between them. Prefabricated lattice slabs provided permanent formwork for an in situ topping slab and a safe working platform during construction. To fully utilize 3DCP, both the primary beams and lattice slab elements were 3D-printed, applying principles of efficient composite construction commonly seen in conventional T-beam designs.

In summary, various computational and experimental design approaches have been employed to optimize material usage and enhance structural efficiency in 3D-printed concrete bridges. Topology optimization (TO) has been extensively used to reduce material waste and improve structural efficiency, but its reliance on predefined constraints and the need for extensive post-processing can limit design flexibility. Parametric design enables greater geometric adaptability and iterative refinement, making it ideal for complex freeform structures, such as the Castilla-La Mancha Park footbridge. Meanwhile, form-finding techniques like Thrust Network Analysis (TNA) focus on generating compression-optimized geometries, which have been crucial in masonry-inspired designs such as the Striatus and Phoenix Bridges. The 'Design by Testing' approach bridges the gap between digital design and physical validation, ensuring real-world feasibility through experimental material characterization and full-scale structural testing, as seen in the world's longest 3D-printed bicycle bridge. Finally, the Minimass technique represents a shift toward efficient axial force-based structures, integrating 3D printing with post-tensioning for improved structural behavior under bending loads.

These methods are not mutually exclusive but can be combined to maximize the advantages of 3D-printed bridge design. For example, TO can be integrated with parametric modeling to refine printability constraints, and form-finding can be complemented by structural testing to validate real-world performance. As computational and experimental methodologies evolve, future 3D-printed bridge designs will likely leverage hybrid approaches, combining the precision of algorithm-driven optimization with performance-based testing to achieve more sustainable and structurally efficient designs.

3. Materials Used in 3D-Printed Concrete Bridges

The concrete material used for printing purposes has the features of self-compacting concrete (i.e., no need to vibrate) and sprayed concrete (i.e., expelling fresh concrete from a nozzle) to conform to the essential requirements of a freeform construction system. Printable concrete should meet certain fresh properties to be printable [27]. The most critical of these properties are extrudability and buildability [28]. Extrudability is the ability of concrete to move through the pipes and nozzles at the printing head and is influenced mainly by the workability (consistency) of the concrete and mix proportions (i.e., cementitious binder–aggregate ratio, water–binder ratio, admixture usage). After passing the extrudability criteria, self-compacting filaments can be produced. The printed filaments should experience minimum deformations under the weight load of subsequent layers. Also, the bottom filaments should bond to the top ones to build monolithic elements. At this stage, the printed concrete needs buildability which refers to the capacity to print a certain number of layers or heights. Buildability is also related to workability and mix proportions, especially the dependency of workability with time, i.e., open time. The open time is defined as the period in which the workability of fresh concrete is at a level where extrudability is preserved and is crucial for maintaining a balance between structural stability and layer adhesion during the printing process [28]. Achieving these properties requires a careful balance of mix design, ensuring that the concrete maintains the necessary flowability and stability during extrusion while providing sufficient strength and durability in its hardened state. This balance is achieved by strategically incorporating various additives and admixtures, which enhance specific properties of the concrete.

One of the critical challenges in 3D concrete printing is the weak interlayer bonding or the occurrence of "cold joints" [29] between layers. In multilayered structures, the hardened properties of the interlayer are typically weaker than those of the bulk material, forming potential weak links in the structure. Unlike conventional cast concrete, these weak interlayer zones introduce mechanical anisotropy, which can significantly impact structural performance and durability [30]. Some studies have investigated this anisotropy by performing mechanical tests such as compression and flexure in different loading directions relative to the printed layers, as shown in Figure 4 [23,24]. According to a review by Wang et al. [31], the anisotropy in compressive strength of 3D-printed concrete (3DPC) is relatively low, regardless of mix design variations. However, flexural and tensile strengths exhibit the highest anisotropy, with significant reductions in strength perpendicular to the printing direction compared to parallel directions (see Figure 4b). This trend was also observed in the work of Salet et al. [23], where flexural strength decreased by approximately 32%, from 1.9 MPa to 1.3 MPa. This reduction in strength is a crucial consideration in the design of 3D-printed concrete bridges, as it directly influences load-bearing capacity and long-term structural integrity.

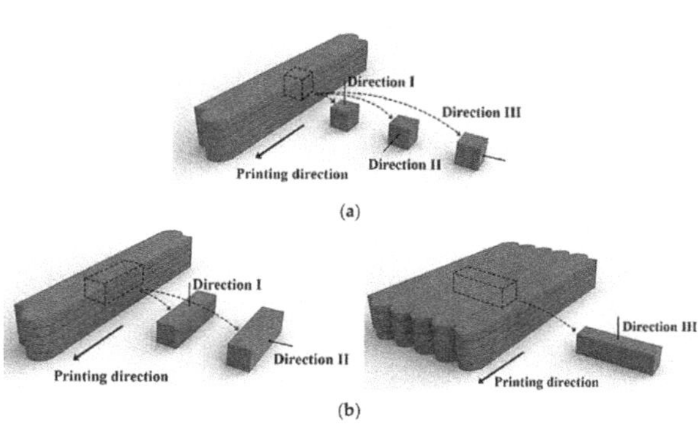

Figure 4. Different cases of loading direction and specimen configuration for (**a**) compressive and (**b**) flexural strength [32].

The pronounced anisotropic behavior is largely attributed to poor interlayer adhesion and the formation of microvoids during the printing process [33]. One of the primary factors influencing interlayer bonding strength is the time interval between layer depositions, commonly referred to as "delay time" [2]. Delay time plays a crucial role in determining the surface moisture of the printed concrete, which directly impacts bond strength [4]. A higher surface moisture content provides an aqueous medium that enhances interlayer adhesion, allowing fresh concrete in successive layers to bond more effectively [2]. Surface moisture is influenced by both printing parameters (e.g., extruder type, nozzle pressure, and deposition speed) and material properties (e.g., mix composition, bleeding rate, and evaporation rate) [34].

In summary, achieving strong interlayer bonding is crucial for minimizing the anisotropic behavior of 3D-printed concrete, thereby enhancing its mechanical performance and structural integrity. Table 1 presents the specific additives and admixtures incorporated in mix designs for 3D-printed concrete bridges, emphasizing their role in optimizing fresh properties, improving interlayer adhesion, and ensuring the successful execution of large-scale 3D printing applications.

Table 1. Materials used in 3D-printed concrete bridges.

Ref.	Additive/Admixture	Purpose	Property Enhanced
[16]	Water retention agent	Improves water retention, provides thixotropic behavior, and prevents pressurized bleeding	Thixotropy
[35]	Superplasticizer	Enhances flowability and reduced water demand	Workability, extrudability
[36]	Superplasticizer	Enhances flowability and reduces water demand	Workability, extrudability
	Deforming agent	Improves workability and prevents segregation	Consistency, stability
	Retarder	Delays setting time to extend the workability window	Open time, buildability
	Viscous agent	Enhances thixotropic behavior	Buildability, layer adhesion
[21]	Polypropylene (PP) fibers	Improves tensile strength and crack resistance	Printability
[17,37]	Limestone powder	Improves workability and reduces shrinkage	Flowability, dimensional stability
	Polycarboxylate ether (PCE)	Enhances flowability and reduces water demand	Pumpability, extrudability
	Hydroxypropyl methylcellulose (HPMC)	Enhances viscosity and thixotropy	Buildability, layer adhesion
	Alkali-free shotcrete accelerator (ACC)	Controls setting time for rapid layer stacking	Buildability
[38]	Water-reducing agent	Enhances flowability and reduces water content	Pumpability, extrudability
	Silica ash and fly ash	Improves workability and durability	Workability, dimensional stability
	Polypropylene (PP) fibers	Enhances interlayer bonding and reduces cracking	Interlayer adhesion, crack resistance
[25]	Silica fume	Enhances strength and durability	Compressive strength, durability
	Metakaolin	Improves thixotropy and extrusion properties	Buildability, extrusion stability
	Superplasticizer	Improves flowability and reduces water demand	Workability, extrudability
	PVA fiber	Enhances tensile strength and crack resistance	Interlayer adhesion, crack resistance

4. Construction Techniques for 3D-Printed Concrete Bridges

This section reviews the printing systems and construction techniques used in 3D-printed concrete bridges, highlighting their applications, advantages and challenges.

4.1. Printing System

Large-scale 3D printers used in bridge construction can be broadly categorized into two primary types: robotic arm printers and gantry printers. Each type offers unique advantages and limitations, making them suitable for different project requirements. Robotic arm printers are known for their mobility and flexibility, enabling the creation of intricate designs due to their six degrees of freedom (see Figure 5b). However, they are constrained by smaller print sizes, the need for fine aggregates, and the requirement for advanced programming skills. In contrast, gantry printers shown in Figure 5a are more suitable for producing larger structures, accommodating coarse aggregates, and achieving longer element lengths with simpler software. However, these systems are bulkier, less mobile, and demand significant on-site setup [39]. The choice between robotic arm and gantry printers is ultimately dictated by the specific needs of a project, including size, complexity, and material requirements. This section reviews the different printing systems used in the construction of 3D-printed concrete bridges, as presented in Table 2. It focuses on the types of printers and nozzle geometries used alongside key innovations in printing technologies that have enhanced these systems' scalability, precision, and efficiency.

(a) (b)

Figure 5. Printer type: (**a**) Gantry printer [40] and (**b**) Robotic arm.

Table 2. Printing systems used in 3D-printed concrete bridges.

Ref.	Printer Type	Nozzle Geometry	Innovations in the Printing System
[16]	6 DOF ABB IRB6650 robotic arm	Round with a diameter of 25 mm	-
[35]	Self-developed System with horizontal and vertical actuators mounted on a raptor track drive	Rectangular (12 inch in length, 1 inch in thickness, and 1.5 inch in height per layer) with side plates to control material flow	Accelerated Heat Curing: Expedites printing with heat curing at 150 °F for rapid compressive strength development Mobile Platform: Raptor track drive enhances scalability and allows for complex element printing
[23]	4-DOF gantry robot	Rectangle measuring 40 mm by 10 mm	Hybrid down/back-flow nozzle: Combining down-flow and back-flow nozzles allows for cable integration while ensuring effective bonding between layers.
[21]	6 DOF KUKA R120–2500 robotic arm	Rectangle measuring 60 mm by 15 mm	Spatial Path Fitting Technology: Adjusted robot posture for spatially curved cavity and uneven stacking. Planar Path Fitting Technology: Controlled nozzle rotation for consistency with travel path tangent vector. Sensor-Based Real-Time Monitoring: Sensors on screw pump and printer for real-time parameter adjustments.
[41]	6 DOF robotic arm	-	Not mentioned
[18]	6 DOF robotic arm	-	Variable-Speed Printing: Modifies nozzle speed to reduce overfilling and ensure uniform printing.
[24]	6 DOF robotic arm	Backflow nozzle-rectangle measuring 60 and 80 mm by 12 mm	-
[25]	6 DOF robotic arm	-	Real-Time Path Generation: Controlled via Rhino and Grasshopper platforms, enabling on-the-fly adjustments.

The review of printing systems for 3D-printed concrete bridges highlights the dominance of robotic arm printers, which offer flexibility and precision for creating intricate designs. These systems, enhanced by innovations such as spatial and planar path fitting technologies and real-time path generation, are well-suited for complex geometries and controlled environments. While gantry printers are better suited for larger structures and coarse aggregates, their use in bridge construction remains limited.

4.2. Construction Techniques

This section reviews the construction methods used in 3D-printed concrete bridges, focusing on prefabrication and on-site assembly. Prefabricated methods involve printing structural components off-site in a controlled environment, allowing for precise quality control and optimized material usage. These components are transported to the construction site and assembled into the final structure.

The Striatus Bridge demonstrates a prefabrication approach involving segmenting a digitally modeled arch into 53 individual voussoirs using a stereotomy process [13,41]. Their technique ensured that contact surfaces experience primarily compressive forces, aligning with the principles of unreinforced masonry. Prefabricated segments were printed off-site and transported to the site for assembly. The construction process included preparing a foundation system with steel footings connected by tension ties supported by ground screws tailored to site constraints. A temporary scaffolding system accommodated the curved profile, supporting the segments during assembly. The segments were placed sequentially, with tension ties pre-tensioned before final positioning. The decentering process transferred structural loads from temporary supports to the ties and concrete segments, creating a stable, self-supporting structure.

Another example of prefabrication is constructing a 4.7-m-long prestressed concrete bridge [25]. The bridge was divided into six equal-length units to facilitate transportation and printing logistics, each weighing approximately 80–90 kg. During assembly, the prefabricated units were supported on a simple brick bracket, and mortar was used to fill the joints between them, ensuring proper force transmission.

The FloatArch Bridge also utilized prefabrication but integrated hybrid construction techniques to achieve a complex free-form design [18]. Segmented into nine components shown in Figure 6a, the design used principal stress trajectories to ensure contact surfaces experienced perpendicular compressive forces. After curing, segments were transported and assembled with temporary and permanent supports. Steel cables served as tensile members, connecting the end segments cast conventionally. The absence of grout or mortar allowed for disassembly and potential reuse, emphasizing sustainability.

Vantyghem et al.'s girder [16], which consisted of 18 prefabricated segments joined on-site with cast end blocks, is another example of a hybrid construction approach (the segments and their assembling are shown in Figure 5b). These segments were printed off-site. Traditional casting was utilized for the two end blocks, essential for anchoring the post-tensioning forces and joining the prefabricated segments. These components were fabricated on-site with conventional formwork, ensuring structural integrity in areas requiring reinforcement. Similarly, the Nijmegen 29-m bridge span was divided into five simply supported parts [24]. Prefabricated elements were transported to the site, glued together, and reinforced with traditionally cast concrete anchor blocks, while the columns utilized 3D-printed concrete "lost formwork." The road bridge in Ukraine was also constructed using both prefabrication and traditional in-site casting [26]. To optimize material usage and structural efficiency, the bridge utilized prefabricated minimass beams and lattice slab concrete panels, topped with a cast–in situ slab. The minimass beams relied on axial tension and compression rather than bending strength, while the lattice slabs served as permanent formwork and a safe working platform during construction.

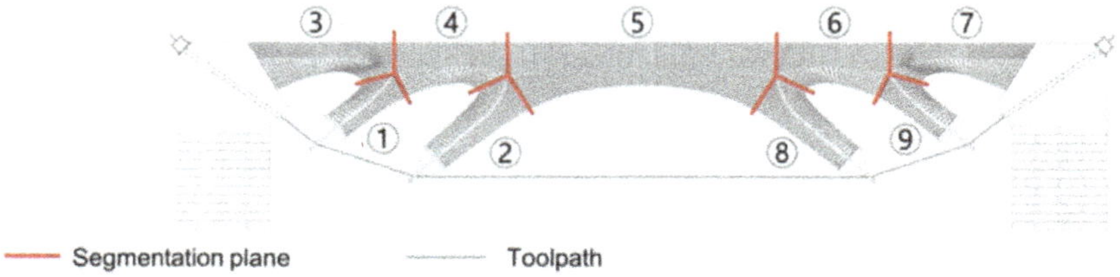

(a)

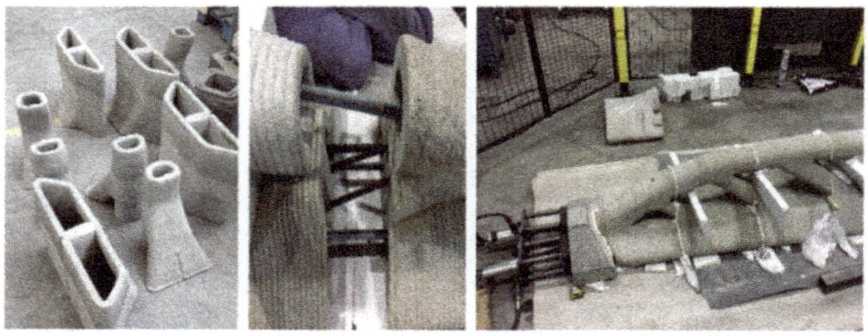

(b)

Figure 6. Examples of segmentation for prefabrication of 3D-printed concrete bridges: (**a**) Toolpath and segmentation planes of FloatArch bridge (the superstructure is divided into 9 segments, denoted from Segment ① to Segment ⑨) [18] and (**b**) 3D-printed segments, reinforcement positioning, and assembly of the girder designed by Vantyghem et al. [16].

The construction of the 3D-printed concrete bicycle bridge in Gemert, Netherlands, further exemplifies this approach [23]. Due to facility limitations, the bridge's 6.5-m span was printed at the TU/e 3D concrete printing facility in six individual elements. The segments were designed to be rotated 90° after printing and assembled on-site. To ensure structural integrity, synthetic epoxy-based interface material was applied between the assembled elements. The post-tensioning system was anchored in traditionally cast and reinforced concrete bulkheads, while the bridge rested on two conventional abutments with pile foundations. Additionally, a steel parapet spanning the full bridge length was supported independently on the foundations.

Lastly, prefabricated and masonry assembly methods were used to construct the "Phoenix" bridge, a 9.5-m-long pedestrian bridge [21]. The bridge's primary arch ring was subdivided into 53 individual blocks of four different types, while the guardrails were divided into 40 pieces of 10 different types. The masonry assembly method was employed on-site, with the prefabricated blocks sequentially laid and bonded using an epoxy-based mortar. Temporary earthwork arch formwork, made of compacted soil, supported the primary arch ring during assembly, beginning at the arch foot and symmetrically progressing towards the arch crown. Each block was rotated 90 degrees during installation to ensure the printed layers were perpendicular to the arch axis. The bridge was further supported by gravity abutments and embedded foundations.

To conclude this section, prefabrication and on-site assembly are the most commonly employed methods for constructing 3D-printed concrete bridges. This approach offers significant benefits while also presenting certain challenges. Below is a summary of the advantages and drawbacks of prefabricated construction techniques:

Advantages:

- Reduced Construction Time: Prefabrication of structural components combined with rapid on-site assembly can reduce construction time by 50% to 80% compared to traditional cast-in-place methods [35,42,43].
- Improved Quality Control: Off-site fabrication of bridge segments in controlled environments ensures consistent conditions, reducing variability and enhancing accuracy and reliability in the final structure [23,35].
- Minimized On-Site Disruptions: Prefabrication reduces on-site activities, minimizing traffic congestion, noise, and environmental impact, which is particularly advantageous in urban or ecologically sensitive areas [19].
- Enhanced Sustainability: Modular prefabricated components allow for disassembly, reuse, and potential recycling, contributing to more sustainable construction practices [13,18].

Disadvantages:

- Transportation and Handling Challenges: The transportation of large, prefabricated segments can be logistically challenging and often requires specialized equipment.
- Connection Complexity: Robust design and execution of connections between prefabricated segments are essential to ensure efficient load transfer, structural integrity, and long-term durability of the bridge.

5. Reinforcement in 3D-Printed Concrete Bridges

Reinforcement is essential for nearly all concrete structural components due to the material's inherent weakness in tension. However, in 3D-printed concrete structures, conventional reinforcement methods, such as rebars, present challenges because of the layered fabrication process. This section reviews the reinforcement methods employed in 3D-printed concrete bridges, as documented in the literature.

Post-tensioning is a widely used method for reinforcing 3D-printed concrete bridge components, addressing the need for enhanced structural capacity and longer spans [16,25,38,44]. This method involves embedding steel tendons or cables within the printed concrete and applying tension after the concrete hardens, inducing compressive stresses to counteract tensile forces. Vantyghem et al. [16] combined post-tensioning with steel rebars to effectively integrate reinforcement into a 3D-printed girder. High-strength steel cables were optimized through topology design to align with the girder's geometry and were anchored in specially designed end blocks to withstand concentrated forces. Steel rebars provided supplementary reinforcement, enhancing tensile strength and ensuring robust force transfer. Their reinforcement positioning is shown in Figure 6b. Zhan et al. [25] used a similar approach, embedding steel bars within cavities during printing and tensioning them post-curing to strengthen the bridge. Mitrović et al. [44] employed post-tensioning to connect printed segments into a single structural element, strategically tensioning steel bars to enhance shear capacity and overall structural integrity.

The "Minimass" bridge design [26] represents another innovative application of post-tensioning, combining reinforced concrete and steel cables. The design features beams acting as trusses, with a reinforced concrete top chord handling compression, a steel cable bottom chord resisting tension, and concrete webs separating the two. The cables were encased in grease-filled ducts, providing corrosion resistance and allowing for re-stressing. Prefabricated lattice slabs and an in situ topping slab ensured seamless load transfer between the printed elements.

Salet et al. [23] also employed post-tensioning in their 3D-printed bicycle bridge using Dywidag-system tendons. These tendons were embedded in the printed elements, stressed, anchored at both ends, and then released. To address shear and torsion, high-strength steel cables were integrated into the concrete filaments during printing. A specialized embedding device was developed to ensure proper integration, maximizing bond strength and structural performance.

Some other projects tackled the challenges of embedding reinforcement during printing by opting for post-printing installation methods. For instance, reinforcement in 3D-printed bridge columns [43] involved installing reinforcement cages into 3D-printed hollow cylindrical formwork. The reinforcement cage comprised steel rebars, ties, and corrugated ducts that were installed within the formwork. The ducts were pathways for energy-dissipating (ED) steel and Fe-SMA rebars, which were inserted and grouted post-assembly. Fe-SMA rebars provided prestressing, with smooth midsections bonded to grout and threaded ends anchored in the footing and top loading block. Once the reinforcement cages were in place, cast concrete was poured into the formwork to encase the steel elements. The segments were stacked over a precast footing, bonded with cement paste, and grouted with the rebars.

The use of fibers within the concrete mix offers another reinforcement strategy for 3D-printed concrete bridges, which can improve tensile strength, ductility, and interlayer bonding. Polypropylene fibers, for example, were used in the 3D printing mortar for a bridge column to enhance crack resistance and interlayer adhesion [38]. Javed et al. [35] employed 2% steel fibers by volume in their mix to increase the strength and ductility of the printed layers, demonstrating the potential of fiber-reinforced concrete as a viable alternative to traditional reinforcement methods. Additionally, Zhan et al. [25] utilized 13 mm polyvinyl alcohol (PVA) fibers in their fiber-reinforced high-strength cementitious mortar to further enhance the structural properties of the printed elements. These examples underscore the versatility and effectiveness of fiber reinforcement in addressing the mechanical challenges associated with 3D-printed concrete structures.

Recent studies have demonstrated the feasibility of embedding FRP grids within printed concrete layers, improving structural integrity while maintaining printability [45–47]. This method has not yet been applied in 3D-printed concrete bridges, but it has potential for future applications, particularly in lightweight, corrosion-resistant structures where conventional reinforcement may be impractical. While FRP-based reinforcement in 3DPC offers numerous benefits, challenges remain, particularly in ensuring adequate bonding between FRP elements and the printed concrete matrix, optimizing grid spacing for uniform load distribution, and addressing long-term durability concerns. Future research should explore hybrid reinforcement techniques, where FRP is combined with steel or discrete fiber reinforcement to maximize performance while maintaining ease of printing.

Hybrid approaches that combine innovative and traditional techniques have also been explored. For example, the Striatus bridge is a 3D-printed masonry arch bridge that eliminates internal reinforcement in the printed blocks by relying on compression-only principles [41,48]. The structural stability of the arch is achieved through its geometry, while the foundation employs reinforced concrete pads to anchor steel footings and ground screws, demonstrating a blend of traditional and modern construction methods.

In conclusion, post-tensioning was the most used reinforcing approach, combining steel cables, rebars, and optimized designs to enhance structural capacity and stability. However, fiber-reinforced polymer (FRP) reinforcement can be a promising alternative, offering a lightweight and corrosion-resistant solution that can be directly integrated into the printing process. Additionally, post-printing methods, such as installing reinforcement cages in printed formwork, provide flexibility for incorporating advanced materials like Fe-SMA rebars. Hybrid approaches, such as the compression-only Striatus bridge, highlight the potential for integrating traditional and modern techniques in 3D-printed bridge construction.

6. Conclusions

The review indicates that 3D-printed concrete bridges have substantial potential to transform construction practices through enhanced material efficiency, design flexibility, and sustainability. The following conclusions and recommendations were drawn from this research:

Design process: The design processes for 3D-printed concrete bridges demonstrate the potential of computational and experimental methodologies in modern construction. Techniques such as topology optimization, parametric design, and thrust network analysis enable efficient material usage, reduce waste, and create innovative forms while leveraging the unique capabilities of 3D printing. Iterative methodologies, like "design by testing", ensure structural integrity and reliability and validate computational models. Moreover, incorporating advanced numerical methods, such as Discrete Element Modeling (DEM) and Finite Element Analysis (FEA), can lead to more sustainable and efficient solutions in bridge construction.

Materials: Printable concrete must maintain a balanced fresh state to ensure successful 3D printing and long-term performance. To achieve this, various admixtures and additives have been incorporated into the mix design of different 3DCP bridges. These include superplasticizers, water retention agents, polypropylene fibers, metakaolin, etc., all of which work to enhance workability, extrudability, buildability, and mechanical strength. Future research should focus on optimizing eco-friendly materials to enhance the sustainability of 3D-printed concrete structures. This includes developing materials with lower carbon footprints, incorporating recycled or renewable components, and improving the efficiency of material usage. Such advancements could significantly reduce the environmental impact of construction while maintaining or even improving structural performance and durability.

Construction techniques: The review of printing systems used for 3DCP bridges revealed a significant preference for robotic arm printers over gantry printers. Known for their flexibility and precision in producing intricate designs, robotic printers can leverage advanced technologies such as spatial and planar path fitting, real-time path generation, optimized nozzle configurations, and real-time monitoring.

Prefabrication and on-site assembly are the most practical and widely accepted methods for constructing 3D-printed concrete bridges. This approach enables components to be manufactured in controlled environments, enhancing precision and consistency and minimizing material waste. However, ensuring the bridge's structural integrity requires meticulous attention during the assembly and connection of individual components. This attention is especially critical for bridges intended to support substantial loads, such as vehicles, as poorly executed connections can create weak points that compromise the overall stability and durability of the structure.

Reinforcement strategies: Post-tensioning has emerged as an effective method for reinforcing 3D-printed concrete bridges. This technique combines steel cables, rebars, and optimized designs to enhance both structural capacity and stability. Practical post-printing approaches, such as embedding reinforcement cages within printed formwork, further extend the applicability of traditional reinforcement strategies. Additionally, fiber reinforcement can offer the potential for enhancing tensile strength, ductility, and interlayer bonding. However, challenges remain in addressing reinforcement perpendicular to layer-to-layer interfaces—an aspect crucial for maintaining structural integrity. This gap presents an exciting opportunity for future innovation. Integrating perpendicular reinforcement during the printing process, possibly through the use of auxiliary robots, expands the structural capabilities of 3D-printed concrete.

7. Recommendations for Future Research

While significant progress has been made in the field of 3D-printed concrete bridges, several key research areas remain underexplored and require further investigation:

- Investigating advanced numerical modeling approaches. Coupled multi-scale models and machine learning-based simulations could optimize structural design under real-world conditions.
- Exploring the role of anisotropy in mechanical performance. Further research is required to quantify the impact of anisotropic behavior and interlayer bonding on long-term structural integrity.
- Developing standardized guidelines and building codes. A lack of regulatory standards hinders the widespread adoption of 3D-printed bridges. Future research should contribute to the establishment of industry-wide guidelines for structural assessment and approval.
- Eco-friendly cementitious materials. Future research should develop low-carbon cement alternatives for 3D-printed concrete bridges.
- Automated quality control and real-time monitoring. Future research should focus on AI-driven defect detection and adaptive extrusion systems to enhance precision during printing.
- Scaling up in situ 3D printing. While most 3D-printed bridges are prefabricated, advancements in mobile printing technologies could enable on-site, large-scale printing for infrastructure projects.
- Developing hybrid reinforcement strategies. To improve structural resilience, research is needed on integrating pre-tensioning, post-tensioning, and embedded reinforcement techniques.
- Optimization of fiber-reinforced polymer reinforcement. Investigating FRP's role in 3D-printed bridges could expand its application in corrosion-resistant, lightweight structures.

Author Contributions: Conceptualization, Z.S.M.; methodology, Z.S.M.; investigation, Z.S.M.; writing—original draft preparation, Z.S.M.; writing—review and editing, Z.S.M., H.B. and M.A.P.; supervision, H.B. and M.A.P.; project administration, Z.S.M. and H.B.; All authors have read and agreed to the published version of the manuscript.

Funding: This research received no external funding.

Conflicts of Interest: The authors declare no conflicts of interest.

Abbreviations

The following abbreviations are used in this manuscript:

3DCP	3D Concrete Printing
AM	Additive Manufacturing
TO	Topology Optimization
FEA	Finite Element Analysis
BESO	Bi-directional Evolutionary Structural Optimization
TNA	Thrust Network Analysis
DEM	Discrete Element Modeling
FRP	Fiber-Reinforced Polymer

References

1. Marchment, T.; Sanjayan, J.; Xia, M. Method of enhancing interlayer bond strength in construction scale 3D printing with mortar by effective bond area amplification. *Mater. Des.* **2019**, *169*, 107684. [CrossRef]
2. Marchment, T.; Xia, M.; Dodd, E.; Sanjayan, J.; Nematollahi, B. Effect of Delay Time on the Mechanical Properties of Extrusion-based 3D Printed Concrete. In Proceedings of the ISARC International Symposium on Automation and Robotics in Construction, Taipei, Taiwan, 28 June–1 July 2017.
3. Delgado Camacho, D.; Clayton, P.; O'Brien, W.J.; Seepersad, C.; Juenger, M.; Ferron, R.; Salamone, S. Applications of additive manufacturing in the construction industry—A forward-looking review. *Autom. Constr.* **2018**, *89*, 110–119. [CrossRef]
4. Sanjayan, J.; Nematollahi, B.; Xia, M.; Marchment, T. Effect of surface moisture on inter-layer strength of 3D printed concrete. *Constr. Build. Mater.* **2018**, *172*, 468–475. [CrossRef]
5. Will 3D Printing Remodel the Construction Industry? Available online: https://www.bcg.com/publications/2018/will-3d-printing-remodel-construction-industry (accessed on 7 August 2022).
6. World's First 3D-Printed Bridge Opens to Cyclists in Netherlands. Available online: https://www.theguardian.com/technology/2017/oct/18/world-first-3d-printed-bridge-cyclists-netherlands (accessed on 12 January 2025).
7. MEED | Case Study: Using an AI-Powered Robot to Construct Buildings. Available online: https://www.meed.com/artificial-intelligence-robot-construct-buildings/ (accessed on 12 January 2025).
8. World's Largest 3D-Printed Concrete Pedestrian Bridge Completed in China | ArchDaily. Available online: https://www.archdaily.com/909534/worlds-largest-3d-printed-concrete-pedestrian-bridge-completed-in-china (accessed on 26 November 2024).
9. World's Longest 3D Printed Concrete Pedestrian Bridge Unveiled in Nijmegen—3D Printing Industry. Available online: https://3dprintingindustry.com/news/worlds-longest-3d-printed-concrete-pedestrian-bridge-unveiled-in-nijmegen-195951/ (accessed on 26 November 2024).
10. First 3D Printed and Unreinforced Concrete Bridge | ETH Zurich. Available online: https://ethz.ch/en/news-and-events/eth-news/news/2021/07/3d-printed-and-unreinforced.html (accessed on 26 November 2024).
11. Phoenix, The New 3D Printed Concrete Bridge—3Dnatives. Available online: https://www.3dnatives.com/en/phoenix-the-new-3d-printed-circular-concrete-bridge-020120245/ (accessed on 26 November 2024).
12. Holcim Launches Phoenix, the First-of-Its-Kind Circular 3D-Printed Concrete Bridge. Available online: https://www.holcim.com/media/company-news/phoenix-circular-3D-printed-concrete-bridge (accessed on 26 November 2024).
13. Dell'Endice, A.; Bouten, S.; Van Mele, T.; Block, P. Structural design and engineering of Striatus, an unreinforced 3D-concrete-printed masonry arch bridge. *Eng. Struct.* **2023**, *292*, 116534. [CrossRef]
14. Dell'Endice, A.; Bodea, S.; Van Mele, T.; Block, P.; Bhooshan, V.; Bhooshan, S.; Eiz, H.; Chen, T.; Lombois-Burger, H.É.; De La Mothe, L.R.; et al. The Phoenix bridge: Improving circularity of 3D-concrete-printed unreinforced masonry structures. In *Fabricate*; UCL Press: London, UK, 2024.
15. Briseghella, B.; Fenu, L.; Lan, C.; Mazzarolo, E.; Zordan, T. Application of Topological Optimization to Bridge Design. *J. Bridge Eng.* **2013**, *18*, 790–800. [CrossRef]
16. Vantyghem, G.; De Corte, W.; Shakour, E.; Amir, O. 3D printing of a post-tensioned concrete girder designed by topology optimization. *Autom. Constr.* **2020**, *112*, 103084. [CrossRef]
17. Ooms, T.; Vantyghem, G.; Tao, Y.; Bekaert, M.; De Schutter, G.; Tittelboom, K.V.; De Corte, W. The Production of a Topology-Optimized 3D-Printed Concrete Bridge. In Proceedings of the Third RILEM International Conference on Concrete and Digital Fabrication, Loughborough, UK, 27–29 June 2022; RILEM Bookseries. Springer: Cham, Switzerland; Volume 37.
18. Li, Y.; Wu, H.; Xie, X.; Zhang, L.; Yuan, P.F.; Xie, Y.M. FloatArch: A cable-supported, unreinforced, and re-assemblable 3D-printed concrete structure designed using multi-material topology optimization. *Addit. Manuf.* **2024**, *81*, 104012. [CrossRef]
19. Pons-Valladares, O.; Casanovas-Rubio, M.M.; Armengou, J.; de la Fuente, A. Approach for sustainability assessment for footbridge construction technologies: Application to the first world D-shape 3D-Printed fiber-reinforced mortar footbridge in Madrid. *J. Clean. Prod.* **2023**, *394*, 136369. [CrossRef]
20. 3D Printed Bridge—IAAC. Available online: https://iaac.net/project/3d-printed-bridge/ (accessed on 29 November 2024).
21. Yang, M.; Li, C.; Liu, H.; Huo, L.; Yao, X.; Wang, B.; Yao, W.; Zhang, Z.; Ding, J.; Zhang, Y.; et al. Exploring the potential for carrying capacity and reusability of 3D printed concrete bridges: Construction, dismantlement and reconstruction of a box arch bridge. *Case Stud. Constr. Mater.* **2024**, *20*, e02938. [CrossRef]
22. Liu, D.; Zhang, Z.; Zhang, X.; Chen, Z. 3D printing concrete structures: State of the art, challenges, and opportunities. *Constr. Build. Mater.* **2023**, *405*, 133364. [CrossRef]
23. Salet, T.A.M.; Ahmed, Z.Y.; Bos, F.P.; Laagland, H.L.M. Design of a 3D printed concrete bridge by testing. *Virtual Phys. Prototyp.* **2018**, *13*, 222–236. [CrossRef]
24. Ahmed, Z.; Wolfs, R.; Bos, F.; Salet, T. A Framework for Large-Scale Structural Applications of 3D Printed Concrete: The Case of a 29 m Bridge in the Netherlands. *Open Conf. Proc.* **2022**, *1*, 5–19. [CrossRef]

25. Zhan, Q.; Zhou, X.; Yuan, P.F. Digital Design and Fabrication of a 3D Concrete Printed Prestressed Bridge. In Proceedings of the 26th International Conference of the Association for Computer-Aided Architectural Design Research in Asia (CAADRIA), Hong Kong, 29 March 2021.
26. Coward, A.; Forsberg, T. Reconstruction of a Ukrainian road bridge by use of 3D printed minimass beams. In Proceedings of the IABSE Symposium, Manchester, UK, 10–12 April 2024.
27. Zhang, J.; Wang, J.; Dong, S.; Yu, X.; Han, B. A review of the current progress and application of 3D printed concrete. *Compos. Part A Appl. Sci. Manuf.* **2019**, *125*, 105533. [CrossRef]
28. Le, T.T.; Austin, S.A.; Lim, S.; Buswell, R.A.; Gibb, A.G.F.; Thorpe, T. Mix design and fresh properties for high-performance printing concrete. *Mater. Struct.* **2012**, *45*, 1221–1232. [CrossRef]
29. Nerella, V.N.; Hempel, S.; Mechtcherine, V. Effects of layer-interface properties on mechanical performance of concrete elements produced by extrusion-based 3D-printing. *Constr. Build. Mater.* **2019**, *205*, 586–601. [CrossRef]
30. Schröfl, C.; Nerella, V.N.; Mechtcherine, V. Capillary water intake by 3D-printed concrete visualised and quantified by neutron radiography. *RILEM Bookseries* **2019**, *19*, 217–224.
31. Wang, C.; Chen, B.; Vo, T.L.; Rezania, M. Mechanical anisotropy, rheology and carbon footprint of 3D printable concrete: A review. *J. Build. Eng.* **2023**, *76*, 107309. [CrossRef]
32. Chen, Y.; Jia, L.; Liu, C.; Zhang, Z.; Ma, L.; Chen, C.; Banthia, N.; Zhang, Y. Mechanical anisotropy evolution of 3D-printed alkali-activated materials with different GGBFS/FA combinations. *J. Build. Eng.* **2022**, *50*, 104126. [CrossRef]
33. van den Heever, M.; du Plessis, A.; Kruger, J.; van Zijl, G. Evaluating the effects of porosity on the mechanical properties of extrusion-based 3D printed concrete. *Cem. Concr. Res.* **2022**, *153*, 106695. [CrossRef]
34. Marchment, T.; Sanjayan, J.G.; Nematollahi, B.; Xia, M. Interlayer strength of 3D printed concrete: Influencing factors and method of enhancing. In *3D Concrete Printing Technology*; Butterworth-Heinemann: Oxford, UK, 2019; pp. 241–264.
35. Javed, A.; Mantawy, I.M.; Azizinamini, A. 3d-printing of ultra-high-performance concrete for robotic bridge construction. *Transp. Res. Rec.* **2021**, *2675*, 307–319. [CrossRef]
36. Kinomura, K.; Murata, S.; Yamamoto, Y.; Obi, H.; Hata, A. Application of 3D Printed Segments Designed by Topology Optimization Analysis to a Practical Scale Prestressed Pedestrian Bridge. *RILEM Bookseries* **2020**, *28*, 658–668.
37. Tao, Y.; Rahul, A.V.; Lesage, K.; Van Tittelboom, K.; Yuan, Y.; De Schutter, G. Mechanical and microstructural properties of 3D printable concrete in the context of the twin-pipe pumping strategy. *Cem. Concr. Compos.* **2021**, *125*, 104324. [CrossRef]
38. Zhang, D.; Ma, G.; Guan, J.; Wang, L.; Wang, Q. Cyclic behavior of unbonded post-tensioned precast segmental concrete columns fabricated by 3D printed concrete permanent formwork. *Eng. Struct.* **2023**, *292*, 116436. [CrossRef]
39. Puzatova, A.; Shakor, P.; Laghi, V.; Dmitrieva, M. Large-Scale 3D Printing for Construction Application by Means of Robotic Arm and Gantry 3D Printer: A Review. *Buildings* **2022**, *12*, 2023. [CrossRef]
40. Main Differences Between 3D Concrete Printing Technologies. Available online: https://cobod.com/robotic-arm-vs-gantry-3d-concrete-printer/ (accessed on 16 December 2024).
41. Bhooshan, S.; Bhooshan, V.; Dell'Endice, A.; Chu, J.; Singer, P.; Megens, J.; Van Mele, T.; Block, P. The Striatus bridge: Computational design and robotic fabrication of an unreinforced, 3D-concrete-printed, masonry arch bridge. *Archit. Struct. Constr.* **2022**, *2*, 521–543. [CrossRef]
42. Mantawy, I.M.; Thonstad, T.; Sanders, D.H.; Stanton, J.F.; Eberhard, M.O. Seismic Performance of Precast, Pretensioned, and Cast-in-Place Bridges: Shake Table Test Comparison. *J. Bridg. Eng.* **2016**, *21*, 04016071. [CrossRef]
43. Raza, S.; Triantafyllidis, Z.; Anton, A.; Dillenburger, B.; Shahverdi, M. Seismic performance of Fe-SMA prestressed segmental bridge columns with 3D printed permanent concrete formwork. *Eng. Struct.* **2024**, *302*, 117423. [CrossRef]
44. Mitrović, S.; Ignjatović, I. Experimental investigation of bearing capacity of 3D printed concrete segmental girder. *Gradjevinski materijali i konstrukcije* **2024**, *67*, 137–146. [CrossRef]
45. Zeng, J.J.; Yan, Z.T.; Jiang, Y.Y.; Li, P.L. 3D printing of FRP grid and bar reinforcement for reinforced concrete plates: Development and effectiveness. *Compos. Struct.* **2024**, *335*, 117946. [CrossRef]
46. Zeng, J.J.; Li, P.L.; Yan, Z.T.; Zhou, J.K.; Quach, W.M.; Zhuge, Y. Behavior of 3D-printed HPC plates with FRP grid reinforcement under bending. *Eng. Struct.* **2023**, *294*, 116578. [CrossRef]
47. Yan, Z.; Zeng, J.J.; Zhuge, Y.; Liao, J.J.; Zhou, J.K.; Ma, G. Compressive behavior of FRP-confined 3D printed ultra-high performance concrete cylinders. *J. Build. Eng.* **2024**, *83*, 108304. [CrossRef]
48. Striatus 3D-Printed Bridge 'Establishes a New Language for Concrete. Available online: https://www.dezeen.com/2021/10/12/striatus-3d-printed-concrete-bridge-holcim-video/ (accessed on 5 June 2022).

applied sciences

MDPI

Review

Transition Effects in Bridge Structures and Their Possible Reduction Using Recycled Materials

Mariusz Spyrowski [1,*], Krzysztof Adam Ostrowski [2,*] and Kazimierz Furtak [2]

[1] Cracow University of Technology, CUT Doctoral School, Faculty of Civil Engineering, 24 Warszawska Str., 31-155 Cracow, Poland

[2] Cracow University of Technology, Faculty of Civil Engineering, 24 Warszawska Str., 31-155 Cracow, Poland

* Correspondence: mariusz.spyrowski@doktorant.pk.edu.pl (M.S.); krzysztof.ostrowski.1@pk.edu.pl (K.A.O.)

Abstract: This article serves as a review of the current challenges in bridge engineering, specifically addressing the transition effect and the utilization of recycled materials. It aims to identify research gaps and propose innovative approaches, paving the way for future experimental studies. As a review article, the authors critically analyze the existing literature on the transition effects in bridge construction, their causes, and their negative impacts. Integral bridges are discussed as a solution designed to work in conjunction with road or rail embankments to transfer loads, minimizing maintenance and construction costs while increasing durability. Particular attention is given to the potential use of modified plastic composites as an alternative material in integral bridge structures. This concept not only addresses the issue of plastic waste but also promotes the long-term use of recycled materials, a key consideration given recycling limitations. This article highlights the importance of the connection between the embankment and the abutment and provides examples of polymer applications in bridge engineering. By outlining the state of the art, this review identifies future development paths in this niche, but promising, field. Almost 240 literature items were analyzed in detail, and works containing 475 different key words contained in about 3500 individual works were used for scientometric analysis. The results of the analysis clearly indicate the novelty of the presented subject matter.

Keywords: civil engineering; integral bridges; transition zone; transition effect; reinforced polymer composite; recycling; waste material; sustainability

1. Introduction

The transition effect in bridge structures refers to the combination of adverse phenomena occurring at the junction of embankments and engineering structures. This complex issue, caused by differences in settlement, stiffness, and dynamic loads, leads to increased wear, damage, and maintenance costs. Addressing this challenge is critical for ensuring the long-term durability and safety of infrastructure.

In traditional bridge designs, expansion joints, bearings, and transition slabs are employed to manage these effects. However, these solutions often fail to provide a gradual transition, resulting in increased dynamic loads and accelerated wear. Integral bridges, which eliminate these elements, offer a promising alternative by enhancing durability and reducing maintenance needs. However, their application is limited due to challenges in managing differential settlements and stiffness transitions.

Simultaneously, the increasing accumulation of plastic waste has become a pressing global issue. Recycling polymer materials into construction applications, particularly in transition zones, presents an innovative approach to addressing both environmental and engineering challenges. Integrating recycled polymers into embankments could enhance their stiffness, mitigate differential settlements, and reduce the overall impact of the transition effect.

This article provides a comprehensive review of the transition effect in bridge structures and explores the potential for integrating recycled materials into embankments. By analyzing current solutions, identifying gaps in research, and proposing future directions, this study aims to pave the way for sustainable and efficient innovations in bridge engineering.

2. Methodology

The methodology used was appropriate for exact sciences. The dominant approach was the use of the quantitative counting method. Based on the collected finite set of bibliographic data, a graphical result of the research was prepared, and conclusions were drawn using specialist software. A graphical method was used for the development. The size of the cluster reflects the frequency of occurrence of the keywords. The thickness of the connections the frequency of co-occurrence of given words. The color distinction links the words with the date in which the texts containing them were published. This was carried out using simple mathematical formulas embedded in the software. On the set containing the elements necessary for the proper performance of systematic reviews and meta-analyses, relationships were assessed. Thanks to this analysis, other authors can obtain a transparent and complete state of knowledge from this research. Before delving into a detailed analysis of the available literature related to the proposed topics in this publication, a scientometric analysis was conducted based on publicly available library catalogs. This type of analysis aims to highlight the major scientific trends and areas requiring in-depth research. It facilitates the creation of visualizations for data analysis and establishes connections between sources, keywords, authors, and articles within a specific research area. Researchers from various fields of science utilize scientometric analysis [1–5].

Currently, scientists have access to a substantial collection of library catalogs promoted by various publishers [5,6]. The analyzed literature was selected based on the widely accessible and most reliable databases Scopus [7] and Web of Science [8]. Thanks to them, it was possible to prepare the search results in the form of Figure 1, which illustrates the frequency of occurrence of keywords, their mutual connections, and the novelty of their appearance. Due to the multifaceted nature of this article, the explored topics covered various fields, including civil engineering, material engineering, chemical engineering, chemical sciences, environmental engineering, and environmental sciences.

To correctly perform a bibliography analysis, it is necessary to determine the minimum number of keywords that must be included in a single work. Often, the overlap of two keywords is far from sufficient. Only with three keywords do single articles begin to share common parts. Data from Scopus and Web of Science databases for the selected searches (three and four keywords) were exported in comma-separated values (CSV) format for importing into the appropriate software tool. Mapping and visualizing the academic network were carried out using the VOSviewer software version 1.6.20 (developed by the Centre for Science and Technology Studies of Leiden University, Leiden, The Netherlands). VOSviewer is a software tool used to construct and visualize bibliometric networks based on citations, bibliographic couplings, co-citations, or co-authorship relations [9].

The visualization of co-occurrence networks of keywords, their relationships, and the density associated with the frequency of their correlations were examined and are presented in Figure 1. The size of the keyword node denotes its frequency, while its location represents its co-occurrence in the publications. The entirety of the connections is shown, where the material scope has been omitted. To identify clusters among over 4500 author keywords contained in nearly 3500 individual works, those appearing a minimum of three times were selected, reducing the number of analyzed words to 475. This demonstrates a significant diversity of topics and areas. Words that were entirely unrelated thematically were excluded from the visualization, leaving around 350. Many weakly connected words imply their infrequent occurrences, being referenced only a few times with other analyzed words from different texts. This indicates the vast fragmentation within the research area. The limitations of this study include the availability of the literature in the databases, the analyzed journals being in English, the limitation of the number of analyzed manuscripts and words, and the need to compensate for the analysis and review in this article.

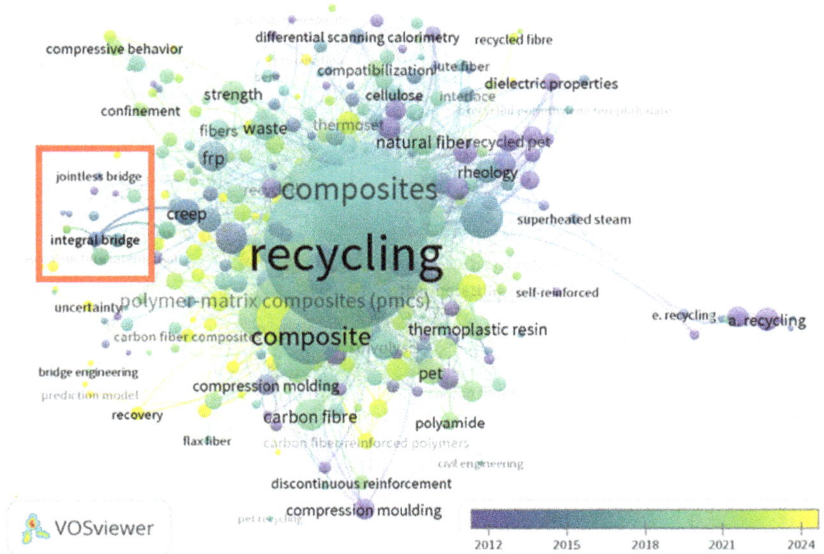

Figure 1. Visualization of the incidence and relationships of individual keywords, created using VOSviewer software version 1.6.20 and additionally showing the date of publication (as of 1 October 2024).

This visualization can not only assist in finding trends but also aid future authors in selecting presented keywords to find published data on specific topics. The results presented in Figure 1 indicate that topics related to bridge engineering have been less frequently addressed in recent years. Popular areas in publications seem to revolve around recycling and the use of reinforced polymers. However, there is a lack of keyword connections between bridge-related topics and the use of reinforced polymer composites, not to mention the transition effect, for which there is a minimal number of keywords. The interesting range in this study is marked with a red square. It is clearly visible that it is not very developed and not very intensive. The analysis presented clearly indicates the necessity for further research in the highlighted subject matter.

3. General Purpose of Research

The aim of this review article is to showcase the current solutions in bridge engineering, focusing on reducing the transition effect and emphasizing the significance of the problem. Additionally, the issue of plastic recycling in the context of utilizing waste plastic as an advanced material base for construction is outlined.

First and foremost, our research introduces a novel approach to addressing the persistent challenge of transition zones in bridge structures. Integral bridges not only eliminate the need for expansion joints and bearings but also enhance durability and reduce maintenance costs—a departure from conventional bridge construction practices.

Furthermore, our study underscores the paramount importance of sustainable development in bridge engineering. We advocate for the utilization of recycled materials, particularly polymers reinforced with fibers, as viable alternatives to traditional construction materials. This emphasis on sustainability aligns with the evolving ethos of modern engineering practices, wherein environmental considerations play an increasingly pivotal role.

Additionally, our research explores the untapped potential of substitute materials processed and shaped to meet the rigorous demands of bridge engineering. By harnessing innovative solutions and materials, we envision a future where construction practices are not only more sustainable but also more cost-effective and ecologically sound.

Moreover, a comprehensive review was conducted on PET (polyethylene terephthalate) as a promising material for integral bridge construction, highlighting its superior strength properties and compatibility as fiber composites. This in-depth analysis lays the groundwork for the further exploration and utilization of PET and similar materials in bridge engineering applications, especially taking into account recycling restrictions. It should be emphasized that the construction of connecting structures forcing cooperation in integrated bridges between the facility itself and the embankment using the indicated polymer material will be a novelty in engineering, eliminating the transition effects.

4. Research Area

The research area focused on assessing transition effects using post-consumer polymer materials is multifaceted. In today's world, solving individual problems in the short term is challenging. When advancing knowledge in one field, it is essential to consider whether materials available in other areas might be useful for solutions. Therefore, construction itself generates a large demand for materials and can effectively utilize materials that are considered waste in other sectors.

This study addresses the following aspects:

- Engineering Aspect: Current projects do not entirely solve the issue of stiffness discontinuity (flexibility) in the subgrade for linear structures approaching bridges. According to the authors, further progress supported by appropriately developed materials, and the spatial structure of the joint construction will undoubtedly enrich contemporary bridge engineering with new design solutions that could reduce the transition effect.
- Environmental Aspect: The new material and its application area offer opportunities to utilize large amounts of waste that accumulate in landfills or are incinerated. Due to their longevity anticipated in engineering constructions, the properties of processed waste will serve future generations. Pursuing the idea of sustainable development involves seeking applications for waste that is considered inefficient, difficult, or costly to process. This fact guarantees a constant demand for solutions enabling recycling.

5. Detailed Overview

5.1. Outline

Bridge construction history dates back several thousand years [10]. Bridges are an integral part of infrastructure and have played a pivotal role in civilization's development. With technological advancements and heightened demands for durability and safety, engineers increasingly seek for superior solutions and construction materials. Bridges stand among the most significant engineering structures, serving as essential components of transportation infrastructure, facilitating access across natural obstacles like rivers, valleys, and artificial barriers, ensuring convenient and safe passage to human settlements. They form a fundamental factor in societal development by enabling the transportation of goods and people [10–15].

Bridge structures must have stability and the necessary load-bearing capacity to meet traffic intensity and load requirements. They must also be durable and safe for users [16,17]. Hence, the construction materials used must meet stringent strength requirements, including resistance to variable and dynamic loads, corrosion, and weather conditions [16,17].

Since the late twentieth century, experts have debated how to repair, renew, and modernize transportation infrastructure in Central Europe [18]. Over the past 25 years, there has been a focus on utilizing modern engineering technologies and decision-making processes to solve typical and regional environmental problems in land transport, especially roads and bridges.

The transition effect occurs at points where different types of permanent way or pavements are connected, laid on various substrates (e.g., soil subgrade, engineering structures) [19,20]. This effect is a complex phenomenon, as illustrated in Figure 2, showing the most significant causes of its occurrence, including permanent settlements at abutments δ_a (usually smaller) and embankments δ_e (usually larger), progressing over time with unequal values for individual elements and with momentary deformations from live loads δ_s (deflection of girders, rails, and substrate deformations), causing the rotation of the span φ_s. Among the surface-related causes, differences in stiffness between the ballast and ballastless tracks, as well as varying stiffness parameters across the transverse direction, can be listed. They also include the primary geometric irregularities of rails and the inaccurate alignment of the gradeline on and outside the facility. Structural causes include primary geometric irregularities, deformations, settlement, and increased stiffness. Causes related to the substrate involve poor sub-ballast compaction, excessive stiffness differences between sub-ballast and substrate, sub-ballast contamination, and improperly chosen geotextiles. Due to different settlement values at the abutment and embankment, a so-called "threshold" is formed, which leads, among other reasons, to additional dynamic loads on the structure and excessive permanent deformations of the surface, amplifying dynamic effects and causing increased wear and tear and damage to pavements, track superstructures, substrates, and structures.

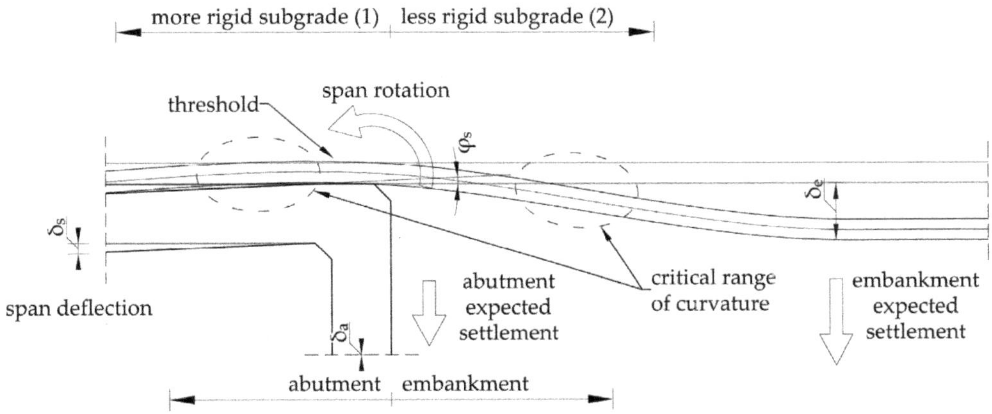

Figure 2. Figure explaining the transition effect.

In order to determine the impact of this effect, the so-called transition zone is distinguished. It is the area where the embankment meets the bridge structure, also referred to as the embankment–abutment interaction zone [21,22]. It is one of the most critical locations along a road or railway line. Due to the fact that the unit stresses acting on the ground beneath the embankment base are usually smaller than those under the abutment foundation slab of the bridge, the settlements of the abutment will be larger, causing unevenness in the grade line [23]. It is impractical to ensure the same settling for the entire embankment over many kilometers as for the rigid bridge abutment. Hence, a limited zone is defined, where a change in stiffness must occur.

5.2. Engineering Aspect

Designing and constructing bridges with their associated embankments require special attention. These structures are often placed in locations where there are poorly compacted, young alluvial sediments and sometimes clayey or peaty soils, which undergo significant settlement [24]. Crossing rivers and watercourses is always associated with the risk of riverbed erosion and changes in ground level directly in contact with the bridge support. The transmission of loads to the subsoil is significantly influenced by repeated, cyclic, and variable loads [25]. The practical application of new soil-strengthening technologies beneath embankments also introduces new risks to the overall stability of the structure.

The transition zone at the embankment–abutment junction is particularly vulnerable. This issue is not new; it was previously addressed in early post-war standards [26,27]. In recent years, this problem has been extensively highlighted in the literature and at numerous conferences [28–38]. It is worth noting that the most significant issues in this area often occur during execution due to gaps in evaluating intermediate schemes during design phases [25]. The issue of ensuring uniform support on the road or railway at the junction of an engineering structure and a deforming earthwork concerns most existing and renovated structures of this kind and, to a lesser extent, new lines, where the issue of non-uniformity can be largely prevented through the use of appropriate designs [30,39]. So far, clear requirements and solutions ensuring a gradual change in the substrate's stiffness for transition zones and reducing the effects of increased rolling stock actions in engineering structure areas have not been defined [30,39].

The embankment adjacent to engineering structures should be protected to equalize settlements at the junction with the support and prevent irregularities in the road pavement or rail track superstructure. Currently, a commonly used solution to mitigate substrate stiffness changes is a transitional slab. Its concept involves softening the stiffness by supporting it unilaterally on a rigid abutment on one side and through resilient embedding in the embankment on the other side The presented solution is demonstrated for integral abutments. The transition slab solution is depicted in Figure 3.

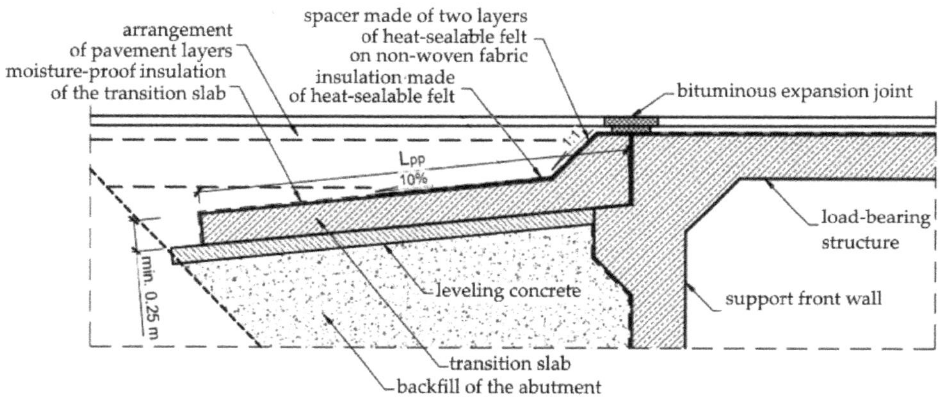

Figure 3. Longitudinal section through an example of an integral abutment.

However, analyzing the most common damages during the "lifetime" of structures, such as road pavement cracks and track superstructure deformations, suggests that this solution might not be an entirely effective solution as it does not ensure a sufficiently gradual change in substrate stiffness (flexibility) [40]. The analysis of the most frequent damages, such as cracks and deformations, suggests that the transition slab is not entirely a suitable solution as it does not provide a sufficiently gradual change in the substrate's stiffness (flexibility) [40]. This statement is vividly illustrated by photographs taken on numerous bridge structures, with selected examples shown in Figure 4 [40,41], and it is also supported by the content presented in publications by several authors [40,42–51]. By reviewing the available literature and observing existing bridge structures, it becomes evident that surface damages occur at the beginning of the transition zone, where the transition slab starts. Road pavement cracks mark the onset of gradual road degradation and can pose a danger to vehicles over time. However, for trains, there are even more risks involved. The track geometry always degrades faster in transition zones than in open tracks, causing significant irregularities (dips) [44]. Such irregularities in geometry can lead to substantial forces, potentially damaging track components, compromising passenger comfort and even causing derailment [31,32,52–60].

(a) (b)

Figure 4. Damage to road pavement (**a**) and rail permanent way (**b**) at the beginning of the transition zone where transition slabs are used.

Very different design solutions for transition zones on railway lines are provided, but they do not include a comprehensive assessment of the transition effect [32,57]. In contrast to roads where reinforced concrete transition slabs are frequently used, they are not popular for railway lines. Instead, the reinforcement of the embankment with geopolymers, draining and vibration-isolating mats, synthetic resins, stabilized soil blocks, gabions, reinforcements using stone columns or micropiles, dogging, chemical stabilization of sub-ballast, extending and widening substructures, and using stiffening rails within track rails are employed. Known solutions also include the compensatory injection method and polyurethane mass-stiffening aggregate.

The use of substitute materials in the integral bridge engineering of embankments behind the abutment is often used to relieve this issue. By reducing the pressure on the abutment, its dimensions can be reduced, thus minimizing the difference in stiffness (compliance) between the surfaces. However, the resulting void between the back wall of the abutment and the embankment creates a discontinuity that can negatively impact operational qualities, increasing the transition effect. This type of solution has been studied, among others, by Horvath [61]. Engineers must therefore balance the selection of solutions to obtain a satisfactory effect.

Road and railway embankments, as well as bridge abutments, can be designed as reinforced soil structures [21,22]. Additional reinforcement at the bottom of the embankment, extending beyond the outline of the transition slab, is often used to minimize settlement differences [23]. The purpose of reinforcing the ground is to increase the soil shear strength and reduce the pressure on the abutment [62]. In such structures, active forces and external loads are partly transferred through friction by the soil and partly through the reinforcement anchored in the ground. Reinforced soil structures can particularly be applied as a soil block separated from the front and side walls of the abutment, bearing the embankment pressure, as a retaining structure forming the abutment walls and as a foundation support for piers. In the case of a significant increase in pressure on native soil, there is a risk of uneven settling or loss of stability. To reduce the differences between the embankment and abutment settlement in the transition zone, often, embankment soil reinforcement is necessary [21–23,45,63–65]. Using a reinforced platform allows for an even distribution of settling and can be controlled [66]. Reinforced soil structures, serving as retaining structures, can be executed using passive systems, where soil reinforcement comprises both structural elements (reinforcement wraps embedded within the reinforced soil block) and auxiliary components (anchoring inserts placed between the structural reinforcements), which anchor the protective elements of the wall. Alternatively, active systems incorporate reinforcing soil inserts anchored in the wall's protective elements (reinforcing–anchoring inserts). Simultaneously, protective elements functioning as formwork for subsequent layers of fill material are placed alongside the formation of the reinforced earth block. The protective wall and reinforced soil block constitute separate structures, enabling their independent construction. For abutment elements, utilizing reinforced soil structures in a passive system is recommended due to the lesser impact of embankment settlements [63–65,67,68]. Flat steel strips, ribbed steel bars, polyethylene geogrids, polymer strips, or strips made of polyester fibers with a polyethylene coating serve as soil reinforcement. The reinforcement connects with the protective elements using systemic connectors.

The use of geosynthetics in constructing transition zones holds significant potential. Reinforcing the subsoil of the embankment and increasing its stiffness will reduce the transition effect.

Implementing engineering structures combined with embankments on weak ground is a complex task. Evaluating the load-bearing capacity and stability of road or railway embankments is the initial step [69–71]. Structures embedded within the embankment are usually supported on piles in challenging conditions [23,72–74].

Placing the structure on deep (pile) foundations results in minimal settlements [23], whereas allowable settlements for adjacent embankments can reach several dozen millimeters [69,71,72,74,75]. If the transition zone is not adequately designed or if the settlements are not uniform, unacceptable irregularities might emerge, which are also unacceptable to users. This aspect is more critical for railway lines [22,69,70].

Engineers often grapple with the durability issue of bridge structures. According to EC 0, bridges fall under category (class) S5, which implies an approximate design service life of at least 100 years. There are also structures over 200 years old that are still in use. Hence, the aspect of the durability and safety of the structure while limiting construction and operation costs is crucial [76].

The elements with the shortest lifespan in bridge structures are expansion joints and bearings. The most advanced ones can operate for only about 30 years [77]. While piers and girders determine a bridge's load-bearing capacity, equipment elements significantly influence its durability and maintenance costs. The immediate cost of equipment element installation constitutes around 15–20% of the bridge's construction costs. Their contribution to maintenance and repairs is often much higher, frequently exceeding 50% [17,62,76].

Ensuring the durability of bridge structures imposes specific requirements on the longevity of individual components. Despite advancements in bearing and expansion joint construction, production, assembly, and maintenance costs continue to rise. Additionally, there are increasing demands for the qualifications of personnel involved in these processes. The growing number of structures adds to maintenance challenges, highlighting the need for simpler solutions. Integral bridges present a viable option in this context.

Frame structures, particularly integral bridges, are not equipped with these expensive solutions [40,62,77–84]. Hence, they are less susceptible to natural and human-induced threats while requiring minimal maintenance throughout their service life. Most conventional bridges are equipped with components compensating for distortions and displacements. Water, salts, and other deicing agents can seep through expansion joints, leading to corrosion in bearings, load-bearing structures, sidewalk parapets, and other bridge elements. The absence of expansion joints in integral bridges reduces repair and maintenance costs throughout their service life [85,86], aligning with the urgent global need for low-maintenance transportation infrastructure [87].

Furthermore, integral bridges, when integrated into highways or railway lines, enhance riding comfort due to the absence of expansion joints and provide better horizontal stiffness perpendicular to the longitudinal axis of the bridge. This reduces the likelihood of rail misalignment [85]. Moreover, modern integral bridges exhibit resilience during earthquakes owing to their monolithic construction [85–93]. Thanks to these advantages, such structures can be appealing to entities managing transportation infrastructure.

These include single- or multi-span structures with a continuous deck connected to the abutment. The connection between the abutment and the deck can be rigid or semi-rigid, depending on the structural solution of the connection. The abutment can be articulated on the foundation or supported on piles. The fundamental structural solutions for these bridges are decks integrated with abutments based on shallow direct foundations, solid abutments, or partially integrated ones.

An integral bridge is a specific type of bridge structure designed to collaborate with an embankment in a way that minimizes the negative effects of the transition effect. Figure 5 shows an example of an integral bridge. In such cases, the embankment soil is often reinforced.

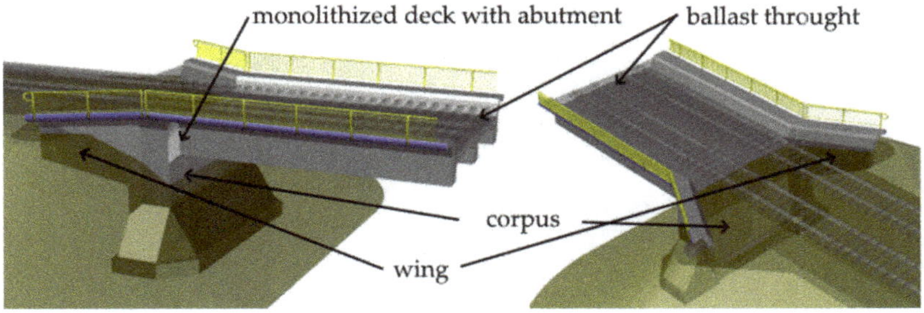

Figure 5. Three-dimensional model of an example integral bridge.

The longitudinal forces arising from the thermal expansion or contraction of the bridge deck are transmitted to the abutments and to the soil behind the abutment [94]. In the design assumption, the abutment does not necessarily have to absorb the entire horizontal forces due to thermal effects, as part of them will be transferred to the embankment, which will be integrated into the cooperation [94].

When calculating the ground reactions behind the abutment, the total length of the bridge is significant, determining the extent of its expansion or contraction due to temperature changes. However, the number of intermediate supports has no effect on the magnitude of the bridge's expansion or contraction, and it does not influence the size of the ground reaction [62].

In integral bridges, forming a slip plane to stimulate movements in the upper layers of the embankment soil is essential to minimize vertical displacements that affect the magnitude of the transition effect [62]. Intermediate supports, typically reinforced concrete, can be articulated on the foundation or based on one or several rows of piles. Spans for structures with integrated supports are usually around 60 m, sometimes reaching 80–100 m. The number of piers can vary, and the lengths of the longest integral bridges can exceed 350 m (Happy Hollow Bridge, Hickman County, TN, USA).

The challenge in evaluating existing structures and applying them lies in the interaction between the bridge and embankment soil. In many cases, this interaction is misunderstood due to the inherent nonlinear behavior of soil during what is known as operational interaction, significantly altering stresses in the embankment soil due to abutment displacements [87]. A step toward a better understanding this phenomenon involves interpreting the support and embankment soil state at the beginning of dynamic excitation based on earlier interactions during operation and assessing the stiffness and strength of the existing cushioning properties of the supports under dynamic loads [87]. The typical geometry of an integral support and typical embankment soil are studied using fully constrained simulations, assuming a viscoelastic–plastic stress model for the soil (coupled approach) under static and dynamic loads [87]. Integral bridges are designed with the assumption that only minor settlements of the end supports of the structure and embankments are permissible for using this solution in a given location. It is essential to note that this assumption significantly restricts the application of integral bridges. The connection of the bridge structure with the embankment, depending on the length of the structure, the shape of the space under the structure, the type of obstacle, and the terrain conditions may be made through different solutions. For example, these may be: massive integral abutments (Figure 6a), integral wall abutments (Figure 6b,f), integral box frame walls abutment (Figure 6c) or pillars support (Figure 6d,e). A semi-integral solution can also be used with span overhang and placement in the embankment (Figure 6e,f) [21,48]. The stability of mutual interactions should be ensured, especially in bridge structures where elastic deformations of supports are used to accommodate the elongation of load-bearing structures [21,22].

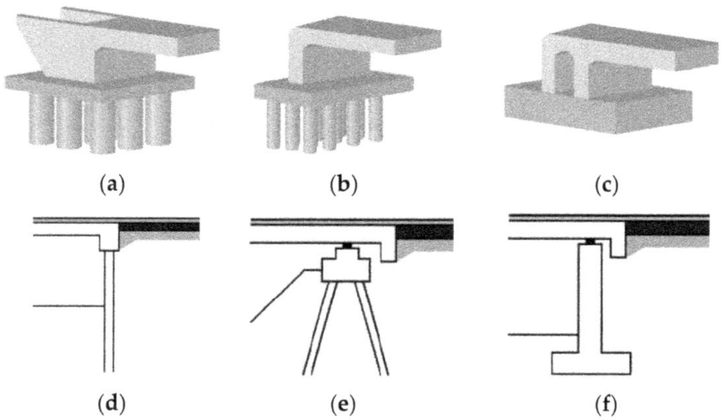

Figure 6. Types of abutments: 3D view of massive integral abutment (**a**), 3D view of integral wall abutment (**b**), ½ 3D view of integral box frame walls abutment (**c**), section of integral pillars support (**d**), section of semi-integral inclined pillars support with frame pillars and with a span overhang (**e**), section of semi-integral wall abutment with span overhang and placement in the embankment (**f**).

When entering a bridge, a vehicle must cross a transition zone. During its design, engineers must consider that the road or rail and the bridge itself may have different mechanical properties, such as stiffness, density, and Young's modulus. If the foundation at the entrance to the bridge is too stiff, it can lead to excessive stress in the bridge structure and increased risk of damage. Conversely, if the foundation is too soft, it may cause excessive settling of the bridge and a change in its geometry, which can result in hazardous situations for users. The mismatch usually occurs between a too compliant embankment and a too rigid bridge structure. Regulations seem to focus separately on the settling of individual elements, overly stiffening the structure (restrictively limiting deformations and settling) while overlooking the fact that soil structures will always settle more than those built from concrete and steel. Hence, it is reasonable to extend the bridge structure into the transition zone and connect it with a system that ensures efficient load transfer and smooth settling changes.

The problem of the transition effect, caused by the varying compliance (stiffness) of the pavement foundation in the transitional section onto the bridge, remains a current challenge for bridge engineers and continues to be researched and analyzed [29–31,33,34,61,80,81,83,84,94–101]. Intensity in addressing this issue has decreased in recent years, yet with increasing speed and tonnage, the problem continues to escalate. In the case of railways, any unevenness in the track profile becomes more important due to the possibility of jolting rolling stock, uneven wear on rail heads, or damage or dislodgement of track beds. For traditional roads, damage occurs to the surface, followed by subsequent structural elements. The issue of transition zones (and the transition effect within) in bridge engineering involves the use of numerous different solutions that are not standardized and have limited impact on reducing the set of adverse phenomena.

Several scientific publications describe the behavior and dynamic response of embankments, including transition zones [42,95,102–104]. Due to the recent increase in train speeds and transported tonnages, many researchers have undertaken studies into their dynamic effects and vibrations transmitted to the engineering object and embankment [105–108].

This aspect is crucial for the safety and durability of bridges, as dynamic forces generated by vehicle movement focus primarily in this area. The complex vehicle–soil–bridge system is where the bridge structure reacts with soil properties such as load-bearing capacity, shock resistance, and displacement [109]. Understanding the interaction between embankment and bridge is essential for the design, construction, and maintenance of bridges.

Presently, bridge design involves complex models encompassing the entire structure, considering mutual influences of parts with different stiffnesses and loads [37]. However, in the initial stages of such projects, simpler computational tools are used to analyze individual structural components. Paradoxically, conservative assumptions are adopted due to significant uncertainty. Consequently, actual settlements are much smaller than calculated, leading to uneven levels among intersecting parts and necessitating costly elimination of the resulting discrepancies. It is worth noting that requirements for limiting settlement differences [21,22,39] and the consequent design work do not consider time-variable settlements such as soil consolidation.

The analysis of deformations and mutual displacements of different parts of the same structure or neighboring structures is crucial. Both excessive and insufficient displacements are undesirable. Phasing of construction and the interaction between structures built at different times must be taken into account. Advancements in knowledge, research capabilities, and computational methods emphasize deformation analysis, displacements, and assessing serviceability limit states, pushing the issue of load-bearing capacity into the background [72,74]. Unlike previous practices, where in some cases only load-bearing capacity was analyzed, settlement matters were hidden in calculations through design assumptions.

The relevance of transition zone themes and the phenomena within them are confirmed by ongoing research [110,111]. The cited articles and their description of the interaction between the embankment soil and the bridge indirectly relate to the collaboration of these elements in response to the transition effect load. An important conclusion is the lack of examination of structures that facilitate and enhance the cooperation between the bridge structure and the embankment.

Despite numerous advantages, designing and constructing integral structures poses a challenge, with primary limitations in widespread application arising from the interaction between the abutment and embankment soil. This interaction results in vertical displacements of embankment soil and increases in passive soil pressure against the abutment [86]. Surprisingly, the literature has not reached a consensus on whether this is a beneficial or detrimental effect. Identified discrepancies point toward a conceptual gap in design and evaluation, requiring further research [86].

5.3. Environmental Aspect

In recent years, escalating environmental pollution and the depletion of natural resources have introduced new challenges to engineers. An effect of the increasing affluence of societies is the intensive growth in the quantity of waste generated. According to a World Bank report [112], currently, approximately 2 billion tons of solid waste are produced each year. Experts also predict that in about 30 years, we might expect nearly a doubling of garbage production. The enormous volume of waste poses a significant burden on the natural environment. Statistics indicate that an average resident of the European Union generates 530 kg of waste per year [113], of which approximately only 48% is processed. The share of polymer waste in recycling is only 25% [114,115].

According to the PlasticsEurope Foundation report [116] from 2022, in 2021, global plastic production increased by 4% to over 390 million tons, indicating a strong and continuous demand for plastics. However, Europe faces many challenges. The latest data show that China's share in global plastic production continues to grow (reaching 32% in 2021), while Europe's share—totaling 57.2 million tons in 2021—continues to decline (reaching 15%). In 2021, the recovery of plastic waste in the 27 EU27+3 countries exceeded 5.5 million tons, and post-consumer plastic waste used in new products and parts accounted for about 10% of plastic recycling and increased by about 20% compared to 2020 [116]. According to the cited report, global plastics production increased from 365.5 million tons in 2018 to 390.7 million tons in 2021. Plastic recyclates accounted for 30 million tons and 32.5 million tons, respectively. A slight increase in the share of recyclates from 8.2% to 8.3% can be observed [117]. In 2022, in the EU27+3 group of countries, out of 17.9 million tons of plastic packaging, 17% was landfilled, 46% was recycled, and 37% was recovered for energy [117]. In two countries, Belgium and the Netherlands, the goal of completely eliminating the landfilling of plastic waste was almost achieved [116]. In Central European countries, including Poland, the plastics industry is an important component of the economy, ranking third among industrial processing sectors in terms of gross added value generated. Employment in the entire industry is estimated at approx. 200,000 employees, and its development over the last dozen or so years has clearly outpaced other industries. The volume of local plastics production, estimated at 1.7 million tons, is insufficient to meet the needs of processors; therefore, a large part of the demand is covered with imported raw materials. Selective waste collection and the use of recycling can significantly improve the situation of the industry, where the collected fraction of PET is 28% [118].

Special attention needs to be paid to waste from high-molecular-weight plastics. Their largest source is packaging. According to data [119], only about 19.5% of plastic waste is sorted. This fact guarantees a constant demand for solutions allowing for recycling.

Growing environmental awareness and the need to reduce the construction industry's impact on the environment encourage the search for alternative materials, such as polymers [120]. The diagram illustrating the global production structure of plastics in 2021 shown in Figure 7 was developed based on data from the PlasticEurope Foundation report [121]. By analyzing the structure of global plastics production, taking into account their frequency [121] and mechanical properties [122–126], we can identify several types of polymers that can be successfully applied in bridge engineering [120].

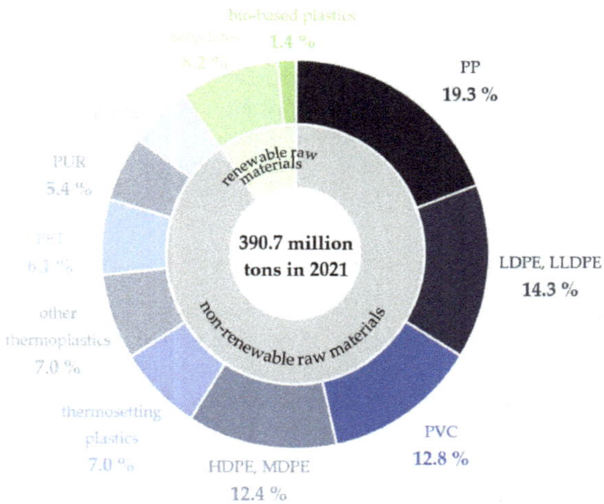

Figure 7. Structure of global plastics production.

Polymers are materials with complex chemical structures, consisting of long chains of molecules called polymers, linked together through the process of polymerization [127–132]. The size and shape of monomers influence polymer properties [127–129]. Due to their exceptional properties, polymers are increasingly being used as structural materials in bridge construction [133,134].

Polymer materials are processed many times, which further reduces the need for limited resources, such as oil or gas, from which they are produced. However, it is important to note that with each subsequent recycling of polymer raw materials, their mechanical properties deteriorate. Essentially, after two cycles of plastic use (use–recycling–use), the material is no longer suitable for rational processing; its strength properties become too low [135]. It is desirable, therefore, to find a long-term use for waste material.

The main challenges hindering plastic recycling are the quality and price of recycled products compared to their virgin counterparts. This is due to the complex process and various difficulties to overcome [136]. This includes sorting technology, specific challenges related to mechanical recycling like thermo-mechanical degradation or degradation over the product's lifespan, and the immiscibility of polymer mixtures. Plastic processors require large amounts of recycled plastics produced to tightly controlled specifications and at competitive prices [137]. Transitioning to a circular economy that retains plastic in its highest-value condition is essential to reducing the environmental impacts, promoting reduction, reuse, and recycling [138]. Mechanical recycling is an essential tool in an environmentally and economically sustainable economy of plastics, but current mechanical recycling processes are limited by cost, degradation of mechanical properties, and inconsistent-quality products [120]. Plastics can be easily customized to meet the functional or aesthetic needs of any manufacturer, complicating the recycling process and increasing its cost while affecting the quality of the final product [137].

A critical look at the recycling process can be found in the Center for Climate Integrity report [139]. Plastic pollution is one of the most serious environmental crises facing the world today. Between 1950 and 2015, over 90% of plastics were landfilled, incinerated, or leaked into the environment [139,140]. Plastic waste is ubiquitous—from our rivers, lakes, and oceans to roadways and coastlines. Underpinning this plastic waste crisis is a decades-long campaign of fraud and deception about the recyclability of plastics. Despite their long-standing knowledge that recycling plastic is neither technically nor economically viable [135,141], petrochemical companies and their trade associations have run misleading campaigns to promote plastic recycling. These efforts have expanded plastic markets and stalled effective regulation. By falsely promising the viability of plastic recycling, they have drastically increased virgin plastic production over the past six decades, worsening the global plastic waste crisis and imposing significant costs on communities [139]. Every product of plastic ever produced still exists today. Since 1950, 2 million tons of plastic have been created worldwide [116]. In studies [140], Roland Geyer et al. estimate that about 79% of all plastic waste ends up in landfills or directly in the natural environment. Ten million tons of plastic enter the oceans every year. The cost of pollution is estimated in billions of USD annually [142]. In 2019, an international research team published data in the Marine Pollution Bulletin stating that each ton of plastic waste in the oceans represents destroyed resources valued at up to USD 33,000. The scientists did not account for the indirect impact on tourism, transportation, and health. Considering these aspects, the social and economic costs of plastic waste in the oceans may be significantly underestimated [142].

Figure 8 shows the changes over time, between 2008 and 2020, in the way plastic waste is managed. We see that the share of landfilled waste has decreased by 47%. There has been an increase in the share of waste that is incinerated, and thanks to this, energy is recovered from it at 77%, and there has been a 117% increase in recycled waste. The charts are based on information contained in [116,117]. The following facts can be added regarding the above figure. In the EU27+3 group of countries, the percentage of waste going to landfills continues to decrease, but progress in this area is slow, especially in countries with a low level of plastic waste recovery. Half of the EU member states recover less than 30%. Every year, 25% of plastic waste in the EU27+3 still goes to landfills [121]. In Europe, 37% of plastic waste is incinerated in modern incinerators [116]. A study by Zero Waste Europe showed that even the most modern incinerators emit dioxins and other harmful pollutants [143,144]. Many of the countries accepting waste perform poorly in plastic waste management rankings. According to the World Bank, in developing countries, "over 90% of waste often ends up in illegal landfills or gets burned [145]". This situation encourages intensified efforts to increase the amount of recycled materials [121].

Green energy still faces challenges, especially in wind energy, where durable, lightweight materials like carbon fibers are needed for producing turbine blades. However, their rapid consumption and the lack of comprehensive solutions for processing these materials pose societal problems [146–157]. From 80% to 90% of wind turbine installations can be recycled (concrete, steel, copper, and silica). The remaining 10% to 20% represents a critical issue in disposing of the materials used to build them [146,147]. This concerns the materials in the blades, known as fiber-reinforced polymers (FRPs), often made of glass, carbon, aramid, or basalt fibers. In the EU, demand for FRPs grew sharply from 5000 tons in 1991 to 346,000 tons in 2015. It is estimated that by 2030, 4 million tons of fiber-reinforced polymers will be used for zero-emission wind energy production in the European Union. The current proposed recycling methods do not solve the rapid increase in the amount of this type of waste [146]. New proposed actions in this area include prevention, involving services and repairs of turbine blades to enable their extended use; reusing them in other sectors of the economy, producing items like small architectural elements, playgrounds, furniture, and similar products; however, low processing capabilities and the dust generated during processing are issues with this; recovery through chemical or thermal treatment, which is currently unprofitable; and disposal through landfilling and incineration, which do not solve the problem and are banned in many countries; additionally, they are not in line with the principles of a circular economy [146]. A major problem is the disposal of used fiber-based components. Their large quantity poses a real challenge to societies [147,149,150,154,157]. Extracting fibers using pyrolysis and using them as dispersed reinforcement in polymer materials will certainly help manage large quantities of this specialized waste to some extent.

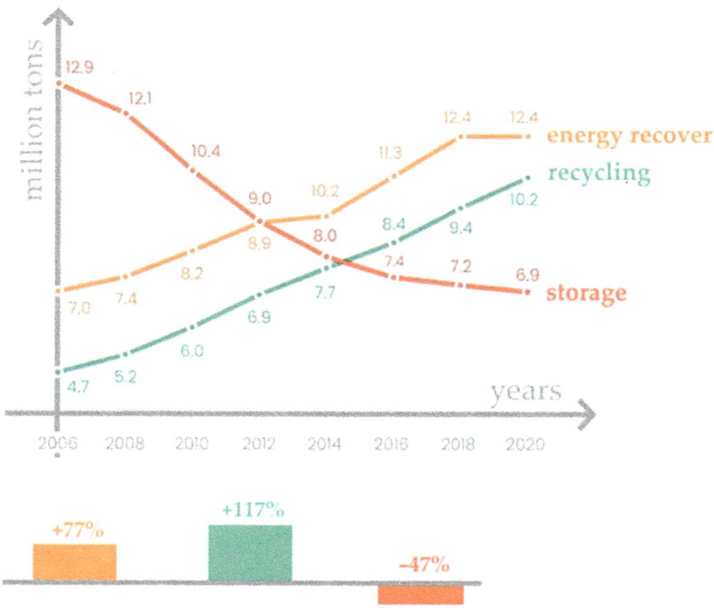

Figure 8. Progress in the management of post-consumer plastic waste (EU27+3).

Currently, the simplest and most effective method of waste disposal appears to be the use of materials in civil engineering as construction materials. Construction accounts for the use of 4 million tons of recyclables in new products in the EU27+3 countries, which constitutes 46% of plastic recyclables [121]. Advances in science and technology allow for the utilization of polymers as structural materials in the construction of bridges and embankments, which may contribute to improved durability, corrosion resistance, and cost-effectiveness. Older structures, built using masonry technology, as well as those made of concrete and steel, require numerous analyses, such as acoustic emission to assess and monitor bridge integrity [93] or well-known non-destructive testing of bridges using acoustic and radar impulses [158–160]. Nowadays, most bridges are built using concrete and steel [16,93], employing well-established computational models and analyses. However, many new designs are being conceptualized as composite systems [161–166] or as concrete structures reinforced with polymer fibers from carbon, glass, aramid, or even basaltic materials [167]. These are structures still being explored, and their behavior in the future remains unknown.

5.4. Properties and Use of Polymers in Bridge Structures

In recent years, polymers have gained popularity due to their light weight, durability, corrosion resistance, flexibility, and ease of shaping [168–170]. The method that gives the greatest strength to polymers is the injection method, where the mass is pressed into a mold at high temperature and pressure [171]. These polymers are relatively easy adapt to the individual needs of a project, making them highly versatile construction materials. Moreover, polymers are readily available and widely used across various industries [116,117,119,121,133,134], making them more cost-effective than traditional construction materials. These are used as insulation material, adhesives, components, surfacing, reinforcement, and barriers [130,132–134]. There has been increasing attention given to thermosetting polymers such as epoxy, polyester, and epoxy resins [172–174], which are especially characterized by their high mechanical strength and corrosion resistance, making them attractive construction materials in bridge building [175,176]. Moreover, these polymers resist environmental factors like UV radiation, contributing to the durability of structures [177,178].

Polymer composites are used in construction [163–165]. They are useful for bridges near water bodies or in high-humidity areas and for underground structural elements due to their better corrosion resistance than steel. For this reason, soil reinforcements are realized using them. Despite numerous advantages, using polymers also has some drawbacks. One of the significant drawbacks is their lower resistance to high temperatures [167]. Polymers are less resistant to heat compared to traditional construction materials like steel or concrete. However, in certain applications, the risk of fire hazards to the structure might be minimal, or the polymer material might be adequately protected by non-combustible materials (e.g., soil or concrete). Another drawback is their reduced resistance to mechanical damage [167]. This aspect can be critical due to the cyclic and long-term dynamic loads that bridge structures experience. Here, in turn, the answer is composite materials, which, thanks to their high strength due to the reinforcement of polymers with fibers, are used as advanced construction materials. A serious threat to composite materials may be UV radiation; however, as engineers have shown, despite this, these materials are often used without additional protection. Protective coatings are also well developed and can effectively support polymer material. Ageing tests carried out on such materials with parameters selected for their use significantly reduce the risk of unexpected destruction. Taking these risks into account, a detailed analysis of the behavior of polymers and their performance in such specific environments should be required depending on the proposed solution. However, the scale at which polymers are used allows us to assume that the material can be appropriately and safely designed [168].

Polymers have been used in bridge construction for many years [133,161–163,166]. One of the initial uses of polymers in construction was the use of saturated polyesters to reinforce wooden bridges [179–181]. Polymers are used in various structural elements of bridges, such as girders, cables, beams, and spans. Presently, polymers are used in the form of laminates reinforced with fiberglass or carbon fiber [179–190]. Another application of polymers is their use as construction materials [161,163,164,166]. Polymers can also be used to increase bridges' resistance to dynamic forces [191–193]. Fiber-reinforced polymer composites (GFRP), can enhance the flexibility and seismic resistance of bridge structures [194].

Another application involves using polymers as anti-slip materials [195,196]. Polymers like polyurethane or rubbers are employed to enhance traction on bridge surfacing, enhancing safety for pedestrians and vehicles. Finally, polymers can be used as fiber-reinforced materials to improve the strength and durability of bridges [180–182,184–190], for instance, epoxy resins reinforced with glass or carbon fibers.

The reinforcement of structures with polymer materials has long been popular [180–182,184–190]. In composite materials, glass, carbon, basalt, or aramid fibers are used as reinforcing fibers [77,164,167,197]. These materials possess excellent mechanical properties like high strength and rigidity, allowing for their usage in various applications [167]. Carbon fibers are among the most commonly used polymer-based materials for reinforcing structures [167,185,197,198]. They exhibit high tensile strength and a high modulus of elasticity. The fibers are highly resistant while remaining lightweight, making them an ideal material for bridge structures. The use of carbon fiber helps to reduce the mass of the structure, subsequently lowering production, transportation, and assembly costs. The process of reinforcing structures with carbon fibers usually involves placing a mat or mesh of carbon fiber on the structure's surface and saturating it with epoxy resin, bonding the carbon fibers to the structure [185,197,198]. This process can be applied to reinforce concrete structural elements such as columns and beams, as well as wooden and metallic elements. Several concepts have been developed so far for creating composite materials using fiber-reinforced polymers [179–181,189,190]. Glass, carbon, aramid, or basalt fibers are used as dispersed fibrous reinforcement in these materials [133]. Research on such fiber-reinforced materials can be found, among others, in the studies [199–205]. When reinforcing structures with glass fibers, the process is similar. While glass fibers have slightly lower strength than carbon fibers, they are more cost-effective and flexible, allowing for their use in reinforcing more complex structural shapes [167,179,180]. Aramid fibers like Kevlar are also popular materials for structural reinforcement [167,206]. These fibers are highly durable and possess outstanding resistance to abrasion and perforation. Reinforcing structures with polymer-based fiber materials can improve their strength in extreme conditions like earthquakes, hurricanes, and tornadoes. Due to their strength and rigidity, these materials are excellent candidates for use in construction projects such as bridges.

Polymers have been extensively used in bridge and embankment construction [66–68,161,163,195,198,207–213]. Their application is known in hydro-insulation and surfacing [214–216]. For reinforcing embankments, geopolymers (geosynthetics) are particularly effective, especially in transition zones [66–68,207–209,211–213,217–223]. Initially, soil reinforcement was carried out using steel bars. The primary drawback of this solution, mainly in terms of durability, was the corrosion of the reinforcement. The use of reinforced concrete layers is also considered more historical. Currently, soil reinforcement primarily involves various forms of geotextiles [62] due to their advantageous properties and application aspects. Geosynthetics are synthetic materials used in geotechnical engineering to improve soil properties and geotechnical structures [62,210,224]. Geotextiles are one type of geosynthetic characterized by a low unit weight and high tensile strength [66].

Materials commonly termed geosynthetics are employed for soil reinforcement [66–68]. There are various types of geosynthetics used based on specific needs, including geogrids, geotextiles, geomembranes, and geocomposites [207–213,217–223]. The effectiveness of soil reinforcement can be increased by using high-strength geosynthetic geogrids, with a tensile strength of up to 1600 kN/m, which vertically absorb, distribute, or dissipate loads onto the soil [66]. Geotextiles for soil reinforcement constitute a kind of mat made from synthetic fibers, usually polypropylene, bonded together in the technological process of their production. It is essential to ensure suitable strength and deformation parameters (ductility) [62]. There are various solutions for constructing embankments using geotextiles. One popular approach involves using geogrids as reinforcements beneath the embankment. This solution increases the soil's bearing capacity and reduces its deformation. Another method involves using geotextiles as a separation layer between different soil layers or as a filtration layer, preventing soil washout from the embankment by water, improving embankment stability and reducing erosion risk. Geotextiles used for embankment reinforcement at abutments cannot act as a diffusion barrier for air and water and must withstand low and high temperatures. They must also resist aggressive compounds (especially bases and acids), fungi, and decay [62]. A comprehensive analysis of geosynthetic reinforcement at the embankment base can be found in [66] and related studies [67,68,207–209,211–213,217–223]. It focuses on geosynthetic reinforcement in geotechnical structures, specifically the use of geogrids in embankment bases. The work presents results from laboratory tests and numerical analyses aimed at evaluating the effectiveness of using geogrids in reinforcing earth embankment bases. The research findings indicate that employing geogrids in reinforcing soil embankment bases enhances the structure's load-bearing capacity and reduces deformation. The author also compares the costs of using geogrids against traditional reinforcement methods, which can be significant from an economic perspective. The study provides a new perspective on geogrid applications in geotechnics and presents research results that contribute to improving the efficiency and durability of soil structures. All these solutions use geotextiles to improve embankment properties and increase embankment durability. The use of geotextiles in embankment construction is extensively described in the scientific literature and engineering practice, and the choice of a specific solution depends on the soil conditions and design requirements.

An interesting application of polymers is their use in epoxy injection for repairing cracks in bridge structures [225]. One of the most intriguing applications can be found in a scientific work concerning an innovative method of reinforcing ground behind abutments using geopolymer injection [226]. This procedure enhances the load-bearing capacity and stiffness, improves embankment and structure cooperation, and enhances durability.

It is crucial to relate the above-mentioned aspect to the transition effect. Ensuring the safety of the structural design itself and appropriately designing the connection between the bridge structure and the embankment are significant aspects in the process of bridge construction. Excessive embankment settlements are one of the main causes of failures in this transition zone [227,228], often leading to numerous damages.

Commonly used embankment reinforcement methods mostly rely on soil replacement, subsoil consolidation, mechanical stabilization, vibratory and dynamic methods, deep-seated reinforcements, or the use of geosynthetics [229–232]. Using typical methods during renovations usually requires dismantling the pavement and conducting extensive earthworks, significantly increasing not only the costs but also hindering the operation of the structure during the works. Geopolymer material injection provides a non-intrusive alternative to traditional substructure work and piling.

For high abutments and abutments founded on piles, as well as ground behind the abutment with a low internal friction angle, horizontal forces play a significant role. These forces can be reduced by using polypropylene blocks resembling honeycomb patterns and expanded polystyrene (EPS) blocks—Styrofoam—for embankment construction [62,233–235]. In engineering, a solution for the embankment behind abutments involves using geosynthetic reinforcement and Styrofoam blocks to simultaneously reduce ground pressure on abutments and control settlement adjacent to the abutments [61]. The use of EPS geofoam as a filling material between viaducts and abutments has been described in several publications [61,62].

Studies on the application of dispersed reinforcement in concrete in the form of fibrous polymers have focused on mechanical properties and durability when used as concrete reinforcement [198,206,236–238]. The results indicate that fibrous polymers increase the strength and durability of concrete. Polymer reinforcement materials are typically provided in the form of mats or meshes, which can be easily cut and adapted to the dimensions of the surface to be reinforced. In the case of concrete, the fibrous polymer reinforcement material is usually placed inside the concrete mass before laying, allowing for better interaction between the materials.

5.4.1. Recycling Polymers

Polymers obtained from waste can be used in the construction of bridge structures [168–170]. One example of utilizing waste polymers is the production of plastics from secondary raw materials like PET bottles, which can be used to create unsaturated resins applied in bridge construction [239,240]. Another instance is the use of PET from recycling, in the form of sheets, to improve the strength parameters of sandy soil [241,242]. Ongoing efforts are also exploring realistic alternatives to classical materials used as reinforcements, such as steel or fiber-reinforced polymers, aiming to produce cheaper and more environmentally friendly structural elements with similar or superior performance during use. PET tapes and polyester yarn tapes could effectively compete with currently used materials. They are among the most commonly used materials in the packaging industry, boasting high tensile strength comparable to steel and being easily recyclable [243,244].

Currently, ongoing research is exploring the impact of reinforcing PET material with carbon fiber on its mechanical and thermal properties. Research has shown this to be the material with the best tensile strength properties among polymers commonly found in waste [116,119]. Due to its good recyclability and simplicity, it is one of the most commonly processed waste polymers [116,119]. A. Brzeski also focused on combining recycling polymers from waste with aggregate, creating a type of synthetic rock. Besides the obvious advantage of recycling, this solution allows for obtaining a material with controlled stiffness that is resistant to weather conditions, fire-resistant, elastic, durable, non-permanently deformable, and environmentally neutral [245]. Nowadays, there is a developing trend in using biopolymer fibers derived from organic organisms reinforced with polyester fibers (PES) in manufacturing construction materials that could be employed in bridge construction [236,237]. These examples clearly indicate that polymers derived from waste hold promise in construction, reaffirming the relevance of addressing this issue.

Numerous scientific articles exist on the recycling of polymer fibers. One such example involves recycling carbon fibers from wind turbine blades [146,147,151,152,156]. The use of post-recycled fibers to reinforce polymers, which are intended to become the building material, will be effective and environmentally justified.

5.4.2. Reducing the Transition Effect

The primary requirement in constructing a bridge abutment along with an embankment is to ensure smooth passage between these entities. Polymers that can be used to make ground reinforcements that reduce the pressure from the fill involve the occurrence of frictional forces between the ground and the reinforcement. Regardless of increasing the soil's shear strength and reducing the pressure on the abutment, reinforced ground provides a more even distribution of deformations. This is particularly crucial in the case of integral bridges, where, due to temperature changes, displacements of the ground behind the abutments alter the direction multiple times [62].

Polymers are actively utilized to reduce the transition effect. These are applied in engineering structures to reinforce the subgrade, increasing its stiffness (while decreasing its susceptibility) and reducing the transition effect, primarily through injection masses.

An example is the use of polyurethane material as an adhesive that gradually stiffens the subgrade along the transition zone. Studies are being conducted on the use of polymer reinforcement in railway subgrades within transition zones at the approaches of engineering structures to mitigate the transition effect [246,247]. The technology involves injecting a polyurethane polymer mass in order to gradually bind the ballast of the track bed along its length. In terms of the vertical displacement of the rail, typical reductions of 25%, 16%, and 3% can be achieved. After static and dynamic testing, we can conclude that performance of polyurethane polymer in the transition zones can achieve the transition of the ballast bed from low stiffness to high stiffness [246]. The maximum reduction occurred with polyurethane alone, and when adding assistant rails and compared to having no settlement, the vertical rail acceleration was reduced by 16.5% and 25.6%, respectively [247]. The research results demonstrated increased subgrade stiffness and a reduced transition effect.

In this scheme, the bonding surface of the aggregate with the polyurethane determined the variation in the subgrade stiffness. This solution was adapted in China for the transition zone of a railway line at the entrance to a tunnel, where the ground changed from flexible to the rigid concrete at the entrance.

Another example is the use of mats from used tires, which is a fully-fledged recycled polymer [248]. In this approach, after numerous analyses, engineers investigated a set of five-plate rubber mats placed under the rails in gradually increasing quantities. Based on the relationship between the rail deflection and rubber mat stiffness, the wheel–rail interaction force was significantly reduced from 120 kN to 85 kN. Rubber mats with different stiffnesses are designed to equalize rail deflection differences. With the proposed design, the smoothness of the transition zone can be significantly improved.

Another solution being studied is the use of dispersed polymer reinforcement for concrete, which can also be recycled [249]. The main difference from standard solutions is the use of a mixture with a continuously changing content and type of reinforcement along the length of the track. It should be noted that changing the rail support in the transition sections, in order to eliminate the transition effect, had a positive effect on the total stiffness of the track system in the transition zone, estimated at up to 15%.

Polymers are used as reinforcing materials in various construction projects, and there is substantial scientific research confirming their strength properties [161,162,198,250]. The aforementioned literature demonstrates that polymers can have numerous applications in the construction of bridge abutments and embankments, both as structural and auxiliary materials. It is worth noting that there is not an abundance of scientific studies linking recycled polymers with bridge construction, particularly with abutments, embankments, and transition zones. This area remains niche and requires interest from scientists. The examples cited clearly demonstrate that polymers possess applicable positive properties. The utilization of recycled polymers in bridge engineering has not yet been thoroughly investigated, creating opportunities for valuable research. Polymers are incredibly significant materials, and their application is crucial for enhancing the safety, durability, and load-bearing capacity of bridge structures.

6. Perspectives and Research Gaps

PET (polyethylene terephthalate or poly(ethylene terephthalate)) is a waste plastic material with excellent strength properties [136,251–257]. It is the most commonly processed material [135–138,244]. However, as mentioned earlier, it is not feasible to rely entirely on recycled raw material because with each processing, the long polymer chains fragmentize, causing a loss in strength properties. Therefore, adding virgin material in the recycling process is necessary to maintain the required strength properties [258]. Consequently, it is evident that there will always be a continuous influx of material for processing. Hence, finding a long-term use for the material to extend its service life is crucial. It appears that PET, as a matrix for fibers, has potential applications [199–205].

Its real strength (strength-to-weight ratio) in Table 1 shows that it could be a more efficient material than steel. The notations 55% and 30% mean the percentage fibers in relation to the weight of the composite. Other composite materials behave similarly. The "strength-to-weight ratio" value in the last column was determined by dividing the tensile strength by the density and converting the units to appropriate ones. It is worth noting that some of these are composite materials made of 100% recycled PET (rPET) with the addition of glass fiber in appropriate amounts. The results come from our research that is currently being conducted [258], research by other authors [199–205], and websites of polymer composite and steel manufacturers [259–262]. In most structures, it is the self-weight that constitutes the majority of the loads, and if this is significantly optimized thanks to innovative, light, and durable materials, engineers will be able to create more advanced structures.

Our experiments evaluated rPET composites reinforced with 20–50% glass fibers (GF) by weight, using standardized methods for mechanical, thermal, and structural analysis. Specimens were prepared via twin-screw extrusion and injection molding, with the process parameters optimized for each composition to ensure material quality. The mechanical tests, including tensile, flexural, impact, and abrasion resistance, followed ISO standards, while the DSC and SEM analyses examined the crystallinity, fiber distribution, and adhesion. Rheological properties were assessed using MFI tests, and the results were compared against theoretical models (Halpin–Tsai, Tsai–Pagano) to evaluate the reinforcement efficiency and predict the performance for construction applications.

Table 1. Comparison of the properties of reinforced polymers to steel (based on [199–205,258–262]).

Material [-]	Tensile Strength [MPa]	Density [g/cm^3]	Strength-to-Weight Ratio [kNm/kg]
PET reinforced with glass fiber in a ratio of 55%	196	1.80	109
PET reinforced with carbon fiber in a ratio of 30%	220	1.45	152
PET reinforced with basalt fiber in a ratio of 30%	113	1.63	69
rPET reinforced with glass fiber in a ratio of 50%	140	1.75	80
rPET reinforced with glass fiber in a ratio of 40%	135	1.66	82
rPET reinforced with glass fiber in a ratio of 30%	122	1.58	77
Steel S235JR	360–510	7.85	46–65
Steel S355	470–630	7.85	60–80

In future research, a critical aspect will be the integration of the abutment with the ground using reinforcing profiles, along with selecting the composite's composition and shaping the geometry of the structure to achieve the required stiffness. The authors are currently exploring the possibility of processing PET polymer (polyethylene terephthalate) with reinforcing fibers. In the recycling process, primarily used for new bottle production, issues arise from contaminated materials (adhesives, labels, chemicals) that cannot be used in products intended for food contact. Consequently, it is essential to find solutions where such limitations do not apply. The planned material's utilization in the construction and modernization of roads, railways, and broadly understood bridge structures could be particularly efficient. Employing polymers as one of the building blocks of engineering structures, ensuring long-term usage, will undoubtedly aid the poorest countries in managing their waste.

Global material consumption in civil engineering is enormous and is still rising, especially in infrastructure [263,264]. The World Bank predicts that the share earmarked for infrastructure construction will be close to 90,000 tonnes per year in 2060 [263]. Using substitute materials in the form of recycled materials for even a small part of the demand will be a great benefit to societies. It is particularly important that these are minimally processed waste materials with a long-term use that will not be an ongoing burden on the recycling industry (also due to material limitations and a finite number of cycles). These materials that do not require expensive and energy-intensive processing and production processes can be defined as low-processed materials or minimally processed materials [265]. These, firstly, are effective, and secondly, they fit into the policy of sustainable development. Post-waste reinforced polymer composites can be successfully considered as such. Material in civil engineering can significantly manage large quantities of waste and ensure their utility for an extended period. The design period, which is often 100 years and is in fact much longer after renovations, also has a positive impact on the possible effectiveness of material recycling.

The construction of transportation infrastructure involves increasingly larger structures. This leads to increased vertical loads on the ground and, in the case of bridges and retaining structures, additional horizontal forces (pressure). The most commonly applied solutions to these problems are costly ground reinforcement techniques (such as piling, various types of columns) or enhancing slope stability and retaining structures (using ground anchors or nails). Moreover, the availability of high-quality natural materials (coarse sands, gravels, aggregates) is often limited, and their transportation over significant distances is uneconomical. An alternative, more cost-effective solution to these issues might involve the use of substitute materials.

In future research, a critical aspect will be the integration of the abutment with the ground using reinforcing bars, along with selecting the composite's composition and shaping the geometry of the bar structure to achieve the required stiffness. It seems that there may be problems with the material's fatigue, its brittle failure pattern, and low ductility. However, the authors provide for the possibility of modifying the composition of the composite, thus giving the material the desired properties.

There is a clear gap in the potential use of composites made from recycled materials in shaping transition zones in integral bridges. Future perspectives aim to propose a new design solution for bridge support structures based on the system of integral bridges, where the abutment will be closely linked to the embankment. In the future, a spatial structure will be designed and analyzed, ensuring cooperation between the abutment and the embankment. Through this, the change in the stiffness (flexibility) of the subgrade along the length of the structure and transition zones will be gradual, predicting a significant reduction in the transition effect.

Another probable outcome of the research will be an expanded scope for the use of integral bridges that will not be significantly restricted by the requirement for minimal settlements under the bridge supports and embankment. As mentioned, elements such as transition slabs, expansions, and bearings are not present in this type of structure. The absence of these elements will have a favorable impact on reducing the construction and maintenance costs of bridge structures.

It is anticipated that the designed structure will be composed of a composite made of polymer reinforced with dispersed fibers, with a direct and clearly defined correlation between the composition and the 'stiffness' of the resulting material. The 'stiffness' of the material and its geometric arrangement will directly influence the stiffness of the designed embankment structure.

The priority selection of components involves utilizing waste in the form of polymers in recyclate form and recycled fibers. The anticipated use of these raw materials will significantly reduce the carbon footprint and effectively utilize the materials (the expected durability of bridge structures is at least 100 years). The scope of the study will, in the near future, encompass a comprehensive exploration of this issue, starting from a theoretical analysis of computational models and concluding with laboratory research and a description of the proposed solution.

Despite significant advancements in the fields of recycling and bridge engineering, the intersection of these disciplines remains underexplored. Existing research on transition zones often lacks a focus on integrating recycled materials as a core component. Moreover, while polymers have been used in various construction contexts, their role in mitigating the transition effect has not been adequately quantified. Future studies should address the development of standardized methodologies for integrating recycled PET in geosynthetic applications; the evaluation of the long-term mechanical properties of recycled polymers under cyclic and dynamic loads; and the design and testing of composite materials specifically tailored for bridge embankments.

At this stage, concepts aligned with sustainable construction trends and the principles of the 6Rs are being formulated. The 6Rs typically refer to a set of principles aimed at promoting sustainability and environmental responsibility. These principles are often used in discussions about waste management and conservation. The 6Rs are rethink, refuse, reduce, reuse, recycle, and rot (or compost).

7. Conclusions

The presented state of the art in the field of bridge engineering and, more specifically, the issue of the transition effect and the field of polymers and their recycling provides several far-reaching facts:

1. The relevance of the transition effect problem in bridge structures and the need for exploring new solutions is still current. The issue of transition zones and the transition effect in bridge engineering is associated with the use of various solutions that lack standardization and have a limited impact on reducing the discussed phenomenon. This topic has been known for quite some time. With the development of transportation, increasing cargo tonnages, and rising speeds, research in this area becomes necessary. The proposed solution related to integral bridges, allowing for the avoidance of expansion joints and bearings, which positively impacts durability and maintenance costs. Transition slabs often do not completely eliminate this and may cause damage.
2. The ever-growing volume of waste poses challenges to humanity. Seeking multifaceted solutions contributing to the common good is the essence of engineers' and scientists' efforts. There is the possibility of using substitute materials that are appropriately processed and shaped, which might find application in bridge engineering, representing a cheaper and more ecological alternative to traditional solutions.
3. There is an opportunity to propose solutions involving the use of processed waste in the construction of transition zones connecting abutments to embankments, where materials made from waste can fulfill their function and contribute to reducing environmental pollution. The potential of recycling in construction and the use of composite materials based on recyclables reinforced with fibers in bridge engineering can contribute to waste reduction and the sustainable utilization of recycled materials. However, it is crucial to emphasize that polymers, including reinforced polymers, especially those made from recyclables, are not widely used in bridge construction, and their impact on reducing the transition effect has not been thoroughly explored.
4. PET (polyethylene terephthalate) has the best strength properties among waste plastics and exhibits good opportunities for creating composites with fibers. Taking into account the fact that the percentage of recycling of this raw material is still not complete, this creates promising applications, including its use as a building material in bridge structures or even as an element of structures reducing the transition effect.

In conclusion, this prospective approach to bridge design, considering technical, durability, and environmental aspects through the utilization of innovative solutions and materials, could lead to more sustainable and efficient construction practices in the future. The integration of recycled polymers in transition zones offers a promising solution to addressing both engineering challenges and environmental concerns. By integrating recycled materials, such as PET composites, into the design of transition zones, it is possible to address both engineering challenges and environmental concerns. While current research provides a solid foundation, significant opportunities exist for future studies to validate these concepts through experimental and field investigations. The findings of this review aim to guide researchers and practitioners in exploring sustainable and cost-effective approaches to bridge construction. The next steps involve validating the proposed materials through computational models and experimental studies, which will be detailed in subsequent publications.

Author Contributions: Conceptualization, M.S.; methodology, K.A.O. and K.F.; investigation, K.A.O.; resources, K.A.O.; writing—original draft preparation, M.S., K.A.O. and K.F.; editing, M.S., K.A.O. and K.F.; supervision, K.A.O. and K.F.; project administration, K.A.O.; validation, M.S., K.A.O. and K.F.; formal analysis, K.A.O. and K.F.; writing—revised draft preparation, M.S. and K.A.O.; visualization, M.S.; funding acquisition, K.A.O. All authors have read and agreed to the published version of the manuscript.

Funding: This research received no external funding.

Institutional Review Board Statement: Not applicable.

Informed Consent Statement: Not applicable.

Conflicts of Interest: The authors declare no conflicts of interest.

References

1. Perron, B.E.; Victor, B.G.; Hodge, D.R.; Salas-Wright, C.P.; Vaughn, M.G.; Taylor, R.J. Laying the Foundations for Scientometric Research: A Data Science Approach. *Res. Soc. Work Pract.* **2017**, *27*, 802–812. [CrossRef]
2. Zitt, M.; Bassecoulard, E. Challenges for scientometric indicators: Data demining, knowledge-flow measurements and diversity issues. *Ethics Sci. Environ. Polit.* **2008**, *8*, 49–60. [CrossRef]
3. Singh, V.K.; Banshal, S.K.; Singhal, K.; Uddin, A. Scientometric mapping of research on 'Big Data'. *Scientometrics* **2015**, *105*, 727–741. [CrossRef]
4. Gile, D. Analyzing Translation studies with scientometric data: From CIRIN to citation analysis. *Perspectives* **2015**, *23*, 240–248. [CrossRef]
5. Falagas, M.E.; Pitsouni, E.I.; Malietzis, G.A.; Pappas, G. Comparison of PubMed, Scopus, Web of Science, and Google Scholar: Strengths and weaknesses. *FASEB J.* **2007**, *22*, 338–342. [CrossRef]
6. Vieira, E.S.; Gomes, J.A.N.F. A comparison of Scopus and Web of science for a typical university. *Scientometrics* **2009**, *81*, 587–600. [CrossRef]
7. Scopus | Abstract and Citation Database | Elsevier. Available online: https://www.elsevier.com/products/scopus (accessed on 18 November 2023).
8. Web of Science Coverage Details—Resources for Librarians—LibGuides at Clarivate Analytics. Available online: https://clarivate.libguides.com/librarianresources/coverage/ (accessed on 18 November 2023).
9. VOSviewer—Visualizing Scientific Landscapes. Available online: https://www.vosviewer.com/ (accessed on 18 November 2023).
10. Brown, D.J. *Bridges: Three Thousand Years of Defying Nature*; Firefly Books: Richmond Hill, ON, Canada, 2005; ISBN 1554070996.
11. Troyano, F.L. *Bridge Engineering: A Global Perspective*; Thomas Telford Pub: London, UK, 2003; ISBN 9780727732156.
12. Leonhardt, F. *Brücken. Asthetik und Gestaltung*; MIT Press: Cambridge, UK, 1984; ISBN 0262121050.
13. Chen, W.-F.; Duan, L. *Bridge Engineering: Substructure Design*, 1st ed.; CRC Press: Boca Raton, FL, USA, 2003; ISBN 9780429214318.
14. Bennett, D. *The Creation Of Bridges—From Vision To Reality, The Ultimate Challenge of Architecture, Design And Distance*, 1st ed.; Aurum Press Ltd.: London, UK, 1999; ISBN 9781854106513.
15. *Bridge Aesthetics Around the World*; Transportation Research Board National Research Council (U.S.): Trenton, NJ, USA, 1991; ISBN 0309050723.
16. Arockiasamy, M.; Butrieng, N.; Sivakumar, M. State-of-the-Art of Integral Abutment Bridges: Design and Practice. *J. Bridg. Eng.* **2004**, *9*, 497–506. [CrossRef]
17. Madaj, A.; Wołowicki, W. *Podstawy Projektowania Budowli Mostowych*, 1st ed.; Furtak, K., Weseli, J., Eds.; Wydawnictwa Komunikacji i Łączności: Warszawa, Poland, 2008.
18. Gutkowski, R.M.; Kmita, J. *Transportation Infrastructure Environmental Challenges in Poland and Neighboring Countries*; Springer: Berlin/Heidelberg, Germany, 1996; Volume 5.
19. Sołkowski, J. Zarys analizy efektu progowego przy łączeniu nawierzchni podsypkowych z innymi typami nawierzchni. *TTS Tech. Transp. Szyn.* **2009**, *15*, 59–65.
20. Sołkowski, J. *Efekt Progowy w Nawierzchniach Szynowych*; Kłosek, K., Sowiński, B., Eds.; Wydawnictwo PK: Kraków, Poland, 2013.
21. Rymsza, J.; Biliszczuk, J.; Onysyk, J.; Toczkiewicz, R. Wytyczne Projektowania Elementów Powiązania Drogowych Obiektów Inżynierskich z Terenem i Drogą. Wzorce i standardy rekomendowane przez Ministra właściwego ds. *Transport* **2020**, 1–73.
22. *Standardy Techniczne—Szczegółowe Warunki Techniczne Dla Modernizacji Lub Budowy Linii Kolejowych 2018–2023*; PKP Polskie Linie Kolejowe: Warszawa, Poland, 2018.
23. Binder, K. Strefa przejściowa pomiędzy obiektem mostowym a nasypem drogowym. *Mosty* **2012**, *4*, 55–57.
24. Cyunel, B.; Kulczycki, B. *Kolejowe Budowle Ziemne: Podręcznik Akademicki. T. 2, Technologia, Organizacja Budowy i Modernizacji*; Wydawnictwa Komunikacji i Łączności: Warszawa, Poland, 1987.
25. Gwizdała, K. Nowy Eurokod 7 w projektowaniu posadowień obiektów mostowych na palach. *Mosty* **2022**, *3*, 28–34.

26. *PN-83/B-02482;* Fundamenty Budowlane—Nośność Pali i Fundamentów Palowych. Wydawnictwa Normalizacyjne „Alfa": Warszawa, Poland, 1983.
27. *PN-69/B-02482;* Fundamenty Budowlane—Nośność Pali i Fundamentów Palowych. Wydawnictwa Normalizacyjne „Alfa": Warszawa, Poland, 1969.
28. Sołkowski, J.; Lisowski, S.; Pawlak-Burakowska, A.; Jamka, M. Zarys metodyki badania sztywności nawierzchni i podtorza w strefach przejściowych do obiektów inżynierskich oraz na obiektach. *Zesz. Nauk. Tech. Stowarzyszenia Inżynierów Tech. Komun. Krakowie* **2012**, *3*, 285–309.
29. Sołkowski, J.; Kudła, D. Analiza niejednorodności mechanicznych nawierzchni i podtorza w obrębie obiektu mostowego. *Zesz. Nauk. Tech. Stowarzyszenia Inżynierów Tech. Komun. Krakowie* **2016**, *2*, 165–175.
30. Skrzyński, E. Podtorze na liniach kolejowych dużych prędkości. *Probl. Kolejnictwa* **2013**, *161*, 87–124.
31. Sasaoka, D. Track Transition Designs for Heavy-Axle-Load Service. *Railw. Track Struct.* **2006**, *102*, 15–18.
32. Sañudo, R.; Dell'Olio, L.; Casado, J.A.; Carrascal, I.A.; Diego, S. Track transitions in railways: A review. *Constr. Build. Mater.* **2016**, *112*, 140–157. [CrossRef]
33. Niemierko, A.; Sołkowski, J. Wzmacnianie stref przejściowych na dojazdach do mostów kolejowych. *Arch. Inst. Inżynierii Lądowej* **2012**, *14*, 173–183.
34. Kłosiński, B.; Niemierko, A. *ERRI D 230.1/RP 3. Bridge Ends; Embankment Structure Transition;* State of the Art Report: Utrecht, The Netherlands, 1999.
35. Kerr, A.D.; Moroney, B.E. Track transition problems and remedies. *Bulletin* **1993**, *94*, 267–298.
36. Jaup, A.; Kempfert, H.-G. Setzungen in Hinterfüllungsbereich von Eisenbahn-brückenwiederlagern. *Bautechnik* **2001**, *9*, 635–640. [CrossRef]
37. Indraratna, B.; Babar Sajjad, M.; Ngo, T.; Gomes Correia, A.; Kelly, R. Improved performance of ballasted tracks at transition zones: A review of experimental and modelling approaches. *Transp. Geotech.* **2019**, *21*, 100260. [CrossRef]
38. Hunt, H.E.M. Winkler Settlement of railway track near bridge abutment. *Proc. Inst. Civ. Eng. Transp.* **1996**, *123*, 68–73. [CrossRef]
39. Skrzyński, E. Osiadania podtorza gruntowego. *Pr. Inst. Kolejnictwa* **2019**, *161*, 28–43.
40. Bień, J. *Uszkodzenia i Diagnostyka Obiektów Mostowych*, 1st ed.; Wydawnictwa Komunikacji i Łączności: Warszawa, Poland, 2010; ISBN 978-83-206-1791-7.
41. Hebda, M. *Materiały Dydaktyczne—Utrzymanie i Remonty Mostów;* Katedra Konstrukcji Mostowych, Metalowych i Drewnianych; Wydział Inżynierii Lądowej; Politechnika Krakowska: Kraków, Poland, 2020.
42. Zauda Coelho, B.E. Dynamics of Railway Transition Zones in Soft Soils. Master's Thesis, University of Porto Portugal, Porto, Portugal, 2011. [CrossRef]
43. Bukowski, M.; Łysiak, P.; Oleszek, R.; Trochymiak, W. Przyczyny powstania uskoku pomiędzy wiaduktem a nasypem na trasie Siekierkowskiej. *Arch. Inst. Inżynierii Lądowej* **2017**, *24*, 39–58. [CrossRef]
44. Wang, H.; Chang, L.; Markine, V. Structural Health Monitoring of Railway Transition Zones Using Satellite Radar Data. *Sensors* **2018**, *18*, 413. [CrossRef]
45. Sołkowski, J.; Kudła, D. Wykonawstwo stref przejściowych do obiektów mostowych i aspekty modelowania numerycznego oddziaływania. *Zesz. Nauk. Tech. Stowarzyszenia Inżynierów Tech. Komun. Krakowie* **2013**, *2*, 275–289.
46. Sołkowski, J.; Jamka, M. Deformacje nawierzchni i podtorza przy obiektach mostowych—Badania i diagnostyka. *Przegląd Komun.* **2016**, *4*, 39–47.
47. Nicks, J.E. The Bump at the End of the Railway Bridge. Ph.D. Thesis, Texas A&M University, College Station, TX, USA, 2009.
48. Krupka, J.; Sobala, D. *Stan Techniczny Mostów Zintegrowanych na Przykładzie Wybranych Obiektów w Polsce;* Praca Magisterska, Politechnika Rzeszowska: Rzeszów, Poland, 2007; pp. 1–73.
49. Germaniuk, K. Uszkodzenia nawierzchni na mostach i w strefie połączenia drogi z mostem. *Mater. Bud.* **2017**, *4*, 82–85. [CrossRef]
50. Dreier, D.; Burdet, O.; Muttoni, A. Transition slabs of integral abutment bridges. *Struct. Eng. Int.* **2011**, *21*, 144–150. [CrossRef]
51. Bukowski, M.; Łysiak, P.; Oleszek, R.; Trochymiak, W. Modeling and analysis of ground settlement between a flyover and reinforced soil embankment. *Arch. Civ. Eng.* **2018**, *64*, 77–100. [CrossRef]
52. Asghari, K.; Sotoudeh, S.; Zakeri, J.A. Numerical evaluation of approach slab influence on transition zone behavior in high-speed railway track. *Transp. Geotech.* **2021**, *28*, 100519. [CrossRef]
53. Dahlberg, T.; Larsson, R. Influence of railway track stiffness variations on wheel/rail contact force. In Proceedings of the Tracks for High-Speed Railways; Faculty of Engineering, University of Porto: Porto, Portugal, 2006; p. 67.
54. Li, D.; Davis, D. Transition of Railroad Bridge Approaches. *J. Geotech. Geoenvironmental Eng.* **2005**, *131*, 1392–1398. [CrossRef]
55. Esveld, C. *Modern Railway Track*, 2nd ed.; Delft University of Technology: Delft, The Netherlands, 2017; ISBN 90-800324-3-3.
56. Banimahd, M.; Woodward, P.K.; Kennedy, J.; Medero, G.M. Behaviour of train–track interaction in stiffness transitions. *Proc. Inst. Civ. Eng. Transp.* **2012**, *165*, 205–214. [CrossRef]
57. Sañudo, R.; Jardí, I.; Martínez, J.C.; Sánchez, F.J.; Miranda, M.; Alonso, B.; Dell'olio, L.; Moura, J.L. Monitoring Track Transition Zones in Railways. *Sensors* **2022**, *22*, 76. [CrossRef] [PubMed]
58. Shi, J.; Burrow, M.P.N.; Chan, A.H.; Wang, Y.J. Measurements and simulation of the dynamic responses of a bridge–embankment transition zone below a heavy haul railway line. *Proc. Inst. Mech. Eng. Part F J. Rail Rapid Transit* **2012**, *227*, 254–268. [CrossRef]
59. Chen, C.; McDowell, G.R. An investigation of the dynamic behaviour of track transition zones using discrete element modelling. *Proc. Inst. Mech. Eng. Part F J. Rail Rapid Transit* **2014**, *230*, 117–128. [CrossRef]

60. Çati, Y.; Gökçeli, S.; Anil, Ö.; Korkmaz, C.S. Experimental and numerical investigation of usp for optimization of transition zone of railway. *Eng. Struct.* **2020**, *209*, 109971. [CrossRef]
61. Horvath, J.S. *Integral-Abutment Bridges: Problems and Innovative Solutions Using EPS Geofoam and Other Geosynthetics*; Manhattan College: New York, NY, USA, 2000.
62. Furtak, K.; Wrana, B. *Mosty Zintegrowane*, 1st ed.; Wydawnictwa Komunikacji i Łączności: Warszawa, Poland, 2005; ISBN 83-206-1550-X.
63. Pisarczyk, S.J. *Geoinżynieria: Metody Modyfikacji Podłoża Gruntowego*; Oficyna Wydawnicza Politechniki Warszawskiej: Warszawa, Poland, 2005; ISBN 8372075530.
64. Łęcki, P.; Różański, M. Wzmacnianie podłoża gruntowego budowli drogowych. *Nowocz. Bud. Inżynieryjne* **2015**, *2*, 46–54.
65. Esmaeili, M.; Gharouni Nik, M.; Khayyer, F. Static and Dynamic Analyses of Micropiles to Reinforce the High Railway Embankments on Loose Beds. *J. Rehabil. Civ. Eng.* **2013**, *1*, 80–89.
66. Duszyńska, A. *Zbrojenie Geosyntetyczne Podstawy Nasypu*, 1st ed.; Ślączka, W., Ed.; Politechnika Gdańska: Gdańsk, Poland, 2016; p. 106.
67. Gajewska, B.; Kłosiński, B.; Rychlewski, P.; Grzegorzewicz, K.; Kraszewski, C. *Zastosowanie Geosyntetyków w Budowlach Ziemnych*; Studium Poznawczo-Techniczne: Warszawa, Poland, 2003.
68. Ajdukiewicz, J. Strome nasypy drogowe zbrojone geosyntetykami efektem wysokospecjalizowanych prac inżynierskich. *Proc. Zesz. Nauk. Bud. Politech. Śląska; Politech. Śląska* **2003**, *97*, 371–386.
69. Grulkowski, S.; Kędra, Z.; Nowakowski, M.J.; Koc, W. *Drogi Szynowe*; Wydawnictwo Politechniki Gdańskiej: Gdańsk, Poland, 2013; ISBN 978-83-7348-511-2.
70. Bednarek, W. Statyczna analiza wybranych imperfekcji podłoża szynowego na dodatkowe ugięcia szyny bezstykowego toru kolejowego. *Przegląd Komun.* **2016**, *11*, 20–26.
71. Kłosek, K.; Sołkowski, J.; Tondera, M. Ocena nośności i stateczności podtorza zgodnie z wymaganiami krajowymi oraz EUROKODU PN-EN 1991:21. *Zesz. Nauk. Tech. Stowarzyszenia Inżynierów Tech. Komun. Krakowie* **2019**, *2*, 93–107.
72. Rychlewski, P. *Przemieszczenia Fundamentów Obiektów Inżynierskich*; Inżynier Budownictwa: Warszawa, Poland, 2017.
73. Waniek, G.; Kwiecień, S. Analiza współpracy przyczółka mostowego z nasypem drogowym posadowionym na podłożu gruntowym o małej sztywności. *Drogownictwo* **2015**, *9*, 302–305.
74. Kłosiński, B. Przemieszczenia poziome przyczółków mostowych. *Mosty* **2012**, *2*, 32–37.
75. England, G.L.; Tsang, N.C.M.; Bush, D.I. *Integral Bridges: A Fundamental Approach to the Time–Temperature Loading Problem*; Emerald Publishing Limited: London, UK, 2000.
76. Madaj, A.; Wołowicki, W. *Budowa i Utrzymanie Mostów*, 3rd ed.; Wydawnictwa Komunikacji i Łączności: Warszawa, Poland, 2007; ISBN 978-83-206-1419-0.
77. Niemierko, A.; Mieczkowski, P.; Gruszczyński, M.; Krasiński, A. *Budownictwo Mostowe Vademecum, 2015 Ed*; Dębińska, A., Małek, A., Bień, P., Mioduszewska, J., Eds.; Wydawnictwo Polskiej Izby Inżynierów Budownictwa Sp. z o.o.: Warszawa, Poland, 2015; ISBN 2353-5261.
78. Musiał, T. Obiekty Mostowe z Przyczółkami Zintegrowanymi; Inżynier Budownictwa, 2012. Available online: https://inzynierbudownictwa.pl/obiekty-mostowe-z-przyczolkami-zintegrowanymi-cz-i/ (accessed on 3 November 2022).
79. Steffen, M.; Chongjie, K. Nowa generacja mostów zintegrowanych na liniach kolei dużych prędkości. *Inżynieria Bud.* **2019**, *75*, 514–518.
80. Copija, M.; Arkadiusz, M. Mosty zintegrowane. *Mater. Bud.* **2008**, *4*, 81–83.
81. Copija, M.; Madaj, A. Doświadczenia z projektowania i eksploatacji mostów zintegrowanych. *Mater. Bud.* **2010**, *4*, 50–53.
82. Arsoy, S.; Barker, R.M.; Duncan, J.M. *The Behavior of Integral Abutment Bridge*; Virginia Polytechnic and State University: Blacksburg, VA, USA, 1999.
83. Burke, J.M.P. *Integral and Semi-Integral Bridges*, 1st ed.; Wiley-Blackwell: Chichester, UK, 2009; ISBN 9781405194181.
84. Burke, J.M.P. *Integral Bridges*; Transportation Research Record; National Research Council: Washington, DC, USA, 1990; pp. 53–61.
85. Dicleli, M. Integral Bridges. In *Innovative Bridge Design Handbook*; Pipinato, A., Ed.; Butterworth-Heinemann: Oxford, UK, 2016; pp. 1–34.
86. Mitoulis, S.A. Challenges and opportunities for the application of integral abutment bridges in earthquake-prone areas: A review. *Soil Dyn. Earthq. Eng.* **2020**, *135*, 106183. [CrossRef]
87. Mitoulis, S.A.; Argyroudis, S.; Kowalsky, M. Evaluation of the stiffness and damping of abutments to extend direct displacement-based design to the design of integral bridges. In Proceedings of the COMPDYN 2015—5th ECCOMAS Thematic Conference on Computational Methods in Structural Dynamics and Earthquake Engineering, Crete Heraklion, Greece, 25–27 May 2015; pp. 1767–1779.
88. Naji, M.; Firoozi, A.A.; Firoozi, A.A. A review: Study of integral abutment bridge with consideration of soil-structure interaction. *Lat. Am. J. Solids Struct.* **2020**, *17*, e252. [CrossRef]
89. Fiorentino, G.; Cengiz, C.; De Luca, F.; Mylonakis, G.; Karamitros, D.; Dietz, M.; Dihoru, L.; Lavorato, D.; Briseghella, B.; Isakovic, T.; et al. Integral abutment bridges: Investigation of seismic soil-structure interaction effects by shaking table testing. *Earthq. Eng. Struct. Dyn.* **2020**, *50*, 1517–1538. [CrossRef]
90. Marchi, A.; Franchin, P. A pushover method for seismic design of Integral Abutment Bridges. In Proceedings of the 14th Fib PhD Symposium in Civil Engineering, Rome, Italy, 5–7 September 2022; pp. 449–457.
91. Liu, X. *Seismic Pushover Analysis of Steel Integral Abutment Bridge*; Fu, C.C., Ed.; ProQuest: College Park, MD, USA, 2018.

92. Kozak, D.L.; Fahnestock, L.A.; LaFave, J.M. *Seismic Behavior and Design of Integral Abutment Bridges in Southern Illinois*; University of Illinois at Urbana-Champaign: Urbana, IL, USA, 2018.
93. Shigeishi, M.; Colombo, S.; Broughton, K.J.; Rutledge, H.; Batchelor, A.J.; Forde, M.C. Acoustic emission to assess and monitor the integrity of bridges. *Constr. Build. Mater.* **2001**, *15*, 35–49. [CrossRef]
94. Jarosz, J. The effect of braking forces on displacement of integral bridges. *Bud. Archit.* **2013**, *12*, 023–030. [CrossRef]
95. Ribeiro, A.C.; Calçada, R.; Delgado, R. Dynamic effects induced by abrupt changes in track stiffness in high speed railway lines. In Proceedings of the Applications of Computational Mechanics in Geotechnical Engineering—5th International Workshop on Applications of Computational Mechanics in Geotechnical Engineering, Guimaraes, Portugal, 1–4 April 2007; Sousa, L.R., Fernandes, M.M., Vargas, E.J., Azevedo, R., Eds.; CRC Press Taylor & Francis Group: Guimaraes, Portugal, 2007; pp. 473–483, ISBN 9780415437899.
96. Jarosz, J. Doświadczalna weryfikacja modelu obliczeniowego zintegrowanego obiektu mostowego. *Bud. Archit.* **2014**, *13*, 103–110. [CrossRef]
97. Eisenmann, J. *Dűrchfűrung von Dauerschwellversuchen An Einer Űbergangkonstruktion*; Technische UniversitatMűnchen: Bavaria, Germany, 1990.
98. Diaz, A. *La Construcción de Terraplenes y Desmontes*; Fundación de los Ferrocarriles Españoles: São Paulo, Brazil, 1993.
99. Davis, D.; Plotkin, D. Track Settlement at Bridge Approaches. *Railw. Track Struct.* **2009**, *105*, 33–38.
100. Davis, D.D.; Otter, D.; Li, D.; Singh, S. Bridge approach performance in revenue service. *Railw. Track Struct.* **2003**, *99*, 18–20.
101. Česká Agentura Pro Standardizaci. Příloha 24. In *ČD S4 Železniči Spodek—Přechodtělesaželezničního Spodku Na Mostníobjekty*; Standardizační Nakladatelství: Praha, Czech Republic, 1998.
102. Sakhare, A.; Farooq, H.; Nimbalkar, S.; Dodagoudar, G.R. Dynamic behavior of the transition zone of an integral abutment bridge. *Sustainability* **2022**, *14*, 4118. [CrossRef]
103. Fröhling, R.D.; Scheffel, H.; Ebersöhn, W. The Vertical Dynamic Response of a Rail Vehicle caused by Track Stiffness Variations along the Track. *Veh. Syst. Dyn.* **1996**, *25*, 175–187. [CrossRef]
104. Sakhare, A.; Punetha, P.; Meena, N.K.; Nimbalkar, S.; Dodagoudar, G.R. Dynamic behavior of integral abutment bridge transition under moving train loads. *Transp. Geotech.* **2023**, *40*, 100989. [CrossRef]
105. Connolly, D.P.; Kouroussis, G.; Woodward, P.K.; Alves Costa, P.; Verlinden, O.; Forde, M.C. Field testing and analysis of high speed rail vibrations. *Soil Dyn. Earthq. Eng.* **2014**, *67*, 102–118. [CrossRef]
106. Connolly, D.P.; Kouroussis, G.; Fan, W.; Percival, M.; Giannopoulos, A.; Woodward, P.K.; Verlinden, O.; Forde, M.C. An experimental analysis of embankment vibrations due to high speed rail. In Proceedings of the 12th International Railway Engineering Conference (Railway Engineering 2013), London, UK, 10–11 July 2013.
107. Kouroussis, G.; Connolly, D.P.; Forde, M.C.; Verlinden, O. An experimental study of embankment conditions on high-speed railway ground vibrations. In Proceedings of the 20th International Congress on Sound and Vibration (ICSV20), Bangkok, Thailand, 7–11 July 2013.
108. Connolly, D.P.; Giannopoulos, A.; Forde, M.C. Numerical modelling of ground borne vibrations from high speed rail lines on embankments. *Soil Dyn. Earthq. Eng.* **2013**, *46*, 13–19. [CrossRef]
109. Idczak, W.; Lewandrowski, T.; Pokropski, D.; Rudnicki, T.; Trzmiel, J. Dynamic Impact of a Rail Vehicle on a Rail Infrastructure with Particular Focus on the Phenomenon of Threshold Effect. *Energies* **2022**, *15*, 2119. [CrossRef]
110. Nabochenko, O.; Sysyn, M.; Gerber, U.; Krumnow, N. Analysis of Track Bending Stiffness and Loading Distribution Effect in Rail Support by Application of Bending Reinforcement Methods. *Urban Rail Transit* **2023**, *9*, 73–91. [CrossRef]
111. Shang, Y.; Nogal, M.; Teixeira, R.; Wolfert, A.R.; Rogier, M. Optimal design of rail level crossings and associated transition zones using adaptive surrogate-assisted optimization. *Eng. Struct.* **2023**, *282*, 115740. [CrossRef]
112. *What a Waste 2.0. A Global Snapshot of Solid Waste Management to 2050*; World Bank: Washington, DC, USA, 2020.
113. *Statistical Data 2021 Luxembourg*; Eurostat European Statistical Office: Luxembourg, 2021.
114. *Statistical Data 2022*; UNESCO: Paris, France, 2022.
115. *Statistical Data 2021*; Główny Urząd Satystyczny Rzeczypospolitej Polskiej: Warszawa, Poland, 2021.
116. Pathways to A Circular, Climate Neutral Plastics System in Europe; Reshaping Plastics; Plastic Europe, 2022. Available online: https://plasticseurope.org/pl/wp-content/uploads/sites/7/2023/04/Tworzywa-Fakty-2022.pdf (accessed on 1 November 2022).
117. *Company Research Data*; Conversio Market & Strategy GmbH: Mainaschaff, Germany, 2022.
118. *Branża Tworzyw Sztucznych 2022*; Raport Fundacji—PlasticsEurope Polska, *Polish Circular Forum*, Warszawa, Poland; 2022. Available online: https://plasticseurope.org/pl/knowledge-hub/branza-tworzyw-sztucznych-2022/ (accessed on 1 November 2022).
119. *Statistical Data 2022*; Plastic Europe: Brussels, Belgium, 2022.
120. Hamad, K.; Kaseem, M.; Deri, F. Recycling of waste from polymer materials: An overview of the recent works. *Polym. Degrad. Stab.* **2013**, *98*, 2801–2812. [CrossRef]
121. *Tworzywa w Obiegu Zamkniętym—Analiza Sytuacji w Europie*; Plastics Europe: Brussels, Belgium, 2021.
122. Deshmukh, K.; Kovářík, T.; Muzaffar, A.; Basheer Ahamed, M.; Khadheer Pasha, S.K. Mechanical analysis of polymers. In *Polymer Science and Innovative Applications: Materials, Techniques, and Future Developments*; Elsevier: Amsterdam, The Netherlands, 2020; pp. 117–152, ISBN 9780128168080.
123. Gościański, M. Struktura i właściwości fizykomechaniczne mieszaniny odpadowych polimerów PA6 +PP—Aspekt ekologiczny i aplikacyjny. *J. Res. Appl. Agric. Eng.* **2002**, *47*, 25–30.

124. Grzesiczak, D.; Postawa, P. Właściwości mechaniczne i termiczne wyprasek po recyklingu odpadów produkcyjnych poduszek powietrznych wykonanych z tworzywa PET. *Przetwórstwo Tworzyw* **2017**, *23*, 535–542.
125. Monnerie, L. Mechanical properties of polymeric materials. In *Statistical Models for the Fracture of Disordered Media A Volume in Random Materials and Processes*; Herrmann, H.J., Roux, S., Eds.; Elsevier: Amsterdam, The Netherlands, 1990; pp. 66–76.
126. Ward, I.M.; Sweeney, J. *Mechanical Properties of Solid Polymers*, 3rd ed.; John Wiley & Sons, Ltd.: Bradford, UK, 2013; ISBN 9781444319507.
127. Florjańczyk, Z.; Penczek, S. *Chemia Polimerów*; Oficyna Wydawnicza Politechniki Warszawskiej: Warszawa, Poland, 2002; ISBN 83-7207-368.
128. Pielichowski, J.; Puszyński, A. *Chemia Polimerów; III*; Wydawnictwo Oświatowe FOSZE: Rzeszów, Poland, 2012; ISBN 978-83-7586-068-9.
129. Stevens, M.P. *Polymer Chemistry. An Introduction*, 3rd ed.; Oxford University Press: Oxford, UK, 1999; ISBN 9780195124446.
130. Plastipedia—The Web's Largest Plastics Encyclopedia. Available online: https://www.bpf.co.uk/plastipedia/Default.aspx#info (accessed on 3 November 2023).
131. Blicharski, M. *Wstęp Do Inżynierii Materiałowej*; Wydawnictwa Naukowo-Techniczne: Warszawa, Poland, 1998; ISBN 8320423015.
132. Hashmi, M.S.J. *Encyclopedia of Materials: Plastics and Polymers*; Hashmi, M.S.J., Ed.; Elsevier: Amsterdam, The Netherlands, 2022; ISBN 978-0-12-823291-0.
133. Furtak, K.; Śliwiński, J. *Materiały Budowlane w Mostownictwie; Wydanie Pi*; Wydawnictwa Komunikacji i Łączności: Warszawa, Poland, 2004; ISBN 83-206-1531-3.
134. Osiecka, E. *Materiały Budowlane: Tworzywa Sztuczne*; Oficyna Wydawnicza Politechniki Warszawskiej: Warszawa, Poland, 2005; ISBN 8372075409.
135. Eriksen, M.K.; Christiansen, J.D.; Daugaard, A.E.; Astrup, T.F. Closing the loop for PET, PE and PP waste from households: Influence of material properties and product design for plastic recycling. *Waste Manag.* **2019**, *96*, 75–85. [CrossRef]
136. Ragaert, K.; Delva, L.; Van Geem, K. Mechanical and chemical recycling of solid plastic waste. *Waste Manag.* **2017**, *69*, 24–58. [CrossRef]
137. Plastic Waste and Recycling in the EU: Facts and Figures I News I European Parliament. Available online: https://www.europarl.europa.eu/news/en/headlines/society/20181212STO21610/plastic-waste-and-recycling-in-the-eu-facts-and-figures (accessed on 3 November 2023).
138. Schyns, Z.O.G.; Shaver, M.P. Mechanical Recycling of Packaging Plastics: A Review. *Macromol. Rapid Commun.* **2021**, *42*, 2000415. [CrossRef]
139. Allen, D.; Spoelman, N.; Linsley, C.; Johl, A. *The Fraud of Plastic Recycling: How Big Oil and the Plastics Industry Deceived the Public for Decades and Caused the Plastic Waste Crisis*; Center for Climate Integrity: Washington, DC, USA, 2024. Available online: https://climateintegrity.org/projects/plastics-fraud (accessed on 1 February 2024).
140. Geyer, R.; Jambeck, J.R.; Law, K.L. Production, use, and fate of all plastics ever made. *Sci. Adv.* **2017**, *3*, e1700782. [CrossRef]
141. Feil, A.; Pretz, T.; Jansen, M.; Thoden Van Velzen, E.U. Separate collection of plastic waste, better than technical sorting from municipal solid waste? *Waste Manag. Res.* **2017**, *35*, 172–180. [CrossRef]
142. *The Guardian*; World Wide Fund for Nature: Gland, Switzerland, 2020.
143. *Statistical Data 2022*; Zero Waste Europe: Brussels, Belgium, 2022.
144. Arkenbout, A. *Hidden Emissions: A Story from the Netherlands Case Study*; Zero Waste Europe: Brussels, Belgium, 2018; pp. 1–11.
145. *Statistical Data 2021*; World Bank: Washington, DC, USA, 2021.
146. Szkudlarek, Ł. *Raport Końcowy—Badanie Ewaluacyjne pn. Recykling Wyeksploatowanych Komponentów Technicznych Odnawialnych źródeł Energii oraz Akumulatorów Pojazdów Elektrycznych jako Element Transformacji w Kierunku Gospodarki o Obiegu Zamkniętym*; Wroconsult Sp. z o.o. oraz Ekovert Łukasz Szkudlarek: Wrocław, Poland, 2019.
147. Drozdowski, W. Łopaty Turbin Wiatrowych Najtrudniejsze do Recyklingu. Available online: https://www.cire.pl/artykuly/opinie/155372-lopaty-turbin-wiatrowych-najtrudniejsze-do-recyklingu (accessed on 30 October 2023).
148. Ziegler, L.; Muskulus, M. Lifetime extension of offshore wind monopiles: Assessment process and relevance of fatigue crack inspection. In Proceedings of the 12th EA WE PhD Seminar on Wind Energy in Europe, Lyngby, Denmark, 23 May 2016; pp. 25–27.
149. Ziegler, L.; Gonzalez, E.; Rubert, T.; Smolka, U.; Melero, J.J. Lifetime extension of onshore wind turbines: A review covering Germany, Spain, Denmark, and the UK. *Renew. Sustain. Energy Rev.* **2018**, *82*, 1261–1271. [CrossRef]
150. Moc zainstalowana (MW)—Stan na 31.12.2022 r. In *Dane Tabelaryczne Dotyczące Poszczególnych Rodzajów Instalacji Odnawialnego Źródła Energii Obejmują Instalacje, Które Uzyskały: Koncesję na Wytwarzanie Energii Elektrycznej*; Urząd Regulacji Energetyki: Warszawa, Poland, 2022.
151. Wronka, K. Pełny Recykling Łopat Turbin Wiatrowych—Przyszłość Czy Złudna Nadzieja? Available online: https://www.cire.pl/artykuly/brak-kategorii/pelny-recykling-lopat-turbin-wiatrowych---przyszlosc-czy-zludna-nadzieja (accessed on 20 November 2023).
152. Albers, H.; Greiner, S.; Seifert, H.; Kuehne, U. Recycling of wind turbine rotor blades. Fact or fiction? *DEWI-Magazin* **2009**, *34*, 32–41.
153. Andersen, P.D.; Borup, M.; Krogh, T. Managing long-term environmental aspects of wind turbines: A prospective case study. *Int. J. Technol. Policy Manag.* **2007**, *7*, 339–354. [CrossRef]
154. Liu, P.; Barlow, C.Y. Wind turbine blade waste in 2050. *Waste Manag.* **2017**, *62*, 229–240. [CrossRef]
155. Liu, P.; Meng, F.; Barlow, C.Y. Wind turbine blade end-of-life options: An eco-audit comparison. *J. Clean. Prod.* **2018**, *212*, 1268–1281. [CrossRef]

156. Lefeuvre, A.; Garnier, S.; Jacquemin, L.; Pillain, B.; Sonnemann, G. Anticipating in-use stocks of carbon fibre reinforced polymers and related waste generated by the wind power sector until 2050. *Resour. Conserv. Recycl.* **2019**, *141*, 30–39. [CrossRef]
157. WindEurope | Wind Energy in Europe 2020 Statistics and the Outlook for 2021–2025. Available online: https://windeurope.org/intelligence-platform/product/wind-energy-in-europe-2020-statistics-and-the-outlook-for-2021-2025/ (accessed on 18 January 2023).
158. Forde, M.C. Sonic and Radar Impulse Non-Destructive Testing of Railroad Bridges. In *Transportation Infrastructure*; Springer: Berlin/Heidelberg, Germany, 1996; pp. 405–412.
159. Forde, M.C. Non-destructive evaluation of bridges: Research in progress. *Proc. US Eur. Bridg. Eng. Work. Tech. Univ. Catalonia Barcelona* **1996**, *26*, 127–152.
160. Nadja, H. Non-Destructive Testing of Concrete Bridges. Master's Thesis, Luleå University of Technology, Luleå, Sweden, 2020; pp. 1–117.
161. Zobel, H.; Karwowski, W.; Sarnowska, J.; Wróbel, M. Nowa generacja mostów—Cz. I, Mosty z kompozytów polimerowych. *Mag. Autostrady* **2004**, *4*, 16–19.
162. Szruba, M. Kompozyty w budownictwie. *Nowocz. Bud. Inżynieryjne* **2020**, *4*, 73–77.
163. Zobel, H.; Karwowski, W. Kompozyty polimerowe w mostownictwie: Pomosty wielowarstwowe. *Geoinżynieria Drog. Most. tunele* **2006**, *2*, 42–49.
164. Szelka, J.; Kamyk, Z. Kompozytowe mosty wojskowe. *Bud. Archit.* **2013**, *12*, 63–70. [CrossRef]
165. Sobczyk, B.; Pyrzowski, Ł.; Chróścielewski, J.; Miśkiewicz, M. Mosty kompozytowe—Pieśń przyszłości czy teraźniejszość? *Mosty* **2022**, *4*, 31–35.
166. Keller, T. *Use of Fibre Reinforced Polymers in Bridge Construction*; International Association for Bridge and Structural Engineering (IABSE): Zurich, Switzerland, 2003.
167. Górski, M.; Kotala, B.; Bialozor, R. Rodzaje i właściwości zbrojenia niemetalicznego. In Proceedings of the XXXIII Warsztaty Pracy Projektanta Konstrukcji, Szczyrk, Poland, 6–9 March 2018.
168. Tamošaitienė, J.; Parham, S.; Sarvari, H.; Chan, D.W.M.; Edwards, D.J. A Review of the Application of Synthetic and Natural Polymers as Construction and Building Materials for Achieving Sustainable Construction. *Buildings* **2024**, *14*, 2569. [CrossRef]
169. Pimentel Real, L.E. *Recycled Materials for Construction Applications: Plastic Products and Composites*; Springer International Publishing: Berlin/Heidelberg, Germany, 2023; ISBN 9783031148729.
170. Lamba, P.; Kaur, D.P.; Raj, S.; Sorout, J. Recycling/reuse of plastic waste as construction material for sustainable development: A review. *Environ. Sci. Pollut. Res.* **2022**, *29*, 86156–86179. [CrossRef]
171. Ornaghi, H.L.; Bolner, A.S.; Fiorio, R.; Zattera, A.J.; Amico, S.C. Mechanical and dynamic mechanical analysis of hybrid composites molded by resin transfer molding. *J. Appl. Polym. Sci.* **2010**, *118*, 887–896. [CrossRef]
172. Ratna, D. *Recent Advances and Applications of Thermoset Resins*, 2nd ed.; Elsevier: Amsterdam, The Netherlands, 2022; ISBN 9780323856645.
173. Sukanto, H.; Raharjo, W.W.; Ariawan, D.; Triyono, J.; Kaavesina, M. Epoxy resins thermosetting for mechanical engineering. *Open Eng.* **2021**, *11*, 797–814. [CrossRef]
174. Kopecky, L.; Machovic, V.; Králík, V.; Bittnar, Z.; Andertová, J.; Šmilauer, V.; Lhotka, M. Surface modification of PET fibers to improve mechanical properties of cement. In Proceedings of the 33rd International Conference on Cement Microscopy, San Francisco, CA, USA, 17–20 April 2011; pp. 355–371.
175. Primachenko, B.M. Effect of Initial Materials and Uncertainty of Production Technology on the Structure and Mechanical Properties of a Thermosetting Polymer. *Fibre Chem.* **2023**, *55*, 106–112. [CrossRef]
176. Malek, A. Thermosetting polymer composites: Manufacturing and properties study. *Rev. Adv. Mater. Sci.* **2023**, *62*, 20230126. [CrossRef]
177. Vaidya, U.K.; Britt, F. Structural Applications Of Reinforced Thermoplastics. In Proceedings of the International Corrosion Conference Series 2007 CORROSION 2007, Nashville, TN, USA, 11–15 March 2007; OnePetro: Nashville, TN, USA, 2007; pp. 075461–075468.
178. Moniruzzaman, M.; Christogianni, P.; Kister, G. Self-Healing in Epoxy Thermoset Polymer Films Triggered by UV Light. *Procedia Eng.* **2016**, *148*, 114–121. [CrossRef]
179. Jankowiak, I. Materiały Kompozytowe w Budownictwie Mostowym; Inżynier Budownictwa: 2012. Available online: https://inzynierbudownictwa.pl/materialy-kompozytowe-w-budownictwie-mostowym/ (accessed on 11 March 2024).
180. Siwowski, T. Wzmacnianie Mostów Materiałami Kompozytowymi; Izbudujemy.pl, 2016. Available online: https://www.izbudujemy.pl/artykuly/564/Wzmacnianie-mostow-materialami-kompozytowymi (accessed on 11 March 2024).
181. Meier, U. Strengthening of structures using carbon fibre/epoxy composites. *Constr. Build. Mater.* **1995**, *9*, 341–351. [CrossRef]
182. Meier, U.; Kotynia, R.; Walendziak, R. Badania Zmęczeniowe Żelbetowych Płyt Wzmocnionych Naprężonymi Taśmami CFRP. In *Problemy Naukowo-badawcze Budownictwa; T.2 Konstrukcje Budowlane i Inżynierskie—Konferencja Naukowa Komitetu Inżynierii Lądowej i Wodnej PAN i Komitetu Nauki PZITB*; Bronowicz, M., Prusiel, J.A., Eds.; Wydawnictwo Politechniki Białostockiej: Białystok, Poland, 2007; pp. 141–148.
183. Łagoda, M. *Wzmacnianie Konstrukcji Mostowych Kompozytami Wstępnie Sprężonymi*; Nowocz. Bud. Inżynieryjne: Kraków, Poland, 2012; pp. 52–55.

184. Łagoda, M. Wzmacnianie mostów przy pomocy materiałów kompozytowych, realizacje i założenia do zaleceń stosowania. In Proceedings of the Zeszyt Pokonferencyjny—Konferencja Naukowo-Techniczna Materiały Kompozytowe w Budownictwie mostowym", Łódź, Poland, 18 December 2000.
185. Kamiński, M.; Wydra, W. Wzmacnianie konstrukcji żelbetowych za pomocą włókien węglowych z uwzględnieniem wymagań ochrony przeciwpożarowej. In Proceedings of the XIII Ogólnopolska Konferencja „Warsztaty Pracy Projektanta Konstrukcji" 1998, Ustroń, Poland, 22 April 1998; pp. 87–98.
186. Radomski, W.; Trochymiak, W. Przegląd wspsółczesnych metod wzmacniania mostów betonowych. In Proceedings of the VIII Seminarium Współczesne Metody Wzmacniania i Przebudowy Mostów, Poznań, Poland, 9–10 July 1998; Volume 8, p. 174.
187. Wołowicki, W. *Projekt wykonawczy wzmocnienia mostu przez rzekę Wartę w Śremie*; PO.MOST ARS Sp. z o.o.: Poznań, Poland, 1997.
188. Siwowski, T.; Radomski, W. Pierwsze krajowe zastosowanie taśm kompozytowych do wzmocnienia mostu. *Inżynieria Bud.* **1998**, *54*, 382–388.
189. *Guide for the Design and Construction of Externally Bonded FRP Systems for Strengthening Concrete Structures*; American Concrete Institute: Washington, DC, USA, 2008.
190. Jankowiak, I. Efektywność Wzmacniania Materiałami Kompozytowymi Żelbetowych Belek Mostowych. Doctoral Thesis, Politechnika Poznańska, Poznań, Poland, 2010.
191. Cheng, Z.; Li, S.; Lu, Y.; Li, W.; Liu, Z. Dynamic compressive properties of lightweight engineered geopolymer composites containing ceramsite (LW-EGC). *Constr. Build. Mater.* **2023**, *370*, 130717. [CrossRef]
192. Gallegos-Calderón, C.; Oliva-Quecedo, J.; Pulido, M.D.G.; Goicolea, J.M. Influence of the Properties of a Glass Fiber Reinforced Polymer Deck on the Dynamic Response of a Road Bridge. *Rev. Politec.* **2023**, *52*, 73–82. [CrossRef]
193. Dai, L.; Zhu, Z.; Zhang, C.; Zhu, D. Experimental study on the influence of glass fiber reinforced concrete isolation layer on the seismic dynamic response of tunnels. *Case Stud. Constr. Mater.* **2023**, *19*, e02303. [CrossRef]
194. Mincigrucci, L.; Civera, M.; Lenticchia, E.; Ceravolo, R.; Rosano, M.; Russo, S. Comparative Structural Analysis of GFRP, Reinforced Concrete, and Steel Frames under Seismic Loads. *Materials* **2023**, *16*, 4908. [CrossRef]
195. Qureshi, J. A Review of Fibre Reinforced Polymer Bridges. *Fibers* **2023**, *11*, 40. [CrossRef]
196. Russell, J.M.; Wei, X.; Živanović, S.; Kruger, C. Vibration serviceability of a GFRP railway crossing due to pedestrians and train excitation. *Eng. Struct.* **2020**, *219*, 110756. [CrossRef]
197. Mayer, P.; Kaczmar, J. Właściwości i zastosowania włókien węglowych i szklanych. *Tworzywa Sztuczne Chem.* **2008**, *6*, 52–56.
198. Mossakowski, P. Pręty z kompozytów polimerowych z włóknami do zbrojenia betonowych konstrukcji inżynierskich. *Drog. Most.* **2006**, *1*, 35–52.
199. Kráčalík, M.; Pospíšil, L.; Šlouf, M.; Mikešová, J.; Sikora, A.; Šimoník, J.; Fortelný, I. Recycled poly(ethylene terephthalate) reinforced with basalt fibres: Rheology, structure, and utility properties. *Polym. Compos.* **2008**, *29*, 437–442. [CrossRef]
200. Katalagarianakis, A.; Van de Voorde, B.; Pien, N.; Polyzos, E.; Duretek, I.; Holzer, C.; Cardon, L.; Bernaerts, K.V.; Van Hemelrijck, D.; Van Vlierberghe, S.; et al. The effect of carbon fiber content on physico-mechanical properties of recycled poly(ethylene terephthalate) composites additively manufactured with fused filament fabrication. *Addit. Manuf.* **2022**, *60*, 103246. [CrossRef]
201. Liang, Z.; Zhuang, W.; Lin, C.; Liang, F.; Fan, X.; Zhu, Y.; Li, Q.; Li, C.; He, J. Glass fiber network reinforced polyethylene terephthalate composites for support insulators. *J. Phys. D Appl. Phys.* **2023**, *56*, 415501. [CrossRef]
202. Giraldi, A.L.F.D.M.; Bartoli, J.R.; Velasco, J.I.; Mei, L.H.I. Glass fibre recycled poly (ethylene terephthalate) composites: Mechanical and thermal properties. *Polym. Test.* **2005**, *24*, 507–512. [CrossRef]
203. Monti, M.; Scrivani, M.T.; Kociolek, I.; Larsen, Å.G.; Olafsen, K.; Lambertini, V. Enhanced Impact Strength of Recycled PET/Glass Fiber Composites. *Polymers* **2021**, *13*, 1471. [CrossRef]
204. Mondadori, N.M.L.; Nunes, R.C.R.; Canto, L.B.; Zattera, A.J. Composites of Recycled PET Reinforced with Short Glass Fiber. *J. Thermoplast. Compos. Mater.* **2012**, *25*, 747–764. [CrossRef]
205. Worku, B.G.; Alemneh Wubieneh, T. Composite material from waste poly (ethylene terephthalate) reinforced with glass fiber and waste window glass filler. *Green Chem. Lett. Rev.* **2023**, *16*, 2169081. [CrossRef]
206. Amalraj, E.F.P.; Ilangovan, P. Experimental Behavior of High-Strength Concrete Reinforced with Aramid Fiber and Polyurethane Resin. *Buildings* **2023**, *13*, 1713. [CrossRef]
207. Koerner, R.M. *Design with Geosythetics*, 2nd ed.; PreNTICe Hall: Hoboken, NJ, USA, 1994.
208. Wesołowski, A.; Krzywosz, Z.; Brandyk, T. *Geosyntetyki w Konstrukcjach Inżynierskich*; Wydawnictwo Szkoły Głównej Gospodarstwa Wiejskiego: Warszawa, Poland, 2000.
209. Wysokiński, L.; Kotlicki, W. *Projektowanie Konstrukcji Oporowych, Stromych Skarp i Nasypów z Gruntu Zbrojonego Geosyntetykami*; Instytut Techniki Budowlanej: Warszawa, Poland, 2008; pp. 1–51.
210. Holtz, R.D. Geosynthetics for Soil Reinforcement. *Struct. Infrastruct. Eng.* **2009**, 1–19.
211. Holtz, R.D.; Christopher, B.R.; Berg, R.R. *Geosynthetic Engineering*; BiTech Publishers: Richmond, BC, Canada, 1997; ISBN 0921095201.
212. Jewell, R.A. *Soil Reinforcement with Geotextiles*; Construction Industry Research and Information Association: London, UK, 1996; ISBN 978-0-86017-425-7.
213. Kawalec, J. *Stabilizacja Podłoża z Wykorzystaniem Georusztów*; ImiG: Philadelphia, PA, USA, 2010; Volume 4.
214. Mieczkowski, P. Hydroizolacja Płyt Pomostów Na Obiektach Mostowych. Available online: https://www.izbudujemy.pl/artykuly/300/Hydroizolacja-plyt-pomostow-na-obiektach-mostowych (accessed on 12 March 2024).

215. Bichajło, L.; Siwowski, T. Nawierzchnie na obiektach mostowych—Doświadczenia z nadzoru nad realizacją. Sympozjum: Nawierzchnie i izolacje na obiektach inżynieryjnych betonowych i stalowych. Konsekwencje i możliwości działań. In Proceedings of the XVII Międzynarodowe Targi Budownictwa Drogowego AUTOSTRADA-POLSKA, Kielce, Poland, 18 May 2011; Nowoczesne Budownictwo Inżynieryjne: Kielce, Poland, 2011.
216. Kilarski, R.; Mąkosa, J.; Germaniuk, K. Analiza i ocena materiałów hydroizolacyjnych na pomosty obiektów mostowych. *Pr. Inst. Badaw. Dróg Most.* **1996**, *3–4*, 105–134.
217. Duszyńska, A.; Sikora, Z. Dobór wyrobów geosyntetycznych do zbrojenia gruntu. *Inżynieria Morska Geotech* **2014**, *nr 5*, 468–473.
218. Duszyńska, A.; Szypulski, P. Projektowanie nasypów komunikacyjnych ze zbrojeniem geosyntetycznym podstawy na słabym podłożu gruntowym. *Przegląd Komun.* **2014**, *5*, 11–18.
219. Duszyńska, A.; Makasewicz-Dzieciniak, M. Nasyp z geosyntetycznym wzmocnieniem podstawy posadowiony na pionowych elementach nośnych. *Inżynieria Morska Geotech.* **2013**, *3*, 217–224.
220. Duszyńska, A.; Szypulski, P. Wymiarowanie wzmocnienia geosyntetycznego podstawy nasypu na słabym podłożu. *Inżynieria Morska Geotech.* **2012**, *3*, 215–223.
221. Duszyńska, A.; Bolt, A. *Współpraca Georusztu i Gruntu w Badaniu na Wyciąganie*; Politechnika Gdańska: Gdańsk, Poland, 2004.
222. Duszyńska, A. Co warto wiedzieć o geosyntetykach? *Inżynieria Morska Geotech* **2010**, *2*, 211–222.
223. Duszyńska, A. Stateczność nasypów drogowych ze wzmocnieniem geosyntetycznym. *Inżynieria Bud.* **2014**, *70*, 256–269.
224. Abdullah, N.H.H.; Ng, K.S.; Jais, I.B.M.; Idrus, J. Use of geosynthetic reinforced soil-integrated bridge system to alleviate settlement problems at bridge approach: A review. *Phys. Chem. Earth Parts A/B/C* **2023**, *129*, 103304. [CrossRef]
225. Wipf, T.J.; Phares, B.; Dahlberg, J.M.; Lu, P. *Investigation and Evaluation of Iowa Department of Transportation Bridge Deck Epoxy Injection Process, February 2019*; Bridge Engineering Center: Iowa, IA, USA, 2019.
226. Banaś, A.; Bałachowski, L.; Kuryłowicz, A.; Poteraj-Oleksiak, A. Innowacyjna metoda wzmocnienia gruntu za przyczółkiem przy użyciu iniekcji geopolimerowych. *Builder* **2020**, *9*, 24–28. [CrossRef]
227. Kokotkiewicz, P. Problemy geotechniczne na styku obiektów inżynierskich z nasypami drogowymi. *Mosty* **2016**, *1*, 14–17.
228. Trojnar, K. Jak projektować nasypy na dojazdach do obiektów mostowych na słabym podłożu? *Mosty* **2016**, *1*, 10–13.
229. Poteraj-Oleksiak, A. Poziomowanie i stabilizacja gruntów—Zalety metody iniekcji geopolimerowych. *Mag. Autostrady* **2018**, *11–12*, 46–49.
230. Gajewska, B.; Kłosiński, B. Rozwój metod wzmacniania podłoża gruntowego. In *Proceedings of the Seminarium IBDiM i PZWFS: Wzmacnianie Podłoża i Fundamentów*; Instytut Badawczy Dróg i Mostów: Warszawa, Poland, 2011; pp. 1–42.
231. Jermołowicz, P. Wzmocnienie nasypów drogowych. Zasady projektowania i wykonawstwa. In *Zeszyty Naukowo-Techniczne*; Stowarzyszenia Inżynierów i Tech. Komun. w Krakowie: Kraków, Poland, 2011.
232. Jermołowicz, P. Wzmacnianie Nasypów Drogowych Wraz z ich Odwadnianiem. Zasady projektowania i wykonastwa. In *Materiały Szkoleniowe*; Warmińsko-Mazurska OIIB: Olsztyn, Poland, 2016; pp. 1–157.
233. Grzegorzewicz, K.; Rychlewski, P. Styropian jako lekkie wypełnienie nasypu przy przeciążonej konstrukcji oporowe. *J. Mater. Bud.* **2003**, *3*, 42.
234. Górecki, Ł.; Grzegorzewicz, K. Keramzyt i Styropian Jako Lekkie Wypełnienia Nasypów Drogowych; Izolacje, 2020, 6. Available online: https://www.izolacje.com.pl/artykul/drogi-mosty/196510,keramzyt-i-styropian-jako-lekkie-wypelnienia-nasypow-drogowych (accessed on 13 March 2024).
235. Gajewska, B.; Grzegorzewicz, K.; Kłosiński, B. Wzmacnianie podłoża gruntowego z zastosowaniem lekkich wypełnień. *Drogownictwo* **2013**, *1*, 9–14.
236. Łysoń, M. Przełom w Budownictwie. Ten BETON Z Biowłóknami Przetrwa Wszystko. Available online: https://www.chip.pl/2024/03/galaxy-a55-5g-dopiero-zadebiutowal-a-juz-kupimy-go-w-plusie (accessed on 10 January 2024).
237. Houshmand Khaneghahi, M.; Kamireddi, D.; Rahmaninezhad, S.A.; Sadighi, A.; Schauer, C.L.; Sales, C.M.; Najafi, A.R.; Cotton, A.; Street, R.; Farnam, Y. (Amir) Development of a nature-inspired polymeric fiber (BioFiber) for advanced delivery of self-healing agents into concrete. *Constr. Build. Mater.* **2023**, *408*, 133765. [CrossRef]
238. Ostrowski, K.A.; Furtak, K. Influence of Carbon Fibre Reinforced Polymer and Recycled Carbon Fibres on the Compressive Behaviour of Self-Compacting High-Performance Fibre-Reinforced Concrete. *Arch. Civ. Eng.* **2023**, *69*, 53–64. [CrossRef]
239. Rebeiz, K.S.; Fowler, D.W.; Paul, D.R. Recycling Plastics in Polymer Concrete for Construction Applications. *J. Mater. Civ. Eng.* **1993**, *5*, 237–248. [CrossRef]
240. Rebeiz, K.S.; Fowler, D.W.; Paul, D.R. Recycling Plastics in Polymer Concrete Systems for Engineering Applications. *Polym. Plast. Technol. Eng.* **1991**, *30*, 809–825. [CrossRef]
241. Salimi, K.; Ghazavi, M. Soil reinforcement and slope stabilisation using recycled waste plastic sheets. *Geomech. Geoengin.* **2021**, *16*, 497–508. [CrossRef]
242. Tao, Y.; Hadigheh, S.A.; Wei, Y. Recycling of glass fibre reinforced polymer (GFRP) composite wastes in concrete: A critical review and cost benefit analysis. *Structures* **2023**, *53*, 1540–1556. [CrossRef]
243. Marusceac, V.; Vlad, M. A Different Approach to Classic Structural Reinforcements Using Recycled Synthetic Materials. *Procedia Eng.* **2017**, *181*, 273–279. [CrossRef]
244. Dahlbo, H.; Poliakova, V.; Mylläri, V.; Sahimaa, O.; Anderson, R. Recycling potential of post-consumer plastic packaging waste in Finland. *Waste Manag.* **2018**, *71*, 52–61. [CrossRef]

245. Gałczyńska, M. Eko-Asfalt z Plastiku i Piasku? Wynalazł Go Łodzianin, Państwo Nie Było Zainteresowane—Wiadomości; Onet, 2016. Available online: https://wiadomosci.onet.pl/lodz/eko-asfalt-z-plastiku-i-piasku-wynalazl-go-lodzianin-panstwo-nie-bylo-zainteresowane/t775jfs (accessed on 13 March 2024).
246. Wei, Q.; Cheng-Hui, L. Performance of Polyurethane Polymer in the Transition Zones of Ballasted and Ballastless Track. In Proceedings of the 1st GeoMEast International Congress and Exhibition, Egypt 2017 on Sustainable Civil Infrastructures, Sharm Elsheikh, Egypt, 15–19 July 2017; Pombo, J., Jing, G., Eds.; Springer: Berlin/Heidelberg, Germany, 2017; pp. 142–152.
247. Zhao, C.; Wang, P.; Yi, Q.; Meng, D. Application of polyurethane polymer and assistant rails to settling the abnormal vehicle-track dynamic effects in transition zone between ballastless and ballasted track. *Shock Vib.* **2015**, *1*, 1–9. [CrossRef]
248. Xin, T.; Ding, Y.; Wang, P.; Gao, L. Application of rubber mats in transition zone between two different slab tracks in high-speed railway. *Constr. Build. Mater.* **2020**, *243*, 118219. [CrossRef]
249. Idczak, W.; Lewandrowski, T.; Pokropski, D.; Rogojsz, G.; Rudnicki, T. The Influence of the Type of Fibers on the Reduction of the Threshold Effect in the Transition Zone of a Railway Track. *Materials* **2022**, *15*, 5730. [CrossRef] [PubMed]
250. Czarnecki, L. Betony polimerowe. *Cem. Wapno Bet.* **2010**, *15/77*, 63–85.
251. Welle, F. Twenty years of PET bottle to bottle recycling—An overview. *Resour. Conserv. Recycl.* **2011**, *55*, 865–875. [CrossRef]
252. Tsochatzis, E.D.; Lopes, J.A.; Corredig, M. Chemical testing of mechanically recycled polyethylene terephthalate for food packaging in the European Union. *Resour. Conserv. Recycl.* **2022**, *179*, 106096. [CrossRef]
253. Barthélémy, E.; Spyropoulos, D.; Milana, M.R.; Pfaff, K.; Gontard, N.; Lampi, E.; Castle, L. Safety evaluation of mechanical recycling processes used to produce polyethylene terephthalate (PET) intended for food contact applications. *Food Addit. Contam. Part A* **2014**, *31*, 490–497. [CrossRef]
254. Panowicz, R.; Konarzewski, M.; Durejko, T.; Szala, M.; Łazińska, M.; Czerwińska, M.; Prasuła, P. Properties of Polyethylene Terephthalate (PET) after Thermo-Oxidative Aging. *Materials* **2021**, *14*, 3833. [CrossRef]
255. Koo, H.J.; Chang, G.S.; Kim, S.H.; Hahm, W.G.; Park, S.Y. Effects of recycling processes on physical, mechanical and degradation properties of PET yarns. *Fibers Polym.* **2013**, *14*, 2083–2087. [CrossRef]
256. Frounchi, M. Studies on degradation of PET in mechanical recycling. *Macromol. Symp.* **1999**, *144*, 465–469. [CrossRef]
257. Lee, J.H.; Lim, K.S.; Hahm, W.G.; Kim, S.H. Properties of recycled and virgin poly(ethylene terephthalate) blend fibers. *J. Appl. Polym. Sci.* **2013**, *128*, 1250–1256. [CrossRef]
258. Ostrowski, K.A.; Spyrowski, M.M.; Romańska, P.; Furtak, K. Evaluation of mechanical properties of recycled polyethylene terephthalate (rPET) reinforced with glass fibers. In Proceedings of the 4th International Conference on Sustainable Development in Civil, Urban and Transportation Engineering 2024, Wroclaw, Poland, 14–17 October 2024; pp. 1–7.
259. S355 Structural Steel Grades for Construction—Constructalia. Available online: https://constructalia.arcelormittal.com/en/steel-grades/s355 (accessed on 20 November 2024).
260. Steel S235JR: Characteristics, Properties, Analogues—Metinvest. Available online: https://metinvestholding.com/en/products/steel-grades (accessed on 20 November 2024).
261. AZO MATERIALS | Polyethylene Terephthalate—PET 55% Glass Fiber Reinforced. Available online: https://www.azom.com/article.aspx?ArticleID=789 (accessed on 13 March 2024).
262. AKRO COMPOUNDS | Polyethylene Terephthalate—PET 30% Carbon Fiber Reinforced. Available online: https://akro-plastic.com/en/product/precite-e-icf-30-black-7429-en (accessed on 13 March 2024).
263. Global Material Consumption: Infrastructure Sector Dominates Consumption—The World Bank Data. Available online: https://www.gihub.org/infrastructure-monitor/insights/infrastructure-consumes-more-than-half-of-the-world-s-materials/ (accessed on 20 November 2024).
264. Zhang, H. *Building Materials in Civil Engineering*, 1st ed.; Haimei, Z., Ed.; Woodhead Publishing: Cambridge, UK, 2011; ISBN 978-1-84569-956-7.
265. Górski, J.; Klimowicz, J.; Nowak, A. Application of Pro-Ecological Building Technologies in Contemporary Architecture. In Proceedings of the XV International Conference on Durability of Building Materials and Components DBMC 2020, Barcelona, Spain, 30 June–3 July 2020; pp. 435–442. [CrossRef]

applied sciences

MDPI

Review

The Seismic Behavior of Rectangular Concrete-Encased Steel Bridge Piers: A Review

Mohammadreza Moradian and Munzer Hassan *

Department of Construction Engineering, École de Technologie Supérieure, University of Quebec, Montreal, QC H3C 1K3, Canada; mohammadreza.moradian.1@ens.etsmtl.ca
* Correspondence: munzer.hassan@etsmtl.ca

Abstract: This paper proposes a review of the previous research work and the representative publications regarding the seismic behavior of the concrete-encased steel (CES) columns. Concrete-encased steel sections are composed of steel sections encased in reinforced concrete members. The research work recently showed increased attention to this type of column due to its advantages compared to conventional reinforced concrete columns. Firstly, the analytical studies of the behavior of the CES columns under axial loads, including comparative studies between different research works, are presented. Then, the behavior of the CES columns under combined axial and flexural loads is also highlighted. An overview of the analytical confinement material models is addressed. In addition, the discussion and summary of the seismic behavior of the CES columns and the important parameters affecting the seismic behavior of these types of columns are included. Although important progress has been made by the previous studies in the CES columns under the axial load and the combination of axial and seismic loads, they fundamentally focused on the building columns, and little attention was paid to the impact of lateral reinforcement and their configuration in bridge piers. Due to the lack of studies on the parameters affecting the seismic behavior of the bridges, more studies should still be made to better understand the behavior of the CES bridge piers. This paper provides a reference for the research and engineering practice of concrete-encased steel bridge piers. It also concludes with suggestions for future studies to integrate the seismic requirement of the CES bridge piers in Canada.

Keywords: bridge seismic behavior; concrete-encased steel pier; ductility; confined concrete; plastic hinge; failure mode

1. Introduction

Engineers have traditionally employed structural components made of steel forms combined with either plain or reinforced concrete. Earlier constructions only focused on the fire and corrosion resistance that the concrete provided for the steel shapes. However, studies on the strength of these members were carried out in the early 1900s, and by 1924, design guidelines had been developed. With the introduction of contemporary composite frame construction in tall structures in recent years, engineers have proposed innovative and practical ways to increase the resistance and ductility of concrete and steel reinforcement bars by using steel sections encased in concrete, also known as concrete-encased steel (CES) sections [1].

The use of concrete-encased columns in structures can reduce the size of the columns and reduce construction costs. In addition, its corrosion and fire resistance are both highly improved. Compared to ordinary columns, steel columns encased in concrete have substantially fewer long-term failures. It has been demonstrated that the ultimate axial load capacity of the CES columns is higher than a reinforced concrete column. This means that the structural steel improved the ultimate load resistance of the CES column [2]. Furthermore, an analytical and experimental study reveals that the axially loaded CES columns can sustain higher axial loads and undergo a smaller axial deformation compared to a conventional reinforced concrete column [3]. The steel and concrete composite building technology combines the flexibility and stiffness of reinforced concrete with the assistance of reinforcing bars to create a cost-effective structure. Generally speaking, the idea that two or more components are combined to make an efficient structure is what gives composite buildings their strength [4]. Nevertheless, the CES columns have certain limitations too. In some cases, it is difficult to cast the concrete since the steel ratio is high and the space between the steel section and the rebars makes it difficult to pour the concrete. During the design phase, the engineers must be extremely mindful of any potential issues with the positioning of the reinforcing steel and congestion as they impact the pier's constructability, just like with any other concrete column with reinforcing steel [1]. Furthermore, the initial cost of construction of a CES column could be higher than a conventional reinforced concrete column. However, the increase in durability of the CES columns can justify a higher initial cost.

Major infrastructure, including railroads and bridges, ought to be built to withstand significant seismic ground motion. In the CES columns, integrated concrete and steel profile sections help to obtain a high load-carrying capacity and high ductility. Figure 1 presents an example of a recent bridge construction using concrete-encased steel columns in China in 2024 [5].

Figure 1. Tian'e Longtan Bridge constructed using CES columns, reprinted with permission from [5], 2024, www.highestbridges.com, accessed on 15 June 2024.

The application of the CES column in buildings is more frequent than in bridges. Most applications on bridges until now have been on the steel–concrete composite high bridge piers. More specifically, in the mountainous regions, where large quantities of rebars are required as well as concentrated stirrups are required for a conventional reinforced concrete column. In addition, a large number of stirrups and reinforcing bars lowers the quality of construction and reduces efficiency while also making construction more challenging.

As illustrated in Figure 2, hybrid hollow high piers, which are a form of CES columns, offer a new approach to reducing the excessive use of steel bars in traditional bridge piers. A portion of the axial direct reinforcing bars installed using the conventional method are replaced by structural steel and stirrups or spiral reinforcement. When compared to typical reinforced concrete piers, the hybrid hollow high pier has a number of advantages, including increased earthquake resistance, increased economy and efficiency during construction, improved quality and aesthetics, and the preservation of the environment [6].

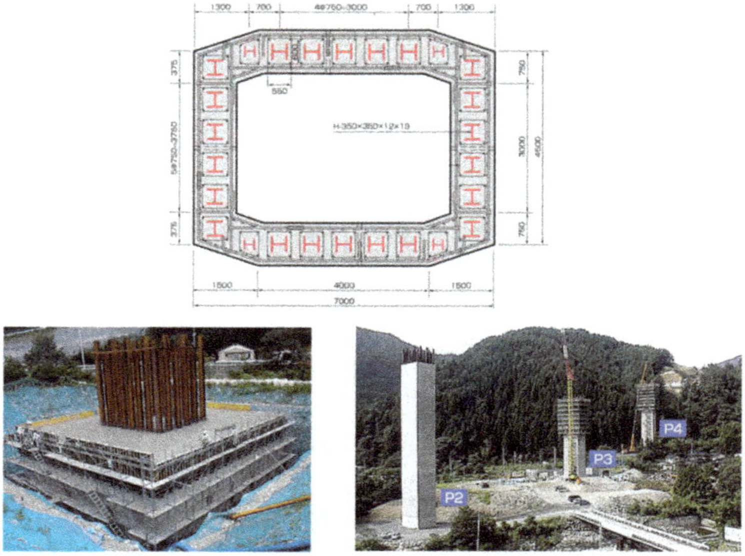

Figure 2. Hybrid hollow high pier of bridge constructed using CES concept; dimensions are in millimeters; reprinted with permission from [7], 2018, www.actec.or.jp.

In the context of the Canadian weather, the CES piers help to increase the durability of the bridges in harsh winter environments. Increasing the durability of the infrastructure becomes one of the most important aspects of the bridge design in view of its large impact on the economy. Reinforcement corrosion directly causes concrete deterioration, specifically in the bridge piers. In order to melt down the ice on the roads, the de-icing salts are spread out on the road during the winter. However, most bridge piers that are near the roads are in the splash zone of the mix of de-icing salt and the melting water caused by passing vehicles. Once the water is absorbed and combined with chloride ions, the corrosion process on the reinforcement rebars begins [8]. With the corrosion of the reinforcement rebars, the expansion causes concrete cover spalling, the reinforcement rebars become more exposed to the watery de-icing salt, and the corrosion process accelerates quickly, causing a significant reduction in the reinforcement and therefore leading to a load capacity reduction of the bridge. In CES piers, since the steel profile is in the center of the column, the possibility of the penetration of the de-icing salt to the steel profile reduces considerably. However, the reinforcing rebars are still exposed, but the risk is less since the central steel profile can carry the vertical load without the contribution of the reinforcing rebars. Thus, the use of CES piers can improve the impact on the traffic during the repair works. In the case of conventional reinforced concrete columns, the repair of the corroded rebars causes a bridge closure, either partially or completely because of the reduction in the capacity of the column during the repair works. However, using the CES piers can allow us to keep the bridge open during the repair of the corroded rebars since the steel profile can assure the capacity of the bridge for the service load.

The CES column has great potential to be widely used in future construction as engineering constructions evolve toward high-rise buildings, heavy-loading structures, and large-span bridges. Given the lack of well-developed Canadian design code specifications, it is crucial to compile the body of knowledge describing the behavior of the CES columns in order to facilitate the codification of the design guide.

This paper aims to present a thorough analysis of different CES column configurations, emphasizing their axial, flexural, and seismic performance as well as analytical modeling techniques. The main objective of this review is to present a thorough, insightful, and useful overview of the field of bridge engineering, with a special emphasis on the concrete-encased steel (CES) column pier section. The goal of this study is to be a useful resource by illuminating the subtleties and complexity of different CES section calculations in bridge performance and analysis.

There are four main sections to this paper. The CES column configuration and their analysis methods are covered in Sections 2 and 3 respectively. The axial and flexural behavior of the CES columns is shown in the Section 4. The seismic and cyclic performance of the CES columns is reviewed in the Section 5, and recommendations for further research are offered in Section 6.

2. Typical Configuration of Concrete-Encased Steel Columns

A concrete-encased column has a concrete-covered steel specimen inside of it. The steel specimen's buckling behavior is improved, and its strength is further increased by the concrete surrounding it. One significant benefit of this encasement is its ability to prevent corrosion and fire.

Some of the typical configurations of the CES columns are presented in Figure 3. It should be noted that the configuration of the steel section, rebar ratio, and transversal reinforcement have an impact on the resistance and the ductility of the CES column.

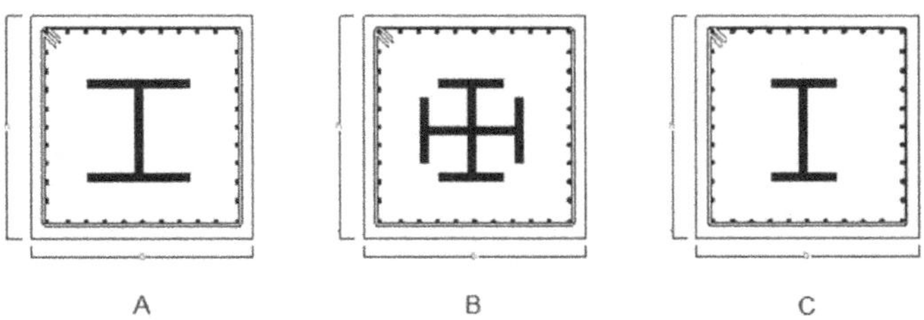

Figure 3. Typical cross section of CES columns. (**A**) H shape section, (**B**) Combined I shape section, (**C**) I shape section.

The CES sections are not presented in the Canadian bridge design code CSA-S6 [9]. However, the American Concrete Institute code ACI 318-8 [10] prescribes that the cross-sectional area of the steel should be at least 1% of the total cross-sectional area. Also, the cross-sectional area of the steel should not exceed 8% of the total cross-sectional area [10]. According to the AISC 360-22 [11] the cross-sectional area of the steel profile should be at least 1% of the total cross-sectional area, and the minimum reinforcement ratio of the longitudinal rebars should be 0.4% of the total cross section.

Since there is no requirement to be found in the Canadian bridge design code CSA-S6 [9] regarding the CES sections, the requirements of the American Institute of Steel Construction code AISC 360-22 [11] are presented in this paper. According to the AISC 360-22, concrete shall have a specified compressive strength, f_c', of not less than 21 MPa nor more than 69 MPa for normal-weight concrete. The specified minimum yield stress of structural steel and the reinforcing rebars used in calculating the strength shall not exceed 525 MPa and 550 MPa, respectively [11].

3. Simulating the Behavior of the Concrete-Encased Steel Columns

For different configurations of the CES columns, the resistance of the confined concrete can be defined accordingly. Moreover, an extensive amount of research has been conducted to gather information regarding past research work on analytical models that consider the effect of confinement and other factors into account in order to simulate the actual mechanism of the CES columns [12].

The CES columns were analyzed using the fiber section model. The section is divided up into "n" fibers, which are not necessarily all of identical sizes. The stresses are integrated into the cross-sectional area in order to determine the straining actions, such as moment or force. Each steel reinforcing bar, structural steel section, or concrete fiber has a specific set of material properties. Fiber stresses are calculated from them using the "plane sections remain plane" theory along with relevant constitutive models. There are several formulas for calculating the strain on the fiber as the loads increase. The section axial strain and curvatures with respect to a fixed reference system were described and they did not have to monitor the evolution of the neutral axis (NA) position [13].

Additionally, the impact of the load on the position of the natural axis (NA) was studied [14]. In order to calculate the stresses of the fibers and modules of elasticity, it should determine a constative model for uniaxial models for steel and concrete including the confinement effect. It was discovered that during an increase in the load, the CES column provided its initial strength against lateral shear. It should be noted that the initial stiffness in the model equations should be considered adequately [15–18].

The impact of flange shear studs for increasing stiffness and bending strength, the concrete's compressive strength, the mechanism of shear resistance of the CES section, and the necessary confinement level to provide appropriate ductility under seismic loading were all explored [16].

In order to account for the confinement impact on the CES columns made of cross-shaped steel and to predict the analytical axial load–displacement relationships of the columns [19,20], we adapted Mander's model. The structural steel segment increases the confining stress in the concrete zone. The confinement regions can be made simpler by transforming the parabolic sections, presented in Figure 4, into rectangular regions [21]. By considering the effect of unstiffened elements of the steel section on the confinement, one-half of the corresponding steel area is taken to calculate the confining stress [22].

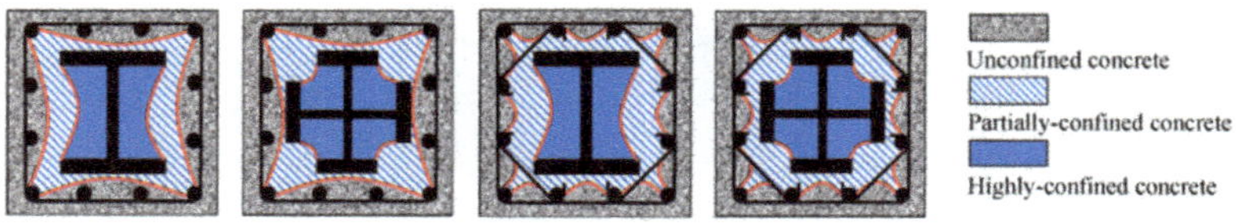

Figure 4. Definition of confinement of CES columns, reprinted with permission from [23], 2023, *Journal of Building Engineering*.

As presented in Figure 4, the concrete section in the CES column's cross section is divided into three sections, unconfined concrete (UC), partially confined concrete (PC), and highly confined concrete (HC). The stress–strain curves for the concrete are presented in Figure 5a. The steel section is also divided into three sections, the steel flange and the steel web as presented in Figure 5b, and the longitudinal reinforcement as presented in Figure 5c. Consequently, six distinct uniaxial stress–strain relationships are presented in this section [19].

Due to the steel flanges' bending stiffness, the steel section of the CES column could contribute to the expanded concrete's lateral confining pressure. The Mander model, first introduced for reinforced concrete [16], can be applied to HC in CES columns as long as it is established that the steel section's effective lateral confining pressure exists. Consequently, by closely examining the lateral interaction between the concrete and steel section, a unified stress–strain relation is provided for HC.

Based on the Mander model, the stress–strain relationship for PC is established. The Mander model predicts that hoops will yield at the maximum strain of confined concrete, but other experimental research work showed that high-strength hoops might not yield when the confinement effect is minimal [24]. As a result, modifications are required to determine the real stress in hoops. Therefore, it is imperative to ascertain the real stress status of the hoops, and the iterative process was used in this instance [25].

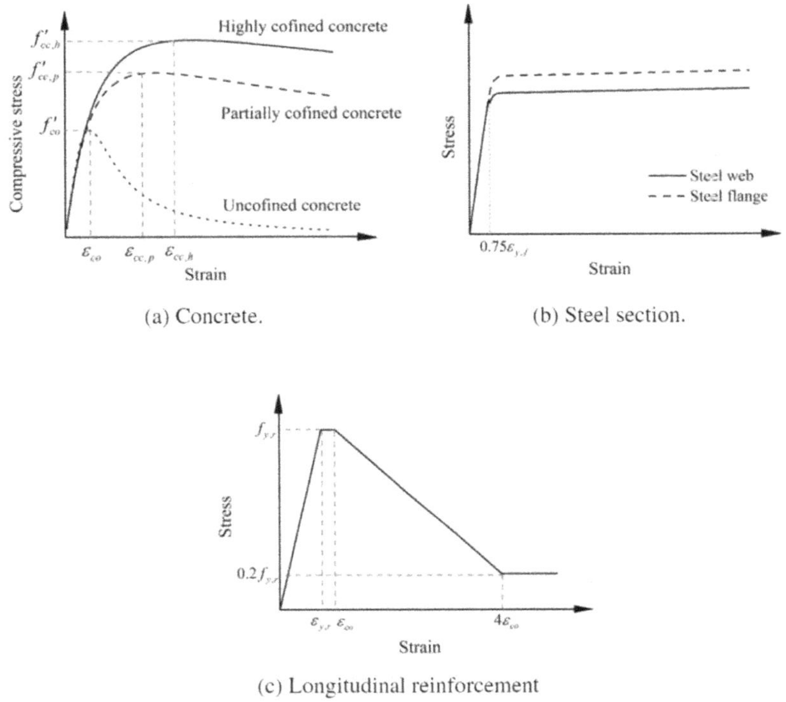

(a) Concrete. (b) Steel section. (c) Longitudinal reinforcement

Figure 5. Stress–strain relationship of materials in CES section, reprinted with permission from [18], 2015, *Journal of Constructional Steel Research.*

It should be noted that the parameters that can impact the resistance of the confined concrete are the distribution of the longitudinal reinforcing bar, the volumetric ratio of the lateral reinforcement, and the spacing of the stirrups [16]. A reduction in the stirrup spacing enhances the post-peak behavior of the axial load–deformation behavior.

Furthermore, in order to investigate the confinement impact for highly and partially confined concrete, the confinement factors (K_p and K_h) have been suggested [22]. The concrete strength f'_{cc} for partially and highly confined concrete is described as presented in Equations (1) and (2), respectively, while f'_{co} is the compressive strength of the unconfined concrete.

$$f'_{cc} = K_p \cdot f'_{co} \quad (1)$$

$$f'_{cc} = K_h \cdot f'_{co} \quad (2)$$

The relations of the tie spacing versus the confinement factor K_p for partially confined concrete for columns are shown in Figure 6 [22]. The effectiveness of confinement by lateral reinforcement can be shown in Figure 6. It can be concluded that reducing stirrup spacing increases the confinement factor K_p.

The form of the structural steel section, which applies confining stress to the core concrete, affects the confinement factor K_h for highly confined concrete. The steel profile form versus confinement factor K_h relations and their impact on the efficiency of confinement are displayed in Figure 7. The structural steel, especially the cross-shaped section, increases the confining effect. The explanation for the change is that in contrast to I- or H-shaped steel sections, which confine the concrete in just one direction, cross-shaped steel sections can produce confining pressure in both directions, resulting in a greater value of the confinement factor K_h [22].

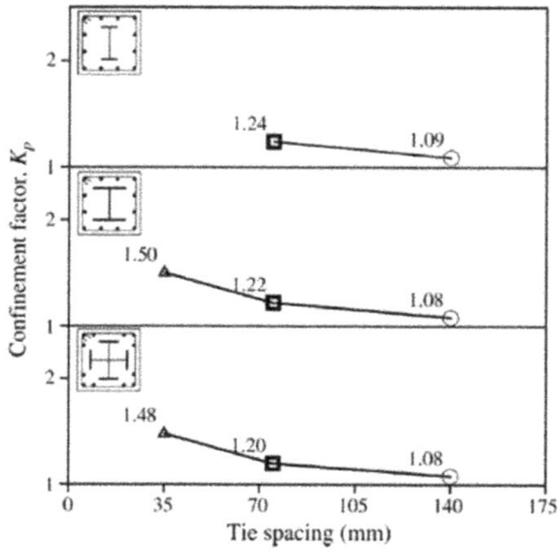

Figure 6. Confinement factor (k_p) for partially confined concrete, reprinted with permission from [22], 2006, *Journal of Constructional Steel Research*.

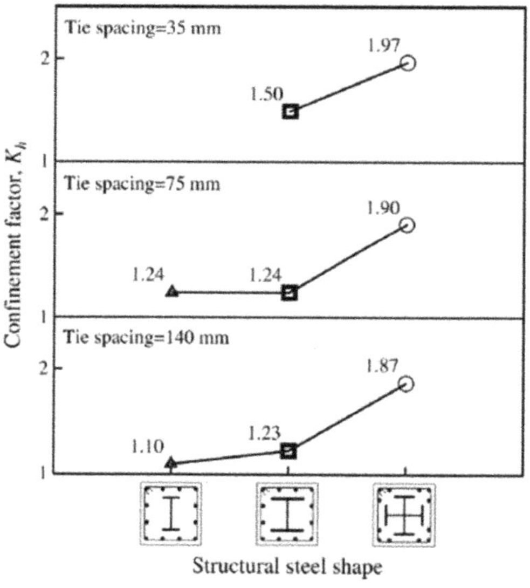

Figure 7. Confinement factor (K_h) for highly confined concrete, reprinted with permission from [22], 2006, *Journal of Constructional Steel Research*.

The effect of concrete confinement improves the ductility of CES columns but has little impact on the axial capacity since it activates after cover spalling. In addition, compared to shear-dominant CES columns, the flexural dominant CES columns benefit more from the confinement effect provided by stirrups [23].

Another research work has developed an analytical model to predict the axial compressive behavior of CES columns [19]. The confinement factors K_p and K_h for PC and HC, respectively, are described as follows:

$$k_p = -1.254 + 2.254\sqrt{1 + \frac{7.94 f'_{le,p}}{f'_{co}}} - 2\frac{f'_{le,p}}{f'_{co}} \quad (3)$$

$$k_h = -1.254 + 2.254\sqrt{1 + \frac{7.94 f'_{le,h}}{f'_{co}}} - 2\frac{f'_{le,h}}{f'_{co}} \tag{4}$$

As presented in Equation (3), $f'_{le,p}$ is the effective confining pressure from stirrups. It can be obtained as follows:

$$f'_{le,p} = \xi f_{rh} \tag{5}$$

where f_{rh} is the real stress in the stirrups, and ξ is the confinement effectiveness factor for the partially confined concrete.

The confinement effectiveness factor for the partially confined concrete can be presented as

$$\xi = \frac{\left(1 - \frac{0.5E_s\varepsilon_{co}}{f_{y,h}}\right) f'_{co}}{14E_s\varepsilon_{co}} \tag{6}$$

The real stress in the stirrups f_{rh} can be obtained as follows:

$$f_{rh} = \frac{0.5\ E_s\varepsilon_{co}}{1 - 14\zeta} \tag{7}$$

where E_s is the elastic modulus for steel, ε_{co} is the peak strain of the unconfined concrete, and the ζ is the coefficient in the simplified method for partially confined concrete presented as follows:

$$\zeta = \frac{\xi E_s\varepsilon_{co}}{f'_{co}} \tag{8}$$

As presented in Equation (4), $f'_{le,h}$ is the effective lateral confining pressure for highly confined concrete. It can be obtained as follows:

$$f'_{le,h} = f'_{le,p} + f'_{le,s} \tag{9}$$

where $f'_{le,p}$ is the effective lateral confining pressure from stirrups presented previously in Equation (5), and $f'_{le,s}$ is the effective lateral confining pressure from the steel section presented as follows:

$$f'_{le,s} = k_{es}k_{ea}f'_{ls} \tag{10}$$

where k_{es} is the stress effectiveness coefficient considering the uneven distribution of confining pressure, k_{ea} is the confinement effectiveness coefficient considering different confining states in highly confined concrete, and $f'_{l,s}$ is the nominal confining pressure between highly confined concrete and steel sections in the CES column which can be obtained as follows:

$$f'_{l,s} = \frac{2}{3}q_u \tag{11}$$

where q_u is the maximum lateral pressure between the highly confined concrete and steel sections in the CES column presented as follows:

$$q_u = \frac{t_f^2}{4b^2}f_{y,f} \tag{12}$$

where t_f is the thickness of the steel flange, b is the cantilevered length of the steel flange, and $f_{y,f}$ is the yield strength of the steel flange.

The stress effectiveness coefficient in Equation (10) is given by

$$k_{es} = \frac{3\lambda}{2\lambda + 1} \tag{13}$$

where λ is the blocking ratio of the steel section presented as follows:

$$\lambda = \frac{b}{h} \quad (14)$$

where b is the cantilevered length of the steel flange, and h is the clear distance between two steel flanges in the steel I section.

The confinement effectiveness coefficient k_{ea} can be obtained as follows:

$$k_{ea} = \frac{1 - \frac{\pi}{4}\left(1 - \frac{\lambda}{3}\right)^2}{1 - \frac{1}{2}(1 - \lambda)^2} \quad (15)$$

In order to better understand the process of deriving the confinement factors (K_p and K_h), Figure 8 presents a flowchart process calculation.

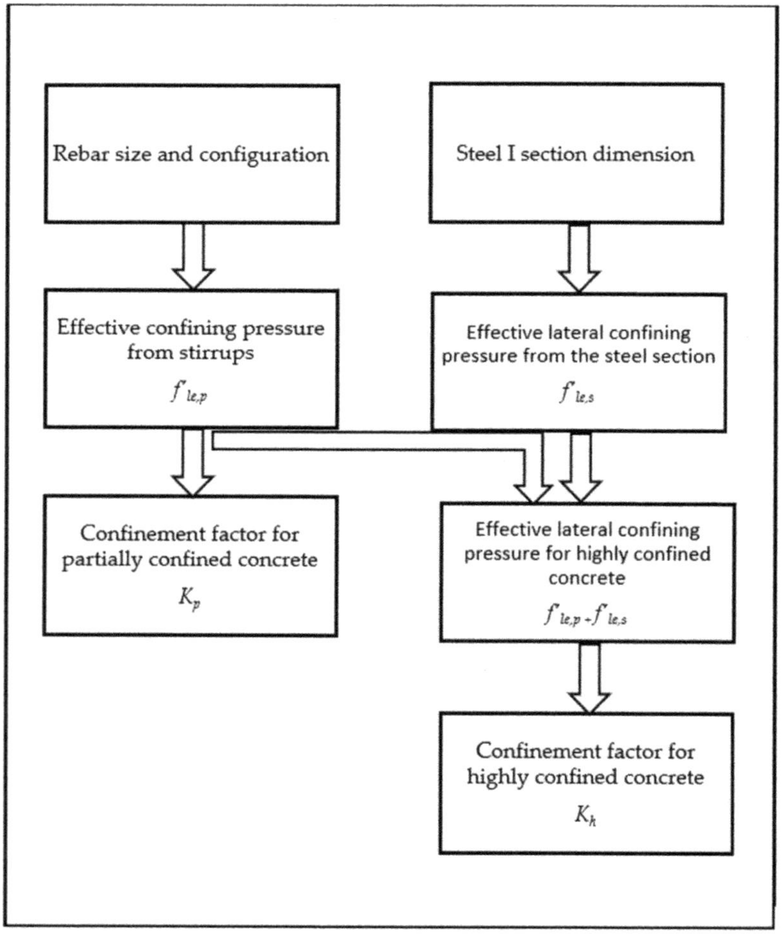

Figure 8. Confinement factor process flowchart.

In this section, two methods are reviewed in order to obtain the confinement factors for partially confined concrete and highly confined concrete (K_p and K_h). The first method presented by [22] is based on the shape of the steel section as well as the longitudinal distance between the stirrups while the second method is based on the shape and the dimension of the steel section as well as the stirrups [19]. The confinement factors calculated from both methods are comparable for a steel I section encased in a rectangular concrete section.

4. Axial–Flexural Behavior of Concrete-Encased Steel Columns

Since axial compression is an uncommon loading scheme in real construction, research on the behavior of the axial–flexural interaction is necessary because the CES bridge piers are typically subjected to a combination of axial force and bending moment. However, there are a few experimental investigations of the eccentric-compressed CES columns. The limited number of experimental investigations reveals the axial–flexural failure mechanism, and Figure 9 illustrates the common failure modes [26]. In other words, tension failure and compression failure are two distinct groups into which axial–flexural failure may fall. Important elements having an impact on the CES column axial behaviors include the concrete strength, the end eccentricity ratio, the slenderness ratio, the structural steel index, and the transverse reinforcement ratio [27].

Figure 9. Failure mode for eccentrically loaded CES column, reprinted with permission from [28], 2021, *Journal of Building Engineering*.

Concrete cracking and the occurrence of the longitudinal bars' yield strength at the tension side are characteristics of tension failure, while concrete cover spalling and crushing, rebar buckling, and stirrup opening are features of compression failure. These are followed by the longitudinal rebar yielding at the tension side [29]. While some specimens experience the cover spalling phenomena, it is not as noticeable as it is in specimens that are concentrically compressed. This could be due to the concrete's enhanced deformability under flexural loading and its non-uniform compressive stress distribution [30]. It is possible to approximate the overall deflection along the height as a half sinusoidal wave, and as there is no visible bond slippage between the steel section and the concrete up until failure, the cross-section analysis can assume a flawless bond [31]. It appears that an increase in structural steel strength has minimal effect on the CES strength for columns that have higher relative slenderness ratios due to the flexural buckling failure mode [32].

Figure 10 presents the axial–flexural behavior of the CES column according to the Architectural Institute of Japan Standard [33].

Axial force–bending moment (N-M) interaction diagrams are typically utilized to illustrate the axial–flexural capacity of CES columns. These diagrams can be condensed into piece-wise diagrams to make calculations easier. The estimated bending moment capacity is decreased by a reduction coefficient in order to account for the potential non-attainment of the full plastic stress distribution. Figure 11 presents the axial–flexural diagram of a CES column according to [34]. Figure 11a presents the cross-section capacity of CES columns under external forces, through seven interaction diagrams from a 3D surface on the coordinate system. Figure 11b shows cutting of the 3D surface in a longitudinal direction while Figure 11c shows that cutting of the 3D surface in a latitudinal direction results in a bundle of curves [34].

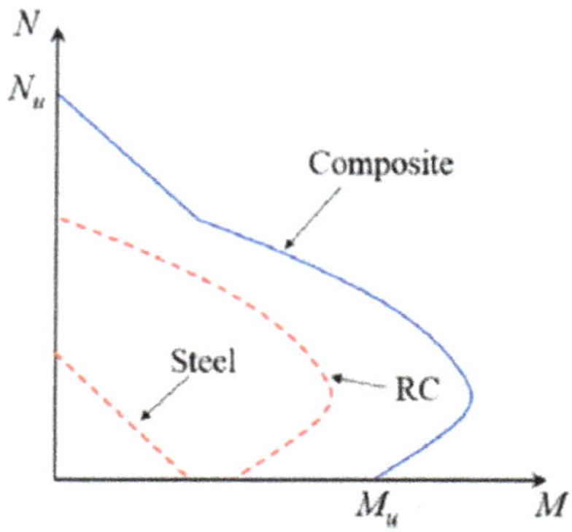

Figure 10. Axial–flexural behavior of CES column according to AIJ-2014 code [33], reprinted with permission from [30], 2019, *Construction and Building Materials*.

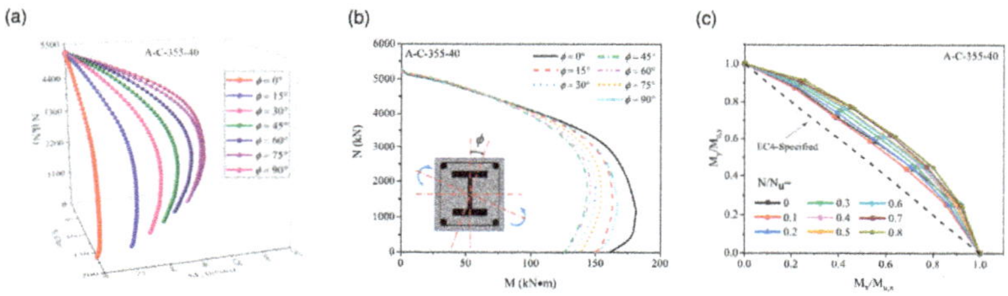

Figure 11. Three-dimensional N-M surface and 2D biaxial N-M curves obtained by fiber section analysis, (**a**) 3D interaction diagrams, (**b**) longitudinal direction cut from 3D interaction diagram, (**c**) latitudinal direction cut from 3D interaction diagram, reprinted with permission from [34], 2022, *Advances in Structural Engineering*.

Another important aspect of the design of the CES columns is the slenderness ration *(L/b)*. The Canadian bridge design code CSA-S6 [9] does not include CES section requirements. However, according to the British code BS.5400-5 [35], the slenderness range for columns *L/b* should be less than or equal to 30 [35].

5. Concrete-Encased Steel Column Behaviors under Cyclic Loads

In this section, a review of the recent study on the seismic and cyclic behavior of the CES columns is presented. In addition, the parameters that can have an impact on the seismic behavior of the CES columns, such as steel ratio, shear span ratio, axial load ratio, embedded depth ratio, steel shear connections, and concrete resistance are highlighted in detail.

A large amount of CES column test data are gathered and reviewed in this paper from a recent decade's worth of research. Based on material composition, this review does not include stainless steel, steel fiber-reinforced concrete, recycled aggregate concrete, ultra-high-performance concrete, or engineered cementitious composite because of differences in material properties.

The primary approaches applied to evaluate the seismic performance of CES columns in laboratory settings are low reversed cyclic loading and constant axial force. In order to simulate a seismic event, the CES column is first loaded axially to reach the target load ratio. After that, it is loaded horizontally with reversed displacement increments. An important performance indicator for evaluating energy dissipation capacity, load-bearing capacity, displacement ductility, stiffness degradation, capacity deterioration, etc., is the hysteresis curve (P-Δ). CES columns typically fail in flexural mode, shear–flexural mode, and shear mode, depending on the primary design parameters.

In seismically active regions, the CES column's exceptional seismic performance is important. Due to this particular advantage, a significant amount of the present review has been conducted to examine the seismic behavior of CES columns using quasi-static loading, in which the column top is subjected to a low reversed cyclic horizontal force and a required axial compressive force. The shear span ratio, axial load ratio, stirrup arrangement, confinement ratio, shear connectors, and material strengths are among the important factors that have an impact on the seismic behavior of the CES columns [12].

According to the research study, when using a sufficient amount of steel, CES column specimens have remarkable cyclic strength and ductility [26]. CES column specimens, in particular, have a high load-bearing capability even after longitudinal bar buckling and spalling of the concrete cover. Based on these tests, the ultimate state and the restorable limit state were determined to be local buckling of the H-shaped steel followed by spalling of the concrete cover.

The seismic behavior of 26 steel concrete composite columns under low cyclic reversed loading has been studied [36]. According to this study, under seismic load, the steel–concrete composite columns exhibit a bending failure mechanism. The deformation capacity is great, and the strength depreciation is modest. It demonstrates that the composite steel–concrete members have advantageous seismic behavior. As presented in Figure 12, all of the studied samples showed signs of bending failure, and the failure mechanism followed the standard procedures for bending failure members.

Figure 12. Flexural failure of the four studied specimens, (**a**,**b**) cover spalling, (**c**,**d**) reinforcement buckling, reprinted with permission from [36], 2014, *Journal of Structural Engineering*.

Five CES columns that were designed to collapse in flexural tension or flexural compression have had their cyclic behavior examined experimentally by [37]. According to current seismic evaluation rules, CES columns with non-seismic detail and low axial loads controlled by flexural tension appear to meet the acceptance criteria. However, the high axial loads controlled by flexural compression and the lack of seismic details in CES columns make them deficient for current seismic assessment acceptance standards [37]. These results show that the presence of the steel section in the concrete column helps to improve the seismic behavior of the CES column even with no additional seismic consideration for the rebars and the stirrups.

It appears that one of the key elements impacting the seismic performance of CES columns is the shear span ratio. While numerous elements influence the damage pattern collectively, an increase in span ratio tends to shift the damage from a shear-dominant failure to a bending-dominant failure. The outcome shows that the shear span ratio and the slenderness ratio have a significant impact on the behavior of CES columns. According to [38], ductility increased, and the failure process slowed down as the ratio of shear span increased. The long columns' carrying capacity decreased as the ratio of slenderness increased, yet the chance of an abrupt collapse increased [39].

Table 1 summarizes some related previous research regarding the seismic behavior of the concrete-encased steel bridge piers. It shows the name of the authors, the year of the study, the type, and the achievement of analysis.

Table 1. Summary of some related previous research.

Author	Type of Analysis	Achievement of Analysis
Hassan et al., 2021 [37]	Seismic performance of steel-reinforced concrete composite columns	CES columns with non-seismic detail and low axial loads controlled by flexural tension appear to meet the acceptance criteria
Dong et al., 2021 [40]	Seismic behavior of large-size encased cross-section steel-reinforced columns	The influence of the steel ratio is more significant under a low axial load ratio
Naito et al., 2011 [26]	Ductility evaluation of concrete-encased steel bridge piers	When using a sufficient amount of steel, CES columns have remarkable cyclic strength and ductility
Chen et al., 2014 [36]	Study on seismic behavior of fully encased steel–concrete composite columns	Under seismic load, the steel–concrete composite columns exhibit a bending failure mechanism
Yang and Li, 2012 [39]	Study on behaviors of steel-reinforced concrete columns	The long columns' carrying capacity decreased as the ratio of slenderness increased
Zhang et al., 2012 [38]	Flexural behavior of SRC columns under axial and bilateral loading	The axial load ratio was the most influencing factor on the ductility of columns
Zhu et al., 2016 [41]	Experimental study on steel reinforced concrete columns under cyclic lateral force	The axial compression ratio has an impact on the deformation capacity and the energy dissipation capacity
Zhang et al., 2019 [6]	Study on ultimate load and ductility of concrete-encased steel composite columns	A larger load ratio is shown to reduce energy dissipation and deformation capacities

One of the other most important factors affecting the cyclic behavior of the CES columns is the axial load ratio. Many researchers have examined the impact of this parameter. The capacity of energy dissipation, ductility, carrying capacity, stiffness degradation, and axial load distribution on the column are all impacted by the axial load ratio. A significant influence on the hysteresis behavior was the ratio of the applied axial compressive load [12]. Because of a significant increase in the moment of secondary bending caused by the axial loads, the columns appeared to have less capacity for energy dissipation and displacement ductility when the applied axial load was 40% of the ultimate load, as opposed to the columns with 20% of the axial ultimate load [22]. The carrying capacity, damage pattern, and deformation ability of the high-resistance CES columns are all impacted by the axial load ratio. The axial load ratio was the most influencing factor on the ductility of columns [38]. The axial compression ratio has an impact on the seismic resistance behavior of CES columns with high-strength concrete [39]. This has a detrimental influence on the deformation capacity and the energy dissipation capacity [41].

Furthermore, a larger load ratio is shown to reduce energy dissipation and deformation capacities [6]. However, a high load ratio increases the longitudinal reinforcing bars' dowel action and prevents the spread of oblique shear cracks, increasing the load-carrying capacity to some degree [41]. Regarding the impact of shear studs, it was found that while they improve the deformation capacity and lessen stiffness degradation, they do not affect the seismic behavior preceding concrete cover spalling [41]. It has been demonstrated that a bigger steel section size results in superior ductility and less stiffness degradation when considering the steel area ratio. It was found that structural steel is more advantageous in the event of a big axial load [41], while another researcher notes that the influence of the steel ratio is more significant under a low axial load ratio [40].

One other factor that has an impact on the seismic behavior of the CES column is the embedded depth ratio. The steel section of the CES column is embedded in the foundation. This length is known as h_b. The height of the steel section is also known as the h_a. The ratio of the embedded length of the steel section to section height is known as the embedded depth ratio, and it presented as h_b/h_a.

To preserve the CES column's strength capability and seismic behavior, an appropriate steel embedding depth is needed [42]. Also, the steel-embedded depth ratio for the CES column subjected to monotonic loading has been studied, and it was found that 1.67 was a suitable value for an ordinary CES column [43]. However, it is recommended that the minimum steel embedded depth ratio for the CES columns should be a minimum of 2.5 in order to take safety precautions into account [36]. For this reason, the Chinese specification [44] prescribes a minimum embedded depth ratio of 3 for the CES columns.

Another parameter that has an impact on the seismic behavior of the CES column is the shear connector. On the CES composite specimens with steel shear connections, numerous analytical and experimental investigations have been carried out to examine the behavior of shear connectors under axial loads and combinations of axial and seismic loads. The findings demonstrated that, in comparison to non-seismic shear capacities, seismic loads cause a 40% decrease in the shear stud's capacity. This reduction happens as a result of seismic fatigue of the weld/connector materials and concrete crushing, which moves the stress distribution up the connector's shank and onto its head. This increases the connector's bending stresses and leads to earlier failures [12]. The shear stud capacity was unaffected by the studs' shear connectors' initial bending. Similar to static capabilities, the concrete strength had an impact on the seismic shear capacity of the studs [45]. As stiffness deterioration decreased, the columns with studs shown in Figure 13 demonstrated a greater capacity for deformation and energy dissipation.

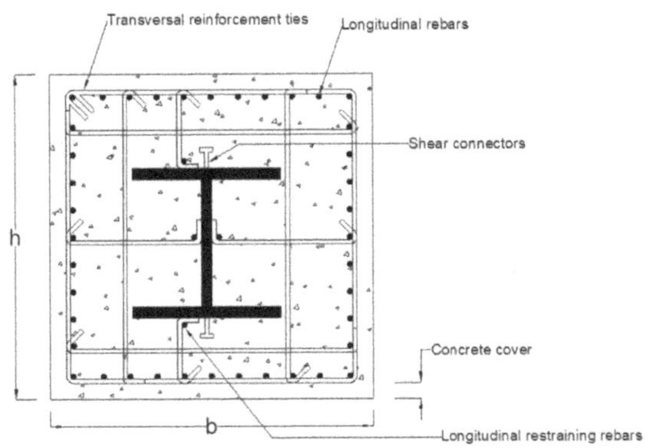

Figure 13. Cross section of CES column with shear connectors.

Many types and strengths of concrete have been used in composite buildings and substantial research in recent years. Concrete with a compressive strength greater than 50 MPa is referred to as high-strength concrete [46]. This type of concrete can be made with premium aggregates and superplasticizers, and its strength can even surpass 100 MPa. The cyclic behavior of CES columns with long webs of cross-shaped steel was more influenced by the concrete strength when the concrete strength rose, increasing the columns' shear capacity and a decrease in their ability to deform [47]. Additionally, as concrete strength grew, so did the capacity of long CES columns [39]. The common failure modes of cyclically loaded high-strength CES columns, shear failure, shear–flexural failure, and flexural failure, are shown in Figure 14.

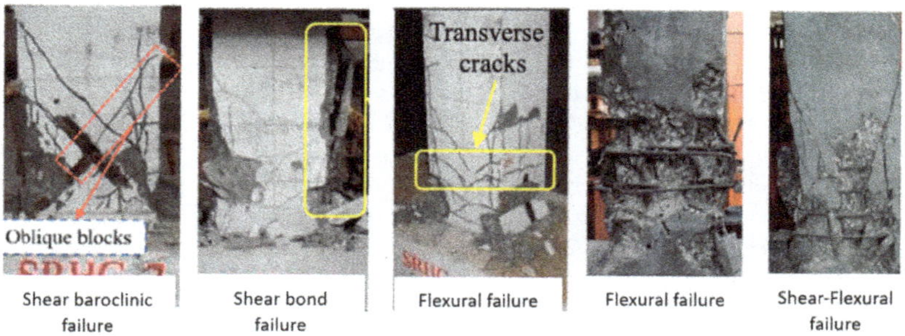

Figure 14. Typical failure mode of cyclically loaded high-strength CES column, reprinted with permission from [23], 2023, *Journal of Building Engineering*.

Based on the lateral displacement capacity of the column and its integration with the curvature distribution, the following equation for the length of the plastic hinge of the CES column is calculated and proposed [26]:

$$L_{rp} = (0.5d + 0.05h)\left\{(1 + 0.04t_{ek})\frac{M_m}{M_{y0}} - 0.25\right\} + 12(D_r - 12) \quad (16)$$

In Equation (16), the M_m is the maximal moment, M_{y0} is the yield moment, D_r is equal to the rebar diameter in mm, and t_{ek} is the ratio of A_s/A_r. Here, A_s is defined as the gross area of the steel section, and A_r is the gross area of the longitudinal rebars. Also, d is considered the effective column depth, and h is the column length. Figure 15 presents the plastic hinge length along the column height [26].

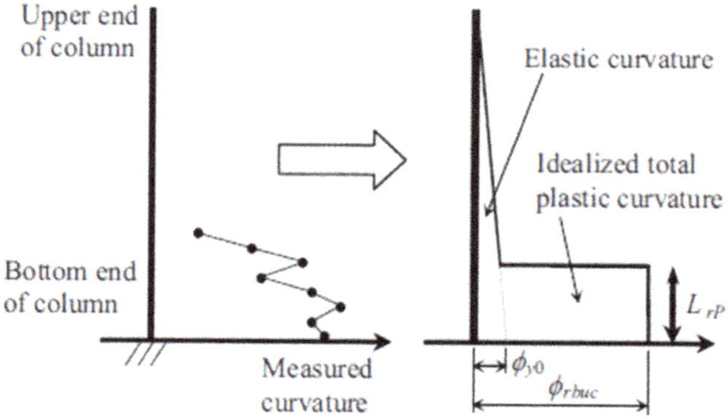

Figure 15. Plastic hinge length along the column, reprinted with permission from [26], 2010, *Journal of Bridge Engineering*.

When assessing seismic performance, the CES column's ductility is an essential metric. As seen in Figure 16, the ductility coefficient is essentially calculated as the ratio of lateral displacement corresponding to ultimate displacement and yield displacement, derived from the specimen's backbone curve. Generally speaking, excellent ductility is considered to be displayed by CES columns with a ductility coefficient greater than 3 [23].

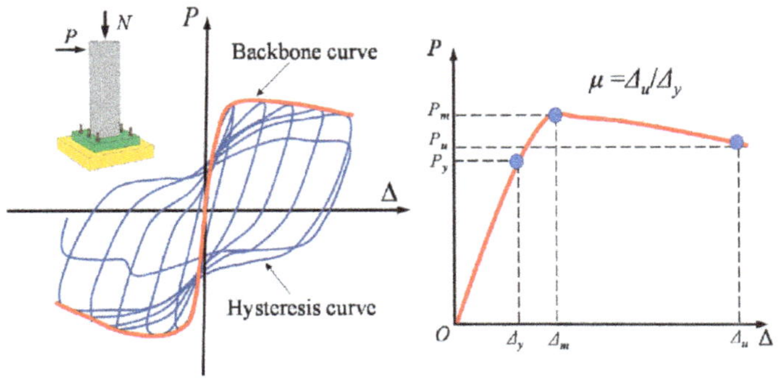

Figure 16. Ductility coefficient calculation on hysteresis curve, reprinted with permission from [23], 2023, *Journal of Building Engineering*.

Given the divergent perspectives, additional experimental research is required to elucidate the impact of steel sections at different load ratios.

Effect of Transverse Reinforcement on Behavior of Concrete-Encased Steel Columns

A large number of research works have been carried out in the past decade on the impact of transversal reinforcement on the behavior of the CES columns. The stirrup effectively confines the core concrete, and the seismic behavior of the CES compression-bending members is directly impacted by the stirrup spacing. It was found that as stirrup spacing increases, bearing capacity and ductility decrease [47]. Additionally, when the stirrup spacing increases from 75 to 100 mm, bearing capacity and ductility are reduced to a higher amount than when the spacing grows from 50 to 75 mm. This indicates that if the stirrup spacing stays below a particular threshold, it does not affect the ductility of CES columns. It suggests that the maximum stirrup spacing interval in the seismic area needs to be restricted. For this reason, the Chinese specification [44] suggests a value of 100 mm as the maximum spacing of stirrups for CES columns [47].

According to the experimental results, the CES column specimens' damage process usually includes the following: the deterioration of the concrete core; local buckling of the H-shaped steel flange; cracking on the buckled steel flange; longitudinal reinforcement bar buckling; and flexural cracks and spalling of the concrete cover at the base of the column. Figure 17 presents the rebar buckling caused by cyclic loading on the specimen of the CES column [26].

The confinement effect provided by stirrups is greatly reduced in short CES columns because the shear deformation allows for the stirrups to sustain a large amount of shear stress. As a result, the confinement stress acting on the concrete is reduced. Previous studies on confinement have shown that the stirrup's ability to confine can resist shear failure, keep the steel bar from buckling, and provide a pressure of confinement for the core concrete, all of which significantly increase the material's capacity and ductility. In addition, the stirrups serve to secure the longitudinal bars during construction and prevent them from buckling outward under pressure. This is particularly important for columns subjected to eccentric stresses, as it prevents spalling of the outer concrete cover, even at low load levels [12].

The stirrup ratio has a significant impact on how steel–concrete composite columns behave under seismic loads [36]. Increasing the stirrup ratio improves the seismic behavior, particularly when the axial compression ratio is high.

Figure 17. Damage to the column under cyclic loading, reprinted with permission from [26], 2010, *Journal of Bridge Engineering*.

6. Critical Views and Recommendations for Future Research

This study is a comprehensive compilation of information on several types of steel columns encased in concrete. A thorough literature review on various steel configurations for use in bridge applications has shown a lack of real-world project experience and a variety of current systems and tests.

In addition to emphasizing areas for development and improving construction techniques, this thorough review is essential for directing future research and promoting these technologies. CES technology opens the door for creative solutions in the industry and helps to navigate the difficulties of constructing bridges. A short list of the analytical parameters is provided, emphasizing how the analysis procedure uses them. The findings can be resumed as follows:

1. Fiber element modeling is a reliable method in order to represent the seismic behavior of the CES columns.
2. Confined concrete could be presented as highly confined concrete or partially confined concrete depending on the position of the concrete in the CES columns.
3. The length of the plastic hinge should be calculated for the CES column considering the presence of the steel profile embedded in the section as well as the rebar ratio.
4. The configuration of the transversal reinforcement also has an impact on the resistance of the CES columns.

Based on the model presented by [22], partially confined concrete could be modeled using the K_p coefficient factor and highly confined concrete could be modeled using the K_h coefficient factor in the CES column. However, based on the configuration and the shape of the steel section, these factors should be different.

The durability and seismic performance of concrete-encased steel bridge piers should be the main topics of future research. Evaluating these construction techniques' sustainability from an economic and environmental standpoint is essential. Additionally, research ought to focus on developing more robust designs, especially for areas with strong seismic activity.

It should be noted that the torsional behavior of the CES bridge piers has not been considered in this paper. There is very little research on the torsional behavior of the

CES columns in the buildings only [48]. Primarily, the torsional mode is not dominant in the seismic behavior of the bridges. However, in irregular or skewed bridges, the torsion effect could be more important. For this reason, future research should focus on the torsional behavior of the CES bridge piers as well as their impact on the seismic behavior of the bridge.

Future designs of concrete-encased steel columns will be greatly influenced by the analytical methodology discussed in this paper. Apart from the suggested configurations, the design parameters are also discussed. Bridges built with concrete-encased steel columns are more durable, leading to higher-quality structures.

7. Conclusions

This study discusses various analytical methods for concrete-encased steel columns. The configuration of the steel section and its seismic performance have been considered the primary subject of the current study since the seismic behavior of the CES columns is one of the most important factors in a successful design. Reviewing recent research, several facets of this issue are covered.

Even while CES columns have been the subject of numerous publications, manuals, and creative studies, there is still much room for more research on this topic. The requirements and needs of new projects might not be met by the commonly offered research studies, particularly in areas with high seismic demand. Developing and testing innovative steel and lateral reinforcement configurations for high seismic zones is one area that could use further improvement and investigation. There should be sufficient ductility and resistance in these CES portions to withstand seismic events.

Furthermore, considering the increasing emphasis on sustainability, future studies and implementations of innovative construction methods must prioritize environmental sustainability over structural soundness.

Author Contributions: Conceptualization, M.M. and M.H.; methodology, M.M. and M.H.; writing—original draft preparation, M.M.; writing—review and editing, M.H.; supervision, M.H. All authors have read and agreed to the published version of the manuscript.

Funding: This research received no external funding.

Institutional Review Board Statement: Not applicable.

Informed Consent Statement: Not applicable.

Data Availability Statement: No new data were created or analyzed in this study. Data sharing is not applicable to this article.

Conflicts of Interest: The authors declare no conflicts of interest.

References

1. American Institute of Steel Construction (AISC). Load and Resistance Factor Design of W-Shapes Encased in Concrete. 2003. Available online: https://www.aisc.org/globalassets/aisc/manual/15th-ed-ref-list/load-and-resistance-factor-design-of-w-shapes-encased-in-concrete.pdf (accessed on 15 June 2024).
2. Kartheek, T.; Venkat Das, T. 3D modelling and analysis of encased steel-concrete composite column using ABAQUS. *Mater. Today Proc.* **2020**, *27*, 1545–1554. [CrossRef]
3. Elbably, A.; Ramadan, O.; Akl, A.; Zenhom, N. Behavior of encased steel-high strength concrete columns against axial and cyclic loading. *J. Constr. Steel Res.* **2022**, *191*, 107161. [CrossRef]
4. Suresh Babu, S.; Priya; Rose Leema, A. State-of-the-Art Review for Concrete-Encased Steel Columns. 2020. Available online: https://rdcu.be/dD2eX (accessed on 15 June 2024).
5. Tian'e Longtan Bridge. Available online: https://www.highestbridges.com/wiki/index php?title=Tian'e_Longtan_Bridge (accessed on 15 June 2024).
6. Zhang, Y.; Liu, Y.; Xin, H.; He, J. Numerical Parametric Study on Ultimate Load and Ductility of Concrete Encased Equal-leg Angle Steel Composite Columns. *Eng. Struct.* **2019**, *200*, 109679. [CrossRef]
7. Hybrid Hollow High Pier. 2018. 20 January 2019. Available online: http://www.actec.or.jp/3h_pier/pdf/3h_pier_pamphlet.pdf (accessed on 15 June 2024).

8. Paeglitis, A.; Gode, K. Concrete Bridge Deterioration Caused by De-Icing Salts in High Traffic Volume Road Environment in Latvia. *Balt. J. Road Bridge Eng.* **2014**, *9*, 200–207. [CrossRef]
9. Canadian Standards Association (CSA). *Canadian Highway Bridge Design Code*; CAN/CSA S6-14: Mississauga, ON, Canada, 2019.
10. *ACI Committee 318, Building Code Requirement for Reinforced Concrete (ACI 318-8) and Commentary (ACI 318R-8)*; American Concrete Institute: Farmington Hills, MI, USA, 2008.
11. *ANSI/AISC 360-16*; Specification for Structural Steel Buildings. American Institute of Steel Construction (AISC): Chicago, IL, USA, 2016.
12. Mostafa, M.A.; Wu, T.; Liu, X.; Fu, B. The Composite Steel Reinforced Concrete Column Under Axial and Seismic Loads: A Review. *Int. J. Steel Struct.* **2019**, *19*, 1969–1987. [CrossRef]
13. Spacone, E.; Filippou, F.C.; Taucer, F.F. Fiber beam–column model for non-linear analysis of R/C frames: Part I. Formulation. *Earthq. Eng. Struct. Dyn.* **1996**, *25*, 711–725. [CrossRef]
14. El-Tawil, S.; Deierlein, G.G. Nonlinear analysis of mixed steel–concrete frames. I: Element formulation. *J. Struct. Eng.* **2001**, *127*, 647–655. [CrossRef]
15. Kent, D.C.; Park, R. Flexural members with confined concrete. *J. Struct. Div.* **1971**, *97*, 1969–1990. [CrossRef]
16. Mander, J.B.; Priestley, M.J.; Park, R. Theoretical stress–strain model for confined concrete. *J. Struct. Eng.* **1988**, *114*, 1804–1826. [CrossRef]
17. Tan, E.L.; Thomas, C.; Siriviatnanon, V. Finite element modeling of nonlinear behaviour of headed stud shear connectors in foamed and lightweight aggregate concrete. In Proceedings of the Fifth International Conference on Advances in Experimental Structural Engineering (I), Taipei, Taiwan, 8–9 November 2013; Volumes 8 and 9.
18. Fang, L.; Zhang, B.; Jin, G.F.; Li, K.W.; Wang, Z.L. Seismic behavior of concrete-encased steel cross-shaped columns. *J. Constr. Steel Res.* **2015**, *109*, 24–33. [CrossRef]
19. Chen, S.; Wu, P. Analytical model for predicting axial compressive behavior of steel reinforced concrete column. *J. Constr. Steel Res.* **2017**, *128*, 649–660. [CrossRef]
20. Liang, C.Y.; Chen, C.; Weng, C.; Yin, Y.; Wang, J. Axial compressive behavior of square composite columns confined by multiple spirals. *J. Constr. Steel Res.* **2014**, *103*, 230–240. [CrossRef]
21. Mirza, S.A.; Skrabek, B. Statistical analysis of slender composite beam–column strength. *J. Struct. Eng.* **1992**, *118*, 1312–1332. [CrossRef]
22. Chen, C.C.; Lin, N.J. Analytical model for predicting axial capacity and behavior of concrete encased steel composite stub columns. *J. Constr. Steel Res.* **2006**, *62*, 424–433. [CrossRef]
23. Lai, B.; Bao, R.; Zhang, M.; Wang, Y.; Liew, J.Y. Evaluation on the static and seismic performance of steel reinforced concrete composite columns with high strength materials. *J. Build. Eng.* **2023**, *79*, 107886. [CrossRef]
24. Sheikh, S.A.; Uzumeri, S. Analytical model for concrete confinement in tied columns. *J. Struct. Div.* **1982**, *108*, 2703–2722. [CrossRef]
25. Cusson, D.; Paultre, P. Stress-strain model for confined high-strength concrete. *J. Struct. Eng.* **1995**, *121*, 468–477. [CrossRef]
26. Naito, H.; Akiyama, M.; Suzuki, M. Ductility Evaluation of Concrete-Encased Steel Bridge Piers Subjected to Lateral Cyclic Loading. *J. Bridge Eng.* **2011**, *16*, 72–81. [CrossRef]
27. Mirza, S.A.; Lacroix, E.A. Comparative Strength Analyses of Concrete-Encased Steel Composite Columns. *J. Struct. Eng.* **2004**, *130*, 1941–1953. [CrossRef]
28. Gautham, A.; Sahoo, D.R. Behavior of steel-reinforced composite concrete columns under combined axial and lateral cyclic loading. *J. Build. Eng.* **2021**, *39*, 102305. [CrossRef]
29. Xu, C.; Cao, P.Z.; Wu, K.; Lin, S.Q.; Yang, D.G. Experimental investigation of the behavior composite steel-concrete composite beams containing different amounts of steel fibres and conventional reinforcement. *Construct. Build. Mater.* **2019**, *202*, 23–36. [CrossRef]
30. Lai, B.; Richard Liew, J.Y.; Xiong, M. Experimental study on high strength concrete encased steel composite short columns. *Constr. Build. Mater.* **2019**, *228*, 116640. [CrossRef]
31. Lai, B.; Richard Liew, J.Y.; Venkateshwaran, A.; Li, S.; Xiong, M.J. Assessment of high-strength concrete encased steel composite columns subject to axial compression. *J. Constr. Steel Res.* **2020**, *164*, 105765. [CrossRef]
32. Ellobody, E.; Young, B. Numerical simulation of concrete encased steel composite columns. *J. Constr. Steel Res.* **2011**, *67*, 211–222. [CrossRef]
33. *AIJ-2014*; AIJ Standard for Structural Calculation of Steel Reinforced Concrete Structures. Architectural Institute of Japan: Tokyo, Japan, 2014. (In Japanese)
34. Lai, B.L.; Tan, W.K.; Feng, Q.T.; Venkateshwaran, A. Numerical parametric study on the uniaxial and biaxial compressive behavior of H-shaped steel reinforced concrete composite beam-columns. *Adv. Struct. Eng.* **2022**, *25*, 2641–2661. [CrossRef]
35. British Standard Institution. BS 5400-05: 1979: Steel, Concrete and Composite Bridges, Part 5, Code of Practice for Design of Composite Bridges. 2002. Available online: https://knowledge.bsigroup.com/products/steel-concrete-and-composite-bridges-code-of-practice-for-design-of-composite-bridges-1?version=standard&tab=preview (accessed on 15 June 2024).
36. Chen, C.; Wang, C.; Sun, H. Experimental Study on Seismic Behavior of Full Encased Steel-Concrete Composite Columns. *J. Struct. Eng.* **2014**, *140*, 04014024. [CrossRef]

37. Hassan, W.; Farag, M. Seismic performance of steel-reinforced concrete composite columns in existing and modern construction. *Soil Dyn. Earthq. Eng.* **2021**, *151*, 106945. [CrossRef]
38. Zhang, S.; Zhao, Z.; He, X. Flexural behavior of SRC columns under axial and bilateral loading. *Appl. Mech. Mater.* **2012**, *166–169*, 3383–3390. [CrossRef]
39. Yang, B.S.; Li, Y.Y. Study on the behaviors of steel reinforced concrete columns. *Adv. Mater. Res.* **2012**, *368–373*, 248–252. [CrossRef]
40. Dong, H.; Liang, X.; Cao, W. Seismic behavior of large-size encased cross-section steel-reinforced high-strength concrete circular columns. *Structures* **2021**, *34*, 1169–1184. [CrossRef]
41. Zhu, W.; Jia, J.; Gao, J.; Zhang, F. Experimental study on steel reinforced high-strength concrete columns under cyclic lateral force and constant axial load. *Eng. Struct.* **2016**, *125*, 191–204. [CrossRef]
42. Azizinamini, A.; Ghosh, S. Steel reinforced concrete structures in 1995 Hyogoken-Nanbu earthquake. *J. Struct. Eng.* **1997**, *123*, 986–992. [CrossRef]
43. Nagata, T.; Fukuchi, Y.; Wakamatsu, S. Experimental study on embedded depth and effective bearing flange width of steel members as bending: Strength of embedded type of steel reinforced concrete column base. In Proceedings of the Annual Meeting Architectural Institute of Japan, Tokyo, Japan, December 1999; Architectural Institute of Japan (AIJ): Tokyo, Japan; pp. 1167–1168.
44. Ministry of Housing and Urban-Rural Development of the People Republic of China. Code for Design of Composite Structures. JGJ 138-2016 (JGJ138-2010). Available online: https://www.chinesestandard.net/Related.aspx/JGJ138-2016 (accessed on 15 June 2024).
45. Civjan, S.A.; Singh, P. Behavior of shear studs subjected to fully reversed cyclic loading. *J. Struct. Eng.* **2003**, *129*, 1466–1474. [CrossRef]
46. Warner, R.; Rangan, B.; Hall, A.; Faulkes, K. *Concrete Structures*; Addison Wesley Longman: Boston, MA, USA, 1998.
47. Wang, Q.; Shi, Q.; Tao, Y. Experimental and numerical studies on the seismic behavior of steel reinforced concrete compression-bending members with new-type section steel. *Adv. Struct. Eng.* **2016**, *19*, 255–269. [CrossRef]
48. Zhou, C.; Xue, J.; Hu, Z.; Liu, Z. Coupled translational-torsional response for SRC frame with special-shaped columns under earthquake. *J. Build. Eng.* **2020**, *31*, 101440. [CrossRef]

applied sciences

MDPI

Review

Retrofitting of Steel Structures with CFRP: Literature Review and Research Needs

Mohamadreza Delzendeh Moghadam, Abbas Fathi and Omar Chaallal *

Department of Construction Engineering, École de Technologie Supérieure, University of Quebec, Montreal, QC H3C 1K3, Canada; mohamadreza.delzendeh-moghadam.1@ens.etsmtl.ca (M.D.M.); abbas.fathi.1@ens.etsmtl.ca (A.F.)

* Correspondence: omar.chaallal@etsmtl.ca

Abstract: The application of the externally bonded (EB) carbon fiber-reinforced polymer (CFRP) technique for retrofitting steel elements offers significant advantages over the conventional method. The high strength-to-weight ratio and corrosion resistance of CFRP materials have made them a viable alternative for retrofitting steel structures. This paper covers a wide range of aspects discussed in the research investigations to date on CFRP bonded steel elements and provides a critical review of the topic under both static and fatigue loading conditions. In the end, research needs and recommendations are presented in this respect.

Keywords: steel structures; retrofitting technique; carbon fiber-reinforced polymer (CFRP); bond–slip relationship; static loading; fatigue loading; finite element analysis; research progress

1. Introduction

Steel structures around the world are susceptible to deterioration over their service life period. This deterioration can reduce the potential strength and stiffness of the steel members due to cracks and corrosion induced by fatigue loading and extreme weather conditions, respectively. Additionally, defects may arise during the design and construction phases. The conventional approach for retrofitting steel structures is by using steel plates that are attached to the structure by weld or bolt [1]. However, this approach presents disadvantages, including the residual stress imposed by welding, which can lead to new damage to the structure [2,3]. Furthermore, the steel plates are susceptible to corrosion and their heavy weight poses challenges during installation [4,5]. Alternatively, the application of externally bonded carbon fiber-reinforced polymer (CFRP) can offer a durable solution to address these challenges [6,7]. The high strength-to-weight ratio and corrosion resistance of CFRP materials have an important role in their selection for retrofitting steel components [8–10].

In recent years, the application of advanced composite materials has been gaining acceptance in retrofitting civil infrastructures. Among these types of materials, CFRP and graphite fiber-reinforced polymer (GFRP) are well established [11]. However, CFRP exhibits superiority over GFRP due to its higher strength. Studies show that CFRP retrofitting systems can effectively enhance the flexural capacity of steel members and prolong their fatigue life [4,12–32]. CFRPs are classified based on their elastic modulus into Low Modulus (LM), Normal Modulus (NM) or Intermediate Modulus (IM), High Modulus (HM), and Ultra-High Modulus (UHM). There is not a unanimous approach to characterize the elastic modulus range of each category. However, it can be expressed relative to the steel elastic modulus as presented in Table 1 [33].

Table 1. CFRP classification in terms of elastic modulus [33].

CFRP Type	CFRP Modulus	CFRP Modulus Relative to Steel
Low Modulus (LM)	<100 GPa	$E_{CFRP} < 0.5\ E_{steel}$
Normal Modulus (NM)	100–200 GPa	$0.5\ E_{steel} \leq ECFRP < E_{steel}$
High Modulus (HM)	200–400 GPa	$E_{steel} \leq ECFRP < 2\ E_{steel}$
Ultra-High Modulus (UHM)	≥400 GPa	$E_{CFRP} \geq 2\ E_{steel}$

The significance of this review lies in its comprehensive and up-to-date examination of the current state of knowledge on steel elements retrofitted with CFRP. This paper integrates findings from multiple dimensions and examines key parameters affecting the performance of CFRP retrofitted components and the bond behavior between CFRP and steel interfaces, including adhesive types, surface preparation, bond length, etc. By synthesizing findings from various investigations, this review aims to guide future research efforts and enhance the application of CFRP in retrofitting steel structures.

This paper is presented in the following sections: (i) a review of studies on the bond behavior between CFRP and steel; (ii) research findings regarding the flexural behavior of the retrofitted steel beams; and (iii) a review of numerical simulations of CFRP/steel retrofitted elements and their bond behavior. Finally, gaps in previous research are identified, and suggestions for further investigations are provided.

2. Bond Behavior between CFRP and Steel

The bond between CFRP and steel plays a major role in the performance of steel components retrofitted with CFRP. The role of the adhesive layer is to carry the tensile forces of the steel substrates and transfer them to the CFRP composites. A comparison between the CFRP/steel and CFRP/concrete bonded interfaces reveals the main contrast in terms of the bond behavior. Indeed, adhesives in the CFRP/steel joints are regarded as the weakest link, whereas the bond behavior of the CFRP/concrete shows that concrete acts as the weakest link [34]. Therefore, to achieve the maximum capacity of the CFRP retrofitting system that follows up with the CFRP rupture, the selection of the most effective adhesives is crucial [35]. Furthermore, many other parameters have been found influential on the bond behavior of the retrofitted elements. These parameters include the CFRP elastic modulus, CFRP bond length, CFRP configuration, adhesive type, adhesive thickness, anchorage system, and surface preparation of the steel substrate. It should be noted that temperature is also considered as one of these influential parameters. There are some comprehensive literature reviews that focus on the thermal and environmental effects [36,37].

2.1. Bond Test Configuration and Failure Modes

To study the bond behavior between the CFRP and steel substrate, two test setups have been mainly adapted, as shown in Figure 1. In the single strap joint configuration (a), two steel plates are attached by CFRP only on one side, while in the double strap joints configuration (b), CFRPs are attached to both sides of steel plates. In both configurations, the loading is applied to the steel plates. It should be noted that in the single strap joint test setup, the adherends are subjected to bending during loading due to load eccentricity. This asymmetrical configuration can lead to rotation of the bond plane, resulting in significant peel stresses at the adhesive layer ends, which may cause premature adhesive failure and consequently lead to an underestimation of the capacity of the CFRP/steel joint [38]. In contrast, the double strap joint is symmetrical about the mid-plane of the specimen. Therefore, given the adherends, the amount of peel stress due to bond rotation is considerably less than a single strap joint configuration [38]. Possible failure modes of the CFRP/steel joints under tensile loading include the following [5]: (a) adhesion failure at CFRP/adhesive interface; (b) adhesive layer failure; (c) adhesion failure at steel/adhesive interface; (d) CFRP delamination; (e) CFRP rupture; (f) steel yielding. Figure 2 presents a schematic view of these failure modes.

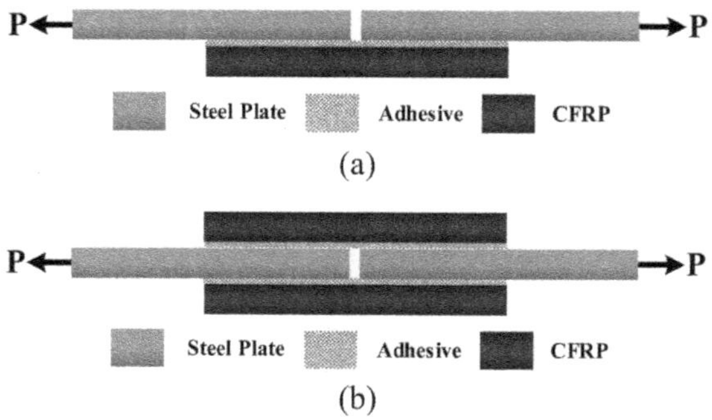

Figure 1. Schematic view of the (**a**) single strap joint; (**b**) double strap joint test setups.

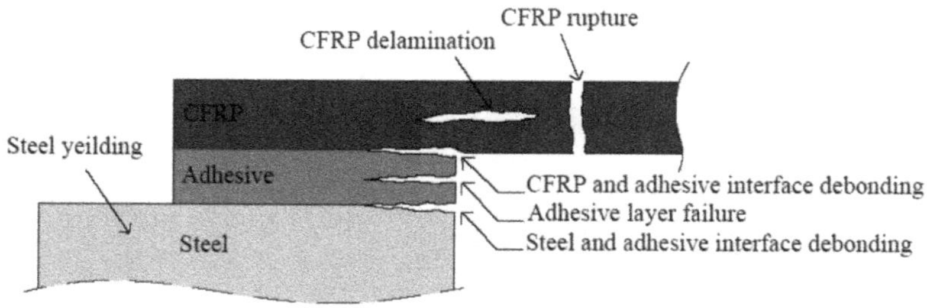

Figure 2. Possible failure modes of CFRP/steel joints under tensile loading [5].

Table 2 presents details of the previous studies on the bond behavior of CFRP/steel joints. It should be noted that in some research studies, the applied CFRP was claimed to be UHM. However, based on the classification presented in Table 1, they should be considered HM or NM CFRP. The elastic modulus of the applied CFRPs in these studies varied from 270 GPa to 640 GPa. However, investigations involving UHM CFRP have not been adequately documented. Based on these studies, it can be inferred that using CFRP with a higher elastic modulus can lead to an improvement in the load-carrying capacity of CFRP/steel joints [39]. Moreover, it has been found that the failure mode is altered as the elastic modulus and type of CFRP change [40,41]. Nevertheless, the influence of fatigue loading on bond behavior was reported to be insignificant, as only a limited bond region adjacent to the joint, known as the fatigue damage zone, was prone to fatigue damage [19]. It was also found that the failure mode changed as the temperature increased [42]. The dominated failure mode at elevated temperatures is cohesion failure within the adhesive layer [43]. In addition, the debonding loads of the CFRP/steel joints are notably decreased at both low and high service temperatures [44].

2.2. Adhesive

The mechanical properties of adhesive have a notable effect on the bond behavior of steel members retrofitted with CFRP [45–48]. As the adhesive is the weakest link in the CFRP and steel joint, selecting a proper adhesive can ensure the maximum capacity of the retrofitting system. In most research studies, commercial adhesives are utilized for retrofitting structures [34,49,50]. Figure 3 shows the stress–strain curves for these adhesives. It should be noted that the mechanical properties of the adhesive may decrease with increasing temperature [43]. Commercial adhesives have been found to possess insufficient strength, often resulting in brittle failure. Consequently, research efforts have been directed towards enhancing adhesive performance, particularly in terms of mechanical properties [45,51–55]. Utilizing enhanced adhesive can change the failure mode of the CFRP/steel joints [56,57].

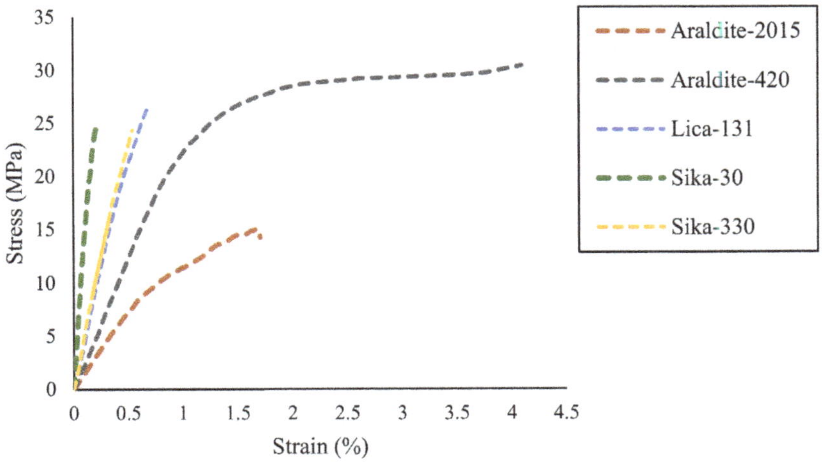

Figure 3. Tensile stress–strain curve of commercial adhesives used in bonding CFRP to steel substrate [45,49].

Among the commercial adhesives, Sikadur and Araldite have been commonly utilized. It was shown that specimens with Araldite 2015 adhesive exhibited a much higher interfacial fracture energy compared to the specimens with Sikadur 30 adhesive [49]. Indeed, specimens retrofitted with nonlinear adhesives having higher strain capacity but lower elastic modulus exhibit higher interfacial fracture energy and thereby a higher bond strength compared to the ones with linear adhesives having similar or even higher elastic modulus [34,58]. The type of adhesive can also significantly influence the bond–slip relationship. Bond–slip can be estimated as a triangular shape for linear adhesive and a trapezoidal shape for nonlinear adhesive [59]. The average effective bond length of specimens bonded with nonlinear adhesive was also found to be greater than that of specimens with linear adhesive, which could be attributable to the higher ductile behavior of nonlinear adhesive compared to the linear one [60]. It is also found that the effective bond length, interfacial fracture energy, and ultimate load increase with the increase in adhesive thickness [49,61]. In addition, increasing the adhesive thickness can change the failure mode from cohesive failure to CFRP delamination [34,62,63]. However, specimens with lower adhesive thickness exhibited better performance in terms of the average peel stresses and failure load [64]. It is worth mentioning that assessing the effect of adhesive thickness on the failure mode is difficult due to the non-uniform thickness of the bond-line [65]. It is suggested to increase the adhesive thickness up to 2 mm to avoid cohesive failure [63,66]. Non-cohesive failure in the bond test has been reported in the literature [34,47,60,67–69]. Increasing the adhesive thickness to 3 mm may result in lower bond strength compared to a thickness of 2 mm [34]. Consequently, it can be concluded that the adhesive's mechanical properties and thickness are the key factors in the bond behavior of CFRP retrofitted steel components [49,55]. Furthermore, the adhesive layer, regardless of the type of adhesive, can help prevent potential galvanic corrosion if the bond quality is strictly controlled [70]. Furthermore, investigations on the impact of bond-line defects on the bond behavior of CFRP steel joints suggest that the size of the defect plays a more significant role in determining the failure mode than the number of defects [71].

Table 2. Experimental details of studies regarding the bond behavior of CFRP/steel joints. Note: (a) adhesive thickness, (b) adhesive type, (c) adhesive elastic modulus, (d) steel stiffeners, (e) CFRP configurations, (f) CFRP thickness, (g) CFRP to steel width ratio, (h) CFRP bond length, (i) CFRP axial rigidity, (j) loading amplitude.

Ref.	f_y (MPa)	E_s (GPa)	f_a (MPa)	E_a (GPa)	t_a (mm)	f_p (MPa)	E_p (GPa)	t_p (mm)	Type of CFRP	Test Set-Up	Steel Plate (mm) ($L_s \times w_s \times t_s$)	Influencing Parameters									
												a	b	c	d	e	f	g	h	i	j
[64]	375	205	14.8 43 32	6.8–7.3 2.1 3.5	0.1 0.5 1	2109.3 1120.8	135.3 270.1	1.2	NM HM	S * D *	225 × 25 × 1.5 225 × 25 × 3 225 × 25 × 6	✔	✔	✔		✔				✔	
[63]	-	-	22.53 20.48 13.89	4.01 10.79 5.43	1 2 3 4	-	165	1.2	NM	S	305 × 118 × 12	✔	✔	✔							
[72]	-	-	37.1	3	1	-	338 460	4 8	HM UHM	D	L_s = 200, 400					✔	✔				
[73]	317.8	-	24.8 30	4.5 3.8	0.8 1	>2800	197	1.4	NM	D	460 × 60 × 6 600 × 60 × 6 1200 × 60 × 60		✔			✔					
[35]	-	200	76	3.1	-	2448 1190	640 340	0.19 1.42	UHM HM	D	128.2 × 25.4 × 3.3 258.3 × 25.4 × 3.3		✔			✔	✔		✔		
[39]	409	200	34.6	3	0.6–0.7	2979 1923	187 514	1.2	NM UHM	D	250 × 50 × 4.8 300 × 50 × 4.8 610 × 50 × 4.8							✔			
[60]	300	200	28.6 24	1.9 9.2	var	1500	460	1.45	UHM	D	300 × 50 × 30		✔						✔		
[34]	-	200	22.34 31.28 14.73 21.46	11.25 4.82 1.75 1.83	0.5 2 3	-	150 235 340	1.2 1.4	NM MM HM	S	L_s = 450 t_s = 30	✔	✔							✔	
[74]	359	200	32	1.9	-	2300	256	-	HM	D	180 × 50 × 5										✔
[19]	487	200.6	28.6	1.90	0.66	1607	478.73	1.45	UHM	D	300 × 50 × 10										✔
[75]	300	-	33.16	11.3	1	1970	185	1.44	NM	S	** 100 × 100 L_s = 250 t_s = 5										
[9]	235	210	35	-	var	-	460	-	UHM	D	450 × 60 × 15 450 × 60 × 10	✔									
[76]	410	-	34.6	3.01	-	1200 2800	450 165	1.2	UHM NM	D	610 × 50 × 5 610 × 50 × 4.8 610 × 50 × 9.5					✔		✔			
[49]	414	198	27.6 15.1	12.2–1.75	var	2760	164	1.4	NM	S	t_s = 20	✔	✔								
[18]	235	-	26–31	11.2	1–2	3100	170	1.2	NM	D	w_s = 75 t_s = 12								✔		
[77]	400	210	21.76	8.4	-	1820 1840	180.5 163.3	1.46 1.26	NM	D	w_s = 55 t_s = 5					✔					

* S = single strap joints/D = double strap joints; ** square steel tubes.

2.3. Surface Preparation

Surface preparation is one of the most crucial parameters of the bond behavior of the CFRP retrofitted elements. The bond capacity can be determined by the cohesion strength of the adhesive as well as the adhesion strength of the interface between the adhesive and the substrate [78]. Adhesion failure can occur at the CFRP/adhesive or the steel/adhesive interface. However, debonding at the steel/adhesive interface is more likely to occur [6,79]. Therefore, the preparation of the steel substrate has received more attention. The adhesion is mainly provided by mechanical interlocking and chemical bonding between the adhesive and the adherend [80]. It is debatable regarding the agent that is responsible for strong bonds. However, it seems that chemical bonding is more important than mechanical bonding [6]. To improve the mechanical bonding, the steel substrate is roughened before the CFRP bonding process. However, applying an improper roughening technique may cause tiny crevices in which the trapped air bubbles or solvent can induce failure in a rigid adhesive by increasing the stress concentrations [81]. It can also reduce the contact between the adhesive and the adherent, which is so-called wetting [82]. Indeed, increasing the contact surface area leads to enhanced wetting and more extensive chemical bonding [83]. The most commonly used mechanical treatments are grit blasting, needle scaler, and sandpaper. In these abrasive techniques, the surface geometry of the substrate is modified and the oxide layer is removed. It is generally accepted that using the grit blasting method is effective [2,6,78]. It also can lead to modifying the surface chemical composition by removing the contaminants such as oil and grease. However, it can introduce grit residues or other contaminants onto the steel substrate [84]. Furthermore, the grit type should be carefully chosen to ensure chemical compatibility with the applied adhesive [78]. Improper surface preparation can lead to premature failure, thereby hindering the achievement of the full capacity of the CFRP retrofitting technique.

2.4. Bond Length

Research findings show that the bond strength, as well as the failure modes of CFRP bonded steel elements, can be affected by the bond length [85]. The bond strength of the CFRP/steel interface tends to increase at greater bond lengths. The strain level experienced by the applied adhesive was seen to drop significantly when a longer bond length was used [85]. Therefore, it is more likely for specimens with shorter bond lengths to undergo debonding failure, whereas steel components retrofitted with greater CFRP bond lengths are more likely to fail due to CFRP rupture. However, increasing the bond length from a certain threshold, the so-called effective bond length, does not yield a further increase in the bond strength [59,86–90]. Moreover, increasing either the CFRP elastic modulus or CFRP thickness results in an increase in CFRP/steel bond strength, although applying CFRP with higher axial stiffness seems to require a greater bond length [90,91]. Furthermore, the type of adhesive has a significant effect on the effective bond length [92]. Applying the nonlinear adhesive can increase the bond strength and the effective bond length compared to the linear adhesive as a result of its ductile behavior and larger elongation at break [60]. It was also found that the rate of impact loading has a trivial effect on the effective bond length; however, it has a notable effect on the bond strength and failure modes [46,90,93–95]. Although the effect of bond length on the stiffness of the retrofitted elements is negligible, it can increase the load-carrying capacity of the components [12]. The proposed formulas to calculate the effective bond length are provided in Table 3. As observed, the ultimate load is not influenced by the adhesive thickness. This issue is also the same for calculating the ultimate load in CFRP/concrete joints [96].

Table 3. Bond strength and effective bond length for CFRP/steel interface.

Ref.	Effective Bond Length	Maximum Shear Stress	Ultimate Load	Remarks
[63]	$\frac{\pi}{2\sqrt{\tau_f/E_pt_p\delta_f}}$	$0.8f_a$	$b_p\sqrt{2G_fE_pt_p}$	-
[97]	$2.77\sqrt{\frac{\beta}{\beta+1}} \times \frac{\sqrt{2G_fE_pt_p}}{\tau_f}$	$\sqrt{\frac{G_aG_f}{2.72t_a}}$	$N\sqrt{\frac{\beta+1}{\beta}} \times b_p\sqrt{2G_fE_pt_p}$	$\beta = \frac{b_st_sE_s}{2b_pt_pE_p}$
[98]	$3.5\sqrt{\frac{E_pt_pt_a}{G_a}}$	$0.8f_a$	$b_p\sqrt{2G_fE_pt_p}$	-
[99]	$a_d + b_e + \frac{1}{\lambda_1}\ln\left(\frac{1+C}{1-C}\right)$	$0.9f_a$	$b_p\sqrt{2G_fE_pt_p}$	$a_d = \frac{1}{\lambda_1}\left[\sqrt{\left(2\frac{\delta_2}{\delta_1} - 1\right)} - 1\right)\right]$ $b_e = \frac{1}{\lambda_2}arcsin\left[\frac{\lambda_2\lambda}{0.97\delta_1\lambda_1{}^2}\left(\delta_f - \delta_2\right)\right]$ $C = \frac{\lambda_2}{\lambda_1\delta_1}\left(\delta_f - \delta_2\right)\cot(\lambda_2 b_e) - \lambda_1 a_d$ $\lambda^2 = \frac{\tau_f}{2G_f}\left(\frac{1}{E_pt_p} + \frac{b_p}{E_st_sb_s}\right)$ $\lambda_1^2 = \frac{2G_f}{\tau_f\delta_1}\lambda^2$ $\lambda_2^2 = \frac{2G_f}{\tau_f(\delta_f-\delta_2)}\lambda^2$
[86]	$\delta_1\sqrt{\frac{2E_pt_p}{G_f(1+2\beta)}}$	-	$b_p\sqrt{2G_fE_pt_p(1+2\alpha)}$	$\alpha = \frac{b_pE_pt_p}{b_sE_st_s}$

2.5. Bond–Slip Models

The bond–slip model presents a formula that predicts the interfacial fracture energy based on the adhesive properties [59]. The effective bond length and bond strength can be achieved by the bond–slip relationship. This relationship can be experimentally determined from strain gauges along the bond length through bonded joint tests [62]. The bond–slip curve for linear adhesive is different from that of nonlinear adhesive. Linear adhesives have an approximately bi-linear shape. However, nonlinear adhesives exhibit an approximately trapezoidal shape, as presented in Table 4. As for the CFRP/concrete joints, the bond–slip curve always has a roughly bi-linear shape due to the brittle behavior of concrete [34]. Therefore, the bond–slip models for CFRP/steel joints are different and should be developed based on the adhesive behavior. A number of bond–slip models have been proposed for CFRP/steel bonded joints, as shown in Table 5.

Table 4. Type of bond–slip models.

Ref.	Type of Model	Bond–Slip Model	Bond–Slip Curves
[100]	Bi-linear	$\begin{cases} \tau = \tau_f\frac{\delta}{\delta_1} & \delta \leq \delta_1 \\ \tau = \tau_f\frac{\delta_f-\delta}{\delta_f-\delta_1} & \delta_1 < \delta \leq \delta_f \\ \tau = 0 & \delta > \delta_f \end{cases}$	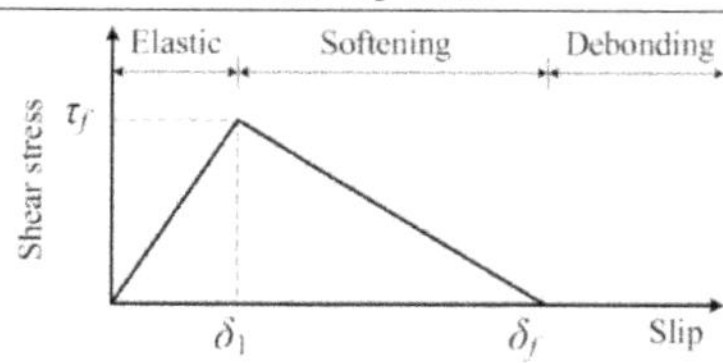
[101]	Simplified	$\begin{cases} \tau = \tau_f\sqrt{\frac{\delta}{\delta_1}} & \delta \leq \delta_1 \\ \tau = \tau_f\exp\left[-\alpha\left(\frac{\delta}{\delta_1} - 1\right)\right] & \delta > \delta_1 \\ \alpha = \frac{3\tau_f\delta_1}{3G_f-2\tau_f\delta_1} \end{cases}$	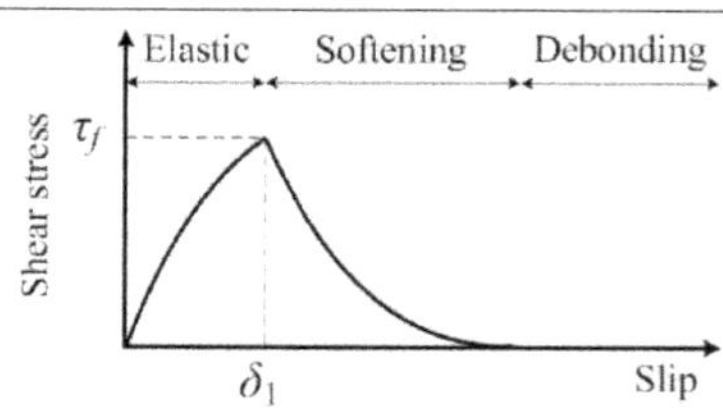

Table 4. *Cont.*

Ref.	Type of Model	Bond–Slip Model	Bond–Slip Curves
[98]	Tri-linear	$\begin{cases} \tau = \tau_f \frac{\delta}{\delta_1} & \delta \leq \delta_1 \\ \tau = \tau_f & \delta_1 < \delta \leq \delta_2 \\ \tau = \tau_f \frac{\delta_f - \delta}{\delta_f - \delta_2} & \delta_2 < \delta \leq \delta_f \\ \tau = 0 & \delta > \delta_f \end{cases}$	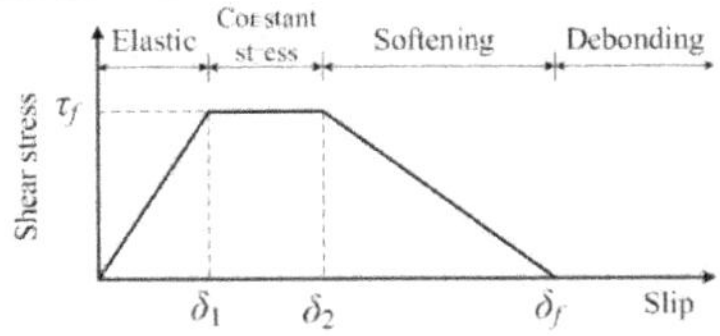

The bond–slip models are categorized as bi-linear, tri-linear, and simplified models. Bi-linear and simplified models were proposed for FRP/concrete and FRP/steel interfaces. The parameters of the bond–slip models for FRP/concrete bonded joints are expressed based on the tensile strength of the concrete, as the concrete is commonly the weakest link of the joint [59]. However, for CFRP/steel joints, these parameters are defined based on the tensile strength of adhesive since the weakest link of the joint is usually adhesive. In addition, the tri-linear model was presented for only CFRP/steel interfaces and utilized for nonlinear adhesive. It should be noted that the area under the curve is representative of the interfacial fracture energy in these models [102,103].

Table 5. Bond–slip models proposed for CFRP/steel interface.

Ref.	Type of Model	τ_f	δ_1	δ_2	δ_f
[63]	Bi-linear	$0.8f_a$	$0.8\frac{t_a}{G_a}f_a$	N/A	$\frac{2G_f}{\tau_f}$
[67]	Bi-linear	f_a	$\frac{t_a}{10}$	N/A	$\begin{cases} \frac{a}{4} & if\ t_a = 0.1–0.5\ \text{mm} \\ 0.125 + \frac{t_a - 0.5}{10} & if\ t_a = 0.5–1\ \text{mm} \end{cases}$
[59]	Bi-linear	$0.9f_a$	$0.3\left(\frac{t_a}{G_a}\right)^{0.65}f_a$	N/A	$\frac{2G_f}{\tau_f}$
[59]	Simplified	$0.8f_a$	$0.3\left(\frac{t_a}{G_a}\right)^{0.65}f_a$	N/A	-
[98]	Tri-linear	$0.8f_a$	$0.8\frac{t_a}{G_a}f_a$	$\frac{\delta_f}{3}$	$\frac{3G_f}{2\tau_f} + \frac{3}{4}\delta_1$
[49]	Bi-linear	$0.9f_a$	$2.61\frac{{t_a}^{0.34}}{G_a}f_a$	N/A	$166.2\frac{{t_a}^{0.4}}{{E_a}^{1.7}}{f_a}^{2.4}$
[49]	Tri-linear	$0.9f_a$	$2.61\frac{{t_a}^{0.34}}{G_a}f_a$	$55.4\frac{{t_a}^{0.4}{f_a}^{2.4}}{{E_a}^{1.7}} + 0.85\frac{{t_a}^{0.34}f_a}{G_a}$	$\frac{2}{3}\left(\frac{2G_f}{\tau_f} + \delta_1\right)$
[104]	Bi-linear	$0.544\tau^{*1.21}$	$1.51\frac{{t_a}^{0.378}}{G_a}\tau^{*1.21}$	N/A	$\frac{2G_f}{\tau_f}$

3. Flexural Retrofitting of Steel Beams

This section reviews the studies related to the flexural behavior of the steel structures retrofitted with CFRP. Various modes of failure in CFRP retrofitted beams were reported, as shown in Figure 4, such as (a) CFRP end-debonding and intermediate-debonding; (b) CFRP delamination or rupture; (c) bending failure in which yielding occurs in beam flanges and web; (d) local bucking, which occurs in compression flange or web; (e) lateral-torsional buckling [18]. To investigate CFRP retrofitting technique efficiency, the lateral and local buckling of steel beams should be controlled or eliminated [105]. Among the aforementioned failure modes, the debonding of the CFRP has been found as the dominant failure mode for the specimens retrofitted with NM CFRP [18,106–109]. CFRP end-debonding is the common failure mode for beams that are retrofitted in flexural yielding [18]. It should be noted that if a longer CFRP plate is applied, the failure mode could turn into a different failure mode, in particular, intermediate-debonding [109]. On the other hand, the CFRP rupture was reported as the common failure mode for HM and UHM CFRP [5,13]. This phenomenon occurs due to the lower rupture strain of CFRP featuring a higher elastic modulus and the reduced stresses of the adhesives at the ends [39]. However, strengthened steel beams using UHM CFRP strips showed debonding failure at strip panel finger joints due to high shear stresses [110]. The most studied parameters in existing research investigations of CFRP retrofitted structures in flexure include adhesive thickness, adhesive type, CFRP configurations, CFRP thickness, and CFRP bond length. Details of previous studies on the flexural behavior of the CFRP-strengthened and rehabilitated steel beams are presented in Table 6.

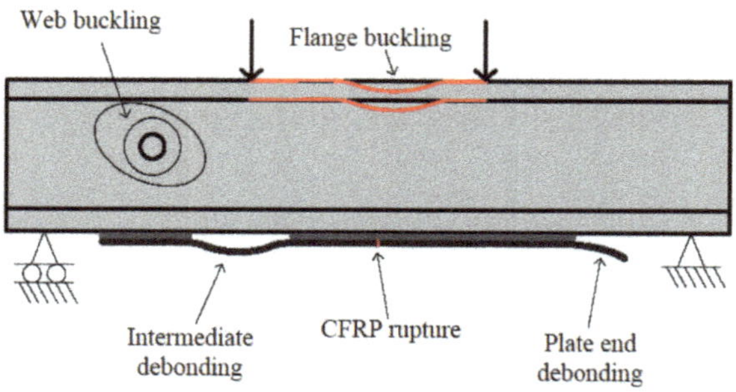

Figure 4. Some of the failure modes of steel beams strengthened with FRP plate [111].

Experimental observations on the effect of CFRP retrofitting systems on steel–concrete composite beams indicated a significant strength increase in specimens retrofitted with CFRP [13,112]. In these cases, concrete crushing and CFRP rupture were reported as the main failure modes. In addition, it was suggested to attach CFRP laminates to beam webs which can lead to a decrease in the interfacial shear stress of the adhesive between the bottom CFRP plate and the steel due to shifting the neutral axis to a lower level [112]. The application of small-diameter strands of CFRP was also found effective in flexural strengthening of steel bridge girder, especially for NM CFRP [113].

The flexural capacity of the steel beams can also be enhanced by using CFRP laminates with higher tensile strength or increased thickness [15]. Furthermore, the CFRP laminates with higher elastic modulus have a better performance in enhancing the ductility of the strengthened beams [112]. Failure analysis and structural behavior of CFRP-strengthened steel beams show that the load-carrying capacity with thicker and longer CFRP plates increases, though excessive thickness could cause premature debonding [114].

In order to prevent debonding failure and use the full capacity of the CFRP, mechanical anchorage systems have been utilized. It was shown that the mechanically fastened CFRP technique outperformed CFRP sheets in flexural strength and ductility [115]. Moreover, the application of the anchorage system to hybrid FRP also resulted in the same results [17]. Furthermore, in this case, prestressing CFRP increases the efficiency of the retrofitting method [116]. In addition, this technique can delay crack propagation and also results in an increase in the fatigue life of specimens [117].

A comparison between the performance of different FRP techniques as well as using steel plates for retrofitting reveal that HM CFRP is superior to the other rehabilitation methods in terms of fatigue behavior [118]. CFRP can significantly extend the fatigue life of the steel specimens and increase the average number of load cycles to failure [4]. Furthermore, CFRP materials can enhance the flexural fatigue behavior of the steel beams in terms of elastic stiffness, yield load, and nominal capacity [119]. The CFRP retrofitting method can also contribute to a reduction in fatigue crack growth in damaged steel beams [20]. In this regard, the elastic modulus of the CFRP is also found an important parameter in enhancing the fatigue life of specimens. Specimens retrofitted with higher elastic modulus outperform those retrofitted with CFRP with lower elastic modulus [120]. Furthermore, the stress intensity factor can be affected by the number of CFRP layers. It was shown that the stress intensity factor in specimens with a double CFRP configuration is significantly less than of the specimens strengthened with a single CFRP layer [24].

Table 6. Experimental flexural tests on CFRP retrofitted steel specimens.

Ref.	Loading Condition	Test Setup	L_{span} (mm)	Steel Cross Section (mm)	E_s (GPa)	f_y (MPa)	f_u (MPa)	CFRP Type	t_p (mm)	E_p (GPa)	f_p (MPa)	Adhesive Type	t_a (mm)	E_a (GPa)	f_a (MPa)	τ_a (MPa)
[112]	Static	F *	3048	** W8 × 15	-	363.4	496.4	NM HM	1.4	152 200	2482 >2482	-	-	-	68.9	-
[4]	Fatigue	F	1300	S127 × 4.5	194.4 (flange) 199.9 (web)	336.4 (flange) 330.9 (web)	-	NM	-	144	2137	-	-	-	-	-
[32]	Static	T *	2743.2	W12 × 14	-	-	-	HM	1.4	>200	>2300	-	-	-	-	-
[121]	Fatigue	F	-	W14 × 68	-	340	-	NM	-	157 114	2600 790	Sikadur 330 Sikadur 30 PLUS 25 DP-460 NS Tyfo TC	-	4.6 - - 1.8 -	41 25 17 35 47	-
[122]	Static Fatigue	F	1800	W100 × 17.5	200	-	-	HM	1.4	300	1800	Sikadur 30	-	2.75	-	-
[119]	Fatigue	F	-	W200 × 19	200	380	-	UHM	-	460	-	-	-	-	-	-
[13]	Static	F	6550	** W310 × 45	-	369–408	-	NM UHM	3.2 2.9 4	229 457	1224 1534	Spabond-345	-	-	-	-
[106]	Static	T	2500	HEA140	-	331	469	NM	1.4	197	>2800	Sikadur 30 Sikadur-330	-	4.5 3.8	24.8 30	24.8 -
[108]	Static	T and F	1200	127 × 76UB13	205	-	-	HM	3 6	212	-	Sikadur 30	1	8	29.7	-
[15]	Static	F	2000	HEA180	212	330	-	NM HM	1.4 1.8 2.4	200 330 165	3300 1500 3100	-	2	7 4.5	25 30	-
[120]	Fatigue	F	-	W310 × 74	200	350	450	NM HM	-	165 210	2520 2805	Sikadur-30	-	-	-	-
[39,110]	Static	F	2743	** W10 × 22 C7 × 9.8	200	407 414	510 531	UHM	1.2	514	1923	Spabond-345	2.5	3	34.6	-
[85]	Static	F	2000	I-section beam: w_b = 100 h_b = 150 t_f = 10 t_w = 6.6	-	250	370	NM	1.2	165	3100	Sikadur-30	1	11.2	24–31	14–19

Table 6. *Cont.*

Ref.	Loading Condition	Test Setup	L_{span} (mm)	Steel Cross Section (mm)	E_s (GPa)	f_y (MPa)	f_u (MPa)	CFRP Type	t_p (mm)	E_p (GPa)	f_p (MPa)	Adhesive Type	t_a (mm)	E_a (GPa)	f_a (MPa)	τ_a (MPa)
[118]	Fatigue	F	-	H350 × 175	-	-	-	UHM NM SW-BFRP	-	436.4 145 108.3	1500 2500 1789	Sikadur-30	-	2.627	31.7	14.4
[115]	Static	F	2000	** UC203 × 203 × 46	200	318.75	459	LM	-	65.364	736.6	Sikadur-330	-	4.5	30	-
[113]	Static/cyclic/fatigue	F	3350	W8 × 13	-	-	-	NM HM	-	134.6 226.3	2288 2218	-	-	-	-	-
[20]	Fatigue	F	1000	IPE120	208	330	444	NM	2.8	195	>2800	Sikadur 30 Sikadur 330	-	>4.5 >3.8	>28.4 >30	-
[22]	Fatigue	F	-	IPE120	199.3	383	462	NM HM UHM	1.4	159 220 440	2800 2800 1200	Araldite AW106	-	-	-	-
[24]	Fatigue	F	1000	IPE120	208	330	444	HM	1.4	205	3200	Sikadur 30 Sikadur 330	-	>4.5 >3.8	>28.4 >30	-
[17]	• Static	T	3000	UB203 × 102 × 23	190	335	429	Hybrid CFRP-GFRP	3.175 6.35	62.19	852	-	-	-	-	-
[123]	Fatigue	F	-	I-shaped	192.8	378.2	519	NM	-	200.4	3022.4	-	-	-	-	-
[105]	Static	F	2000	H-section	185	210.4	332.1	NM	1.2	167.3	2398.3	Sikadur-30	-	11.3	22.3	-
[123]	Fatigue	F	-	I-shaped	192.8	378.2	519	HM	-	200.4	3022.4	Araldite 420 Sikadur 30	-	1.495 11.2	29 30	-
[27]	Fatigue	F	-	I-shaped	192.8	378.2	519	HM	-	200.4	3022.4	Araldite 420 Sikadur 30	-	1.495 11.2	29 30	-
[124]	Fatigue	F	1300	H-section	197.3	158.8	-	NM	2	183.2	2239.5	Lica-131	-	5.7	39.2	-

* Test setup: T = Three-point bending/F = Four-point bending; ** Steel/concrete composite beam.

As for the type of adhesives, it was found that adhesives with relatively high ductility can effectively redistribute stress within the adhesive layer during increased loading [121]. Consequently, it is suggested to use nonlinear adhesives for retrofitting steel structures as the specimens bonded with linear adhesives are more susceptible to interfacial debonding [123].

Prestressing CFRP sheets can effectively increase the fatigue life of cracked steel elements as compressive forces are applied to the crack edges, thereby hindering crack growth in steel structures [21]. It was also shown that the prestressed CFRP can decrease the interfacial stress at the notch location and delay interfacial debonding [124]. Also, the level of the prestressing of the CFRP has a significant effect on extending the fatigue life of strengthened specimens [120]. The minimum level for prestressing the CFRP needed to increase the fatigue life of steel beams was determined in the literature [125].

4. Numerical Analysis

Computational methods present a cost-effective tool for a better understanding of the performance of the retrofitted components. In this regard, many numerical studies have been proposed to investigate the bond behavior of the retrofitted elements with CFRP.

Peiris [39] investigated a numerical study on the bond behavior of CFRP/steel joints using ANSYS (2009) software for NM and UHM CFRP by analyzing double-strap joint and doubly reinforced steel plates. An eight-node element of SOLID45 was used due to its large deflection and strain capabilities. The CFRP layer and adhesive were modeled as a single layer of elements. The results showed that the numerical element strains are in good agreement with experimental results except at the gap of the joints. Furthermore, the tensile stress obtained from finite element (FE) analysis is less than the experimental ones for ultra-high modulus CFRP. In this study, steel beams strengthened with UHM CFRP were also modeled. Contrary to CFRP/steel joint models, 4-node SHELL181 elements were utilized to simulate beams to reduce the computational costs. However, the proposed model could not predict the failure of the beam or the laminate for strip panel configuration as a result of the simplified assumptions of the SHELL181 element.

Wu and Zhao [60] modeled CFRP/steel joints using ABAQUS software, utilizing CPE4R elements for the CFRP and steel adherents and COH2D4 elements for the adhesive layer. The tie constraint, used in ABAQUS to establish perfect interface connection, was applied to the CFRP/adhesive and steel/adhesive surfaces. The numerical findings aligned well with experimental results in terms of the ultimate load, and bond–slip relationships. Al-Mosawe et al. [46] presented a numerical investigation of the effect of high load rates on the bond behavior of CFRP/steel double strap joints using ABAQUS for LM and NM CFRP. The C3D8R, SC8R, and COH3D8 elements were utilized to model steel plates, CFRP laminate, and the adhesive layer, respectively. The tie constraint was also applied to the adhesive layer with steel and CFRP. It was found that the effective bond length for specimens subjected to high loading rates was shorter compared to the specimens tested under quasi-static loading.

The flexural behavior of steel beams strengthened with CFRP sheets using ANSYS was studied by Elkhabeery et al. [18]. SOLID186 elements were selected to model steel and CFRP sheets. To model the epoxy layer, the INTER204 element was used. The adhesive layer was defined as linear elastic material and the mixed-mode bi-linear cohesive zone model (CZM) was selected to model the bond between steel and CFRP. The parametric study showed that the CFRP system is very efficient in strengthening compact mono-symmetric sections, whereas its effect is very negligible for non-compact sections. The CFRP/steel double overlapped bonded joints were simulated by Yang et al. [77]. It was revealed that the initial stiffness of the load versus slip relationship increases with the elastic modulus of CFRP laminate. Colombi et al. [106] carried out a numerical study on the static behavior of the steel beams strengthened with CFRP strips. A standard two nodes beam element was applied to model the beam and a standard eight-node plane stress element was selected to model CFRP strips as well as the adhesive layer. To ensure the compatibility of the deformations of the beams and strips, multipoint constraints were imposed between the beam nodes and the corresponding adhesive nodes. The adhesive stresses estimated by the FE model were in good agreement with their counterparts obtained from the analytical approach. A 2D FE model was proposed by Lenwari [122] to investigate the debonding strength of steel beams strengthened with CFRP. To this end, the eight-node element with two degrees of freedom at each node was utilized. Furthermore, it was assumed that the CFRP material is isotropic. The results indicated that the adhesive modulus, the CFRP thickness, and the CFRP modulus significantly affect the debonding strength. Fernando [59] studied the prediction of the debonding failure in RHS steel tubes strengthened with CFRP using a bond–slip model under an end-bearing load. C3D8, S4R, and COH3D8 elements were used to model the bearing plate, CFRP plate, and adhesive layer. The tie constraint was applied to adhesive surfaces connected to the CFRP plate and tub web. The results indicated that the debonding process in the FE model is much more gradual which can lead to higher stiffness in the load–displacement curve prior to reaching the ultimate load. A numerical investigation on the interfacial behavior of the bond between CFRP laminate and steel beam was presented by Linghoff et al. [126]. In this study, all parts were modeled using C3D20R solid elements and the common nodes at the interfaces were merged. It is found that interfacial shear stress as well as peel stress are not uniformly distributed over the width of the adhesive layer. Furthermore, the distribution of the peeling stresses along the width of the bond-line is not the same for different strengthened beams. This variation is attributed to the higher axial stiffness of the CFRP laminate as a major parameter. Hmidan et al. [16] investigated the flexural behavior of the steel beams repaired with CFRP sheets with various initial crack configurations by simulating a three-dimensional model in ANSYS. SOLID45 and LINK8 elements were utilized to model steel beams and CFRP sheets, respectively. These elements can be connected using interface elements as they have the same degree of freedom. The COMIN39 interface element was used to predefine crack propagation at the midspan of the beam. Results showed that the influence of the initial damage level on the failure mode of the repaired beam is negligible, whereas the damage level can have an effect on the web fracture rate of the beams. Moreover, the level of the initial damage determines the initiation of CFRP debonding. Deng [127] studied the rehabilitation of notched steel beams using CFRP plates. In this regard, a mixed-mode cohesive law was used to model notched retrofitted steel beams. The findings show that increasing the CFRP elastic modulus and thickness can enhance the bearing capacity while reducing the ductility and leading to premature debonding failure. Wang et al. [128] explored the effectiveness of using externally bonded CFRP using ductile adhesive. A trapezoidal mixed-mode CZM was used to simulate the debonding behavior of the CFRP-strengthened steel beam. The findings revealed that thicker or shorter CFRP laminates resulted in higher interfacial stresses, leading to earlier debonding, whereas longer CFRP laminates delayed debonding by changing the stress transfer path.

Table 7 provides a summary of the numerical studies presented in the literature. The realistic representation of stress distribution and failure modes in the rehabilitated elements with CFRP demands high computational costs. However, simplifications of the assumptions may not capture all real-world behaviors. Therefore, finding a middle ground that takes these factors into account is really important.

Table 7. Summary of numerical studies.

Study	Software	Steel Element Type	CFRP Element Type	Adhesive Element Type	Interactions
[39]	ANSYS (2009)	8-node SOLID45	8-node SOLID45	8-node SOLID45	Perfect interface
[60]	ABAQUS (V6.8)	CPE4R	CPE4R	COH2D4	Quadratic traction damage initiation criterion
[46]	ABAQUS (V6.13)	C3D8R	SC8R	COH3D8	Tie constraint
[18]	ANSYS (V17)	SOLID186	SOLID186	INTER204	Mixed-mode bi-linear CZM
[77]	ABAQUS	T2D2	T2D2	COH2D4	Bi-linear bond–slip derived from experimental data
[106]	ABAQUS	2-node beam element	8-node plane stress element	8-node plane stress element	Multipoint constraints
[122]	Not specified	8-node element with 2 DOF per node	8-node element with 2 DOF per node	Not specified	Reciprocal work contour integral method
[59]	ABAQUS (2004)	C3D8	S4R	COH3D8	Tie constraint
[126]	ABAQUS (V6.4.1)	C3D20R	C3D20R	C3D20R	Common nodes merged
[16]	ANSYS	SOLID45	LINK8	COMBIN39	Bi-linear bond–slip model
[127]	ABAQUS	C3D8I	COH3D8	COH3D8	Mixed-mode cohesive law
[128]	ABAQUS	C3D8I	C3D8R	COH3D8	Trapezoidal mixed-mode CZM

5. Research Needs and Recommendations

A review of all the investigations regarding the application of CFRP materials for the retrofitting of steel members suggests that research in this area is rather limited. The number of parameters affecting the behavior of CFRP retrofitted steel members increases the complexity of their behavior. Accordingly, many experimental programs are needed to investigate the effect of these parameters. Furthermore, there are limited numerical investigations that can accurately predict the behavior of CFRP retrofitted steel components. Hence, special attention needs to be devoted to developing reliable numerical modeling. Although the applied CFRP in the literature includes various elastic moduli, investigations involving UHM CFRP have not been adequately documented. Therefore, further research is needed in this regard. Further investigations on the behavior of steel components retrofitted with CFRP could be conducted in the following recommended research areas:

- Further investigation is needed to develop adhesives with enhanced mechanical properties to improve bond strength and durability.
- Further research should be conducted to investigate the effect of interrelated parameters on the bond behavior of the CFRP/steel interface to propose an optimal retrofitting system. The finite element modeling can be considered a cost-effective solution in this regard.
- To better analyze the impact of the CFRP elastic modulus on the performance of the retrofitted steel elements, it is advisable to use CFRPs with approximately the same tensile capacity but varying elastic modulus.
- More investigation is required to develop bond–slip models at the CFRP/steel interface under fatigue loading by considering the influencing variables that are representative of conditions in practice.
- More research is needed to investigate the effect of fatigue loading on the effective bond length. Also, the effect of shear combined with flexure on the bond length is to be clarified.
- More research could be conducted to investigate the effect of galvanic corrosion, especially in the CFRP retrofitting method utilizing a steel anchorage system. The long-term effect of galvanic corrosion has not been properly investigated.

6. Conclusions

A state of knowledge on the application of the CFRP in retrofitting steel elements as well as the influencing parameters is presented. Based on the findings obtained from the available literature, the following conclusions can be drawn:

- Using CFRP with higher elastic modulus results in an increase in CFRP/steel bond strength and contributes to an improvement of the performance of retrofitted steel components by increasing structural load-carrying capacity and flexural strength.
- Applying adhesives with higher tensile modulus generally results in enhanced bond strength of the steel/CFRP interface. Nevertheless, adhesives with nonlinear properties can yield higher failure loads than linear adhesives with an even higher tensile modulus.
- As for bond–slip models, studies have shown that in linear adhesive materials, triangular bond–slip curves are obtained, whereas in nonlinear adhesives, the bond–slip relationship tends to follow a trapezoidal curve.
- Proper surface preparation of steel substrates is crucial for achieving a strong bond strength. Mechanical treatments like grit blasting improve surface roughness and chemical bonding.
- Increasing the elastic modulus of CFRP reinforcement could lead to an improvement in the fatigue life of specimens. Indeed, it has been found that the fatigue life of a steel structure can be enhanced by either applying prestressing to the steel details or by increasing beam stiffness by using UHM CFRP or adding CFRP layers.
- Regarding the performance of prestressed CFRP, experimental results indicate that prestressing can reduce the stress intensity factor and confine crack growth by applying compressive forces to the edges of cracks in notched steel elements. Therefore, the use of prestressed CFRP patches could enhance the effectiveness of CFRP rehabilitation systems.
- Finally, experimental studies of anchorage systems have shown that using anchorage techniques can help delay crack propagation and thereby prolong fatigue life in strengthened steel specimens. The crack mouth opening displacement could also be reduced as a result of using anchorage systems.

Funding: This research received no external funding.

Data Availability Statement: Not applicable.

Conflicts of Interest: The authors declare no conflict of interest.

Notations

b_p	Width of CFRP plate
b_s	Width of steel plate
E_a	Young's modulus of adhesive
E_p	Young's modulus of CFRP plate
E_s	Young's modulus of steel plate
h_b	Height of beam
f_a	Tensile strength of adhesive
f_p	CFRP tensile strength
f_y	Steel yield stress
f_u	Steel ultimate stress
G_a	Shear modulus of adhesive
G_f	Interfacial fracture energy
L_e	Effective bond length of CFRP plate
$L_{span,b}$	Span length of steel beam
N	Number of interfaces working in parallel
P_u	Ultimate load (bond strength)
t_a	Thickness of adhesive layer

t_p	Thickness of CFRP plate
t_s	Thickness of steel plate
W_b	Width of beam
W_s	Steel plate width
τ_f	Peak interfacial shear stress
τ^*	Interlaminar shear strength of the CFRP plate
δ_1	Relative slip corresponding to the peak interfacial stress
δ_2	Relative slip when the shear stress begins to decrease in the tri-linear model
δ_f	Maximum slip

References

1. Dexter, R.J.; Ocel, J.M. *Manual for Repair and Retrofit of Fatigue Cracks in Steel Bridges*; Federal Highway Administration: Ashburn, VA, USA, 2013.
2. Schnerch, D.; Dawood, M.; Rizkalla, S.; Sumner, E. Proposed design guidelines for strengthening of steel bridges with FRP materials. *Constr. Build. Mater.* **2007**, *21*, 1001–1010. [CrossRef]
3. Fisher, J.W. Evolution of fatigue-resistant steel bridges. *Transp. Res. Rec. J. Transp. Res. Board* **1997**, *1594*, 5–17. [CrossRef]
4. Tavakkolizadeh, M.; Saadatmanesh, H. Fatigue strength of steel girders strengthened with carbon fiber reinforced polymer patch. *J. Struct. Eng.* **2003**, *129*, 186–196. [CrossRef]
5. Zhao, X.-L.; Zhang, L. State-of-the-art review on FRP strengthened steel structures. *Eng. Struct.* **2007**, *29*, 1808–1823. [CrossRef]
6. Hollaway, L.C.; Cadei, J. Progress in the technique of upgrading metallic structures with advanced polymer composites. *Prog. Struct. Eng. Mater.* **2002**, *4*, 131–148. [CrossRef]
7. Kamruzzaman, M.; Jumaat, M.Z.; Sulong, N.H.R.; Islam, A.B.M.S. A review on strengthening steel beams using FRP under fatigue. *Sci. World J.* **2014**, *2014*, 702537. [CrossRef] [PubMed]
8. Oudah, F.; El-Hacha, R. Research progress on the fatigue performance of RC beams strengthened in flexure using Fiber Reinforced Polymers. *Compos. Part B Eng.* **2013**, *47*, 82–95. [CrossRef]
9. Chataigner, S.; Benzarti, K.; Foret, G.; Caron, J.; Gemignani, G.; Brugiolo, M.; Calderon, I.; Piñero, I.; Birtel, V.; Lehmann, F. Design and testing of an adhesively bonded CFRP strengthening system for steel structures. *Eng. Struct.* **2018**, *177*, 556–565. [CrossRef]
10. Hu, B.; Li, Y.; Jiang, Y.-T.; Tang, H.-Z. Bond behavior of hybrid FRP-to-steel joints. *Compos. Struct.* **2020**, *237*, 111936. [CrossRef]
11. Jagtap, P.; Pore, S.; Prakash, V. Necessity of strengthening of steel structures with FRP composites: A review. *Int. J. Latest Trends Eng. Technol. (IJLTET)* **2015**, *5*, 390–394.
12. Shaat, A.; Fam, A. Repair of cracked steel girders connected to concrete slabs using carbon-fiber-reinforced polymer sheets. *J. Compos. Constr.* **2008**, *12*, 650–659. [CrossRef]
13. Schnerch, D.; Rizkalla, S. Flexural strengthening of steel bridges with high modulus CFRP strips. *J. Bridg. Eng.* **2008**, *13*, 192–201. [CrossRef]
14. Fam, A.; MacDougall, C.; Shaat, A. Upgrading steel–concrete composite girders and repair of damaged steel beams using bonded CFRP laminates. *Thin-Walled Struct.* **2009**, *47*, 1122–1135. [CrossRef]
15. Linghoff, D.; Al-Emrani, M.; Kliger, R. Performance of steel beams strengthened with CFRP laminate—Part 1: Laboratory tests. *Compos. Part B Eng.* **2010**, *41*, 509–515. [CrossRef]
16. Hmidan, A.; Kim, Y.J.; Yazdani, S. CFRP Repair of Steel Beams with Various Initial Crack Configurations. *J. Compos. Constr.* **2011**, *15*, 952–962. [CrossRef]
17. Sweedan, A.M.; Alhadid, M.M.; El-Sawy, K.M. Experimental study of the flexural response of steel beams strengthened with anchored hybrid composites. *Thin-Walled Struct.* **2016**, *99*, 1–11. [CrossRef]
18. Elkhabeery, O.H.; Safar, S.S.; Mourad, S.A. Flexural strength of steel I-beams reinforced with CFRP sheets at tension flange. *J. Constr. Steel Res.* **2018**, *148*, 572–588. [CrossRef]
19. Wu, C.; Zhao, X.L.; Chiu, W.K.; Al-Mahaidi, R.; Duan, W.H. Effect of fatigue loading on the bond behaviour between UHM CFRP plates and steel plates. *Compos. Part B Eng.* **2013**, *50*, 344–353. [CrossRef]
20. Colombi, P.; Fava, G. Experimental study on the fatigue behaviour of cracked steel beams repaired with CFRP plates. *Eng. Fract. Mech.* **2015**, *145*, 128–142. [CrossRef]
21. Emdad, R.; Al-Mahaidi, R. Effect of prestressed CFRP patches on crack growth of centre-notched steel plates. *Compos. Struct.* **2015**, *123*, 109–122. [CrossRef]
22. Ghafoori, E. Fatigue Strengthening of Metallic Members Using Un-Bonded and Bonded CFRP Laminates. Ph.D. Thesis, ETH Zurich, Zurich, Switzerland, 2015.
23. Aljabar, N.; Zhao, X.; Al-Mahaidi, R.; Ghafoori, E.; Motavalli, M.; Powers, N. Effect of crack orientation on fatigue behavior of CFRP-strengthened steel plates. *Compos. Struct.* **2016**, *152*, 295–305. [CrossRef]
24. Colombi, P.; Fava, G. Fatigue crack growth in steel beams strengthened by CFRP strips. *Theor. Appl. Fract. Mech.* **2016**, *85*, 173–182. [CrossRef]
25. Hu, L.L.; Zhao, X.L.; Feng, P. Fatigue Behavior of Cracked High-Strength Steel Plates Strengthened by CFRP Sheets. *J. Compos. Constr.* **2016**, *20*, 04016043. [CrossRef]

26. Aljabar, N.; Zhao, X.; Al-Mahaidi, R.; Ghafoori, E.; Motavalli, M.; Koay, Y. Fatigue tests on UHM-CFRP strengthened steel plates with central inclined cracks under different damage levels. *Compos. Struct.* **2017**, *160*, 995–1006. [CrossRef]
27. Yu, Q.-Q.; Wu, Y.-F. Fatigue retrofitting of cracked steel beams with CFRP laminates. *Compos. Struct.* **2018**, *192*, 232–244. [CrossRef]
28. Tafsirojjaman, T.; Fawzia, S.; Thambiratnam, D.P.; Zhao, X.-L. FRP strengthened SHS beam-column connection under monotonic and large-deformation cyclic loading. *Thin-Walled Struct.* **2021**, *161*, 107518. [CrossRef]
29. Tong, L.; Xu, G.; Zhao, X.-L.; Yan, Y. Fatigue tests and design of CFRP-strengthened CHS gap K-joints. *Thin-Walled Struct.* **2021**, *163*, 107694. [CrossRef]
30. Wang, H.-T.; Wu, G.; Pang, Y.-Y.; Shi, J.-W.; Zakari, H.M. Experimental study on the bond behavior between CFRP plates and steel substrates under fatigue loading. *Compos. Part B Eng.* **2019**, *176*, 107266. [CrossRef]
31. Ke, L.; Li, C.; He, J.; Shen, Q.; Liu, Y.; Jiao, Y. Enhancing fatigue performance of damaged metallic structures by bonded CFRP patches considering temperature effects. *Mater. Des.* **2020**, *192*, 108731. [CrossRef]
32. Liu, X. *Rehabilitation of Steel Bridge Members with FRP Composite Materials*; University of Missouri-Rolla: Rolla, MO, USA, 2003.
33. Ghafoori, E.; Motavalli, M. Normal, high and ultra-high modulus carbon fiber-reinforced polymer laminates for bonded and un-bonded strengthening of steel beams. *Mater. Des.* **2015**, *67*, 232–243. [CrossRef]
34. Yu, T.; Fernando, D.; Teng, J.; Zhao, X. Experimental study on CFRP-to-steel bonded interfaces. *Compos. Part B Eng.* **2012**, *43*, 2279–2289. [CrossRef]
35. Stanford, K.A. Strengthening of Steel Structures with High Modulus Carbon Fiber Reinforced Polymers (CRRP) Materials: Bond and Development Length Study. Master's Thesis, North Carolina State University, Raleigh, NC, USA, 2009.
36. Gholami, M.; Sam, A.R.M.; Yatim, J.M.; Tahir, M.M. A review on steel/CFRP strengthening systems focusing environmental performance. *Constr. Build. Mater.* **2013**, *47*, 301–310. [CrossRef]
37. Borrie, D.; Al-Saadi, S.; Zhao, X.-L.; Raman, R.K.S.; Bai, Y. Bonded CFRP/Steel Systems, Remedies of Bond Degradation and Behaviour of CFRP Repaired Steel: An Overview. *Polymers* **2021**, *13*, 1533. [CrossRef]
38. Duncan, B. Developments in testing adhesive joints. In *Advances in Structural Adhesive Bonding*; Elsevier: Amsterdam, The Netherlands, 2010; pp. 389–436.
39. Peiris, N.A. *Steel Beams Strengthened with Ultra High Modulus CFRP Laminates*; University of Kentucky: Lexington, KY, USA, 2011.
40. Jiao, H.; Zhao, X.-L. CFRP strengthened butt-welded very high strength (VHS) circular steel tubes. *Thin-Walled Struct.* **2004**, *42*, 963–978. [CrossRef]
41. Fawzia, S.; Zhao, X.; Al-Mahaidi, R.; Rizkalla, S. Bond characteristics between CFRP and steel plates in double strap joints. *Adv. Steel Constr.* **2005**, *1*, 17–27.
42. He, J.; Xian, G.; Zhang, Y. Effect of moderately elevated temperatures on bond behaviour of CFRP-to-steel bonded joints using different adhesives. *Constr. Build. Mater.* **2020**, *241*, 118057. [CrossRef]
43. Zhou, H.; Fernando, D.; Torero, J.L.; Torres, J.P.; Maluk, C.; Emberley, R. Bond Behavior of CFRP-to-Steel Bonded Joints at Mild Temperatures: Experimental Study. *J. Compos. Constr.* **2020**, *24*, 04020070. [CrossRef]
44. Guo, X.; Wu, Z.; Yang, Y.; Bai, J.; Zhou, Q. A Study on the Bond–Slip Relationship of the CFRP–Steel Interface of CFRP Strengthened Steel. *Materials* **2022**, *15*, 4187. [CrossRef]
45. Li, Y.; Li, C.; He, J.; Gao, Y.; Hu, Z. Effect of functionalized nano-SiO_2 addition on bond behavior of adhesively bonded CFRP-steel double-lap joint. *Constr. Build. Mater.* **2020**, *244*, 118400. [CrossRef]
46. Al-Mosawe, A.; Al-Mahaidi, R.; Zhao, X.-L. Bond behaviour between CFRP laminates and steel members under different loading rates. *Compos. Struct.* **2016**, *148*, 236–251. [CrossRef]
47. He, J.; Xian, G. Debonding of CFRP-to-steel joints with CFRP delamination. *Compos. Struct.* **2016**, *153*, 12–20. [CrossRef]
48. Al-Mosawe, A.; Al-Mahaidi, R. Performance of CFRP-steel joints enhanced with bi-directional CFRP fabric. *Constr. Build. Mater.* **2019**, *197*, 72–82. [CrossRef]
49. Wang, H.-T.; Wu, G. Bond-slip models for CFRP plates externally bonded to steel substrates. *Compos. Struct.* **2018**, *184*, 1204–1214. [CrossRef]
50. Li, C.; Ke, L.; He, J.; Chen, Z.; Jiao, Y. Effects of mechanical properties of adhesive and CFRP on the bond behavior in CFRP-strengthened steel structures. *Compos. Struct.* **2019**, *211*, 163–174. [CrossRef]
51. Zhou, H.; Liu, H.Y.; Zhou, H.; Zhang, Y.; Gao, X.; Mai, Y.W. On adhesive properties of nano-silica/epoxy bonded single-lap joints. *Mater. Des.* **2016**, *95*, 212–218. [CrossRef]
52. Sousa, J.M.; Correia, J.R.; Cabral-Fonseca, S. Durability of an epoxy adhesive used in civil structural applications. *Constr. Build. Mater.* **2018**, *161*, 618–633. [CrossRef]
53. Ferreira, F.V.; Brito, F.S.; Franceschi, W.; Simonetti, E.A.N.; Cividanes, L.S.; Chipara, M.; Lozano, K. Functionalized graphene oxide as reinforcement in epoxy based nanocomposites. *Surf. Interfac.* **2018**, *10*, 100–109. [CrossRef]
54. Wan, Y.-J.; Gong, L.-X.; Tang, L.-C.; Wu, L.-B.; Jiang, J.-X. Mechanical properties of epoxy composites filled with silane-functionalized graphene oxide. *Compos. Part A Appl. Sci. Manuf.* **2014**, *64*, 79–89. [CrossRef]
55. Xie, G.-H.; You, Y.; Tao, Z.-A.; Li, S.-Q. Experimental and theoretical study on mechanical behaviors of CFRP-steel interface. *Thin-Walled Struct.* **2023**, *182*, 110208. [CrossRef]
56. Razavi, N.; Ayatollahi, M.R.; Giv, A.N.; Khoramishad, H. Single lap joints bonded with structural adhesives reinforced with a mixture of silica nanoparticles and multi walled carbon nanotubes. *Int. J. Adhes. Adhes.* **2018**, *80*, 76–86. [CrossRef]

57. Saraç, I.; Adin, H.; Temiz, Ş. Experimental determination of the static and fatigue strength of the adhesive joints bonded by epoxy adhesive including different particles. *Compos. Part B Eng.* **2018**, *155*, 92–103. [CrossRef]
58. Mohabeddine, A.; Malik, G.; Correia, J.; Silva, F.; De Jesus, A.; Fantuzzi, N.; Castro, J.M. Experimental parametric investigation on the behavior of adhesively bonded CFRP/steel joints. *Compos. Struct.* **2023**, *307*, 116598. [CrossRef]
59. Fernando, N.D. Bond Behaviour and Debonding Failures in CFRP-Strengthened Steel Members. Ph.D. Thesis, The Hong Kong Polytechnic University, Hong Kong, China, 2010.
60. Wu, C.; Zhao, X.; Duan, W.H.; Al-Mahaidi, R. Bond characteristics between ultra high modulus CFRP laminates and steel. *Thin-Walled Struct.* **2012**, *51*, 147–157. [CrossRef]
61. Wang, H.-T.; Wu, G.; Dai, Y.-T.; He, X.-Y. Experimental Study on Bond Behavior between CFRP Plates and Steel Substrates Using Digital Image Correlation. *J. Compos. Constr.* **2016**, *20*, 04016054. [CrossRef]
62. Wang, H.-T.; Wu, G.; Dai, Y.-T.; He, X.-Y. Determination of the bond–slip behavior of CFRP-to-steel bonded interfaces using digital image correlation. *J. Reinf. Plast. Compos.* **2016**, *35*, 1353–1367. [CrossRef]
63. Xia, S.; Teng, J. Behaviour of FRP-to-steel bonded joints. In *International Symposium on Bond Behaviour of FRP in Structures, BBFS 2005*; International Institute for FRP in Construction (IIFC): Hong Kong, China, 2005.
64. Photiou, N.K. *Rehabilitation of Steel Members Utilising Hybrid FRP Composite Material Systems*; University of Surrey (United Kingdom): Guildford, UK, 2005.
65. Fawzia, S. Bond Characteristics between Steel and Carbon Fibre Reinforced Polymer (CFRP) Composites. Ph.D. Thesis, Monash University, Clayton, Australia, 2007.
66. The Institution of Structural Engineers (ISE). *The Structural Use of Adhesives*; The Institution of Structural Engineers (ISE): London, UK, 1999.
67. Fawzia, S.; Zhao, X.-L.; Al-Mahaidi, R. Bond–slip models for double strap joints strengthened by CFRP. *Compos. Struct.* **2010**, *92*, 2137–2145. [CrossRef]
68. Heshmati, M.; Haghani, R.; Al-Emrani, M.; André, A. On the strength prediction of adhesively bonded FRP-steel joints using cohesive zone modelling. *Theor. Appl. Fract. Mech.* **2018**, *93*, 64–78. [CrossRef]
69. Fawzia, S. Evaluation of shear stress and slip relationship of composite lap joints. *Compos. Struct.* **2013**, *100*, 548–553. [CrossRef]
70. Wu, C.; Yu, Y.-Z.; Tam, L.-H.; Orr, J.; He, L. Effect of glass fiber sheet in adhesive on the bond and galvanic corrosion behaviours of CFRP-Steel bonded system. *Compos. Struct.* **2021**, *259*, 113218. [CrossRef]
71. Wu, C.; Yu, Y.-Z.; Tam, L.-H.; He, L. Effects of bondline defects on the bond behaviour of CFRP-steel double strap joints. *Compos. Struct.* **2023**, *308*, 116682. [CrossRef]
72. Schnerch, D.; Dawood, M.; Rizkalla, S.; Sumner, E.; Stanford, K. Bond behavior of CFRP strengthened steel structures. *Adv. Struct. Eng.* **2006**, *9*, 805–817. [CrossRef]
73. Colombi, P.; Poggi, C. Strengthening of tensile steel members and bolted joints using adhesively bonded CFRP plates. *Constr. Build. Mater.* **2006**, *20*, 22–33. [CrossRef]
74. Nguyen, T.-C.; Bai, Y.; Zhao, X.-L.; Al-Mahaidi, R. Curing effects on steel/CFRP double strap joints under combined mechanical load, temperature and humidity. *Constr. Build. Mater.* **2013**, *40*, 899–907. [CrossRef]
75. He, J.; Xian, G. Bond-slip behavior of fiber reinforced polymer strips-steel interface. *Constr. Build. Mater.* **2017**, *155*, 250–258. [CrossRef]
76. Peiris, A.; Harik, I. FRP-steel bond study of IM and UHM CFRP strips. *Constr. Build. Mater.* **2018**, *185*, 628–637. [CrossRef]
77. Yang, Y.; Zhao, J.; Zhang, S.; Chastre, C.; Biscaia, H. Effect of mechanical anchorage on the bond performance of double overlapped CFRP-to-steel joints. *Compos. Struct.* **2021**, *267*, 113902. [CrossRef]
78. Fernando, D.; Teng, J.G.; Yu, T.; Zhao, X.L. Preparation and characterization of steel surfaces for adhesive bonding. *J. Compos. Constr.* **2013**, *17*, 04013012. [CrossRef]
79. Schnerch, D.A. Strengthening of Steel Structures with High Modulus Carbon Fiber Reinforced Polymer (CFRP) Materials. Ph.D. Thesis, North Carolina State University, Raleigh, NC, USA, 2005.
80. Lorenzo, M.A. Experimental Methods for Evaluating Epoxy Coating Adhesion to Steel Reinforcement. Master's Thesis, University of Texas at Austin, Austin, TX, USA, 1997.
81. Baldan, A. Adhesively-bonded joints and repairs in metallic alloys, polymers and composite materials: Adhesives, adhesion theories and surface pretreatment. *J. Mater. Sci.* **2004**, *39*, 1–49. [CrossRef]
82. Ebnesajjad, S.; Ebnesajjad, C. *Surface Treatment of Materials for Adhesive Bonding*; William Andrew: Norwich, NY, USA, 2013.
83. Ebnesajjad, S.; Landrock, A.H. *Adhesives Technology Handbook*, 2nd ed.; William Andrew: New York, NY, USA, 2014.
84. Islam, M.; Tong, L.; Falzon, P. Influence of metal surface preparation on its surface profile, contact angle, surface energy and adhesion with glass fibre prepreg. *Int. J. Adhes. Adhes.* **2014**, *51*, 32–41. [CrossRef]
85. Narmashiri, K.; Jumaat, M.Z.; Sulong, N.H.R. Strengthening of steel I-beams using CFRP strips: An investigation on CFRP bond length. *Adv. Struct. Eng.* **2012**, *15*, 2191–2204. [CrossRef]
86. Yang, Y.; Biscaia, H.; Chastre, C.; Silva, M.A. Bond characteristics of CFRP-to-steel joints. *J. Constr. Steel Res.* **2017**, *138*, 401–419. [CrossRef]
87. Fawzia, S.; Al-Mahaidi, R.; Zhao, X.-L. Experimental and finite element analysis of a double strap joint between steel plates and normal modulus CFRP. *Compos. Struct.* **2006**, *75*, 156–162. [CrossRef]

88. Bocciarelli, M.; Colombi, P.; Fava, G.; Poggi, C. Prediction of debonding strength of tensile steel/CFRP joints using fracture mechanics and stress based criteria. *Eng. Fract. Mech.* **2009**, *76*, 299–313. [CrossRef]
89. Chiew, S.; Yu, Y.; Lee, C. Bond failure of steel beams strengthened with FRP laminates—Part 1: Model development. *Compos. Part B Eng.* **2011**, *42*, 1114–1121. [CrossRef]
90. Al-Zubaidy, H.; Al-Mahaidi, R.; Zhao, X.-L. Experimental investigation of bond characteristics between CFRP fabrics and steel plate joints under impact tensile loads. *Compos. Struct.* **2012**, *94*, 510–518. [CrossRef]
91. Al-Zubaidy, H.; Al-Mahaidi, R.; Zhao, X.-L. Finite element modelling of CFRP/steel double strap joints subjected to dynamic tensile loadings. *Compos. Struct.* **2013**, *99*, 48–61. [CrossRef]
92. Korayem, A.H.; Li, C.Y.; Zhang, Q.H.; Zhao, X.L.; Duan, W.H. Effect of carbon nanotube modified epoxy adhesive on CFRP-to-steel interface. *Compos. Part B Eng.* **2015**, *79*, 95–104. [CrossRef]
93. Al-Zubaidy, H.A.; Zhao, X.-L.; Al-Mahaidi, R. Dynamic bond strength between CFRP sheet and steel. *Compos. Struct.* **2012**, *94*, 3258–3270. [CrossRef]
94. Huo, J.; Zhang, X.; Yang, J.; Xiao, Y. Experimental study on dynamic behavior of CFRP-to-steel interface. *Structures* **2019**, *20*, 465–475. [CrossRef]
95. Al-Zubaidy, H.A.; Zhao, X.-L.; Al-Mahaidi, R. Experimental evaluation of the dynamic bond strength between CFRP sheets and steel under direct tensile loads. *Int. J. Adhes. Adhes.* **2013**, *40*, 89–102. [CrossRef]
96. Chen, J.F.; Teng, J.G. Anchorage strength models for FRP and steel plates bonded to concrete. *J. Struct. Eng.* **2001**, *127*, 784–791. [CrossRef]
97. Bocciarelli, M.; Colombi, P.; Fava, G.; Poggi, C. Interaction of interface delamination and plasticity in tensile steel members reinforced by CFRP plates. *Int. J. Fract.* **2007**, *146*, 79–92. [CrossRef]
98. Dehghani, E.; Daneshjoo, F.; Aghakouchak, A.; Khaji, N. A new bond-slip model for adhesive in CFRP–steel composite systems. *Eng. Struct.* **2012**, *34*, 447–454. [CrossRef]
99. Fernando, D.; Yu, T.; Teng, J.G. Behavior of CFRP Laminates Bonded to a Steel Substrate Using a Ductile Adhesive. *J. Compos. Constr.* **2014**, *18*, 04013040. [CrossRef]
100. Monti, G.; Renzelli, M.; Luciani, P. FRP adhesion in uncracked and cracked concrete zones. In *Fibre-Reinforced Polymer Reinforcement for Concrete Structures: (In 2 Volumes)*; World Scientific: Singapore, 2003; pp. 183–192.
101. Lu, X.Z.; Teng, J.G.; Ye, L.P.; Jiang, J.J. Bond–slip models for FRP sheets/plates bonded to concrete. *Eng. Struct.* **2005**, *27*, 920–937. [CrossRef]
102. Pang, Y.-Y.; Wu, G.; Wang, H.-T.; Su, Z.-L.; He, X.-Y. Experimental study on the bond behavior of the CFRP-steel interface under the freeze–thaw cycles. *J. Compos. Mater.* **2019**, *54*, 13–29. [CrossRef]
103. Wang, H.-T.; Liu, S.-S.; Liu, Q.-L.; Pang, Y.-Y.; Shi, J.-W. Influences of the joint and epoxy adhesive type on the CFRP-steel interfacial behavior. *J. Build. Eng.* **2021**, *43*, 103167. [CrossRef]
104. Pang, Y.; Wu, G.; Wang, H.; Gao, D.; Zhang, P. Bond-slip model of the CFRP-steel interface with the CFRP delamination failure. *Compos. Struct.* **2021**, *256*, 113015. [CrossRef]
105. Zeng, J.-J.; Gao, W.-Y.; Liu, F. Interfacial behavior and debonding failures of full-scale CFRP-strengthened H-section steel beams. *Compos. Struct.* **2018**, *201*, 540–552. [CrossRef]
106. Colombi, P.; Poggi, C. An experimental, analytical and numerical study of the static behavior of steel beams reinforced by pultruded CFRP strips. *Compos. Part B Eng.* **2006**, *37*, 64–73. [CrossRef]
107. Sallam, H.E.M.; Ahmad, S.S.E.; Badawy, A.A.M.; Mamdouh, W. Evaluation of steel I-Beams strengthened by various plating methods. *Adv. Struct. Eng.* **2006**, *9*, 535–544. [CrossRef]
108. Deng, J.; Lee, M.M. Behaviour under static loading of metallic beams reinforced with a bonded CFRP plate. *Compos. Struct.* **2007**, *78*, 232–242. [CrossRef]
109. Teng, J.; Fernando, D.; Yu, T. Finite element modelling of debonding failures in steel beams flexurally strengthened with CFRP laminates. *Eng. Struct.* **2015**, *86*, 213–224. [CrossRef]
110. Peiris, A.; Harik, I. Steel beam strengthening with UHM CFRP strip panels. *Eng. Struct.* **2021**, *226*, 111395. [CrossRef]
111. Teng, J.; Yu, T.; Fernando, D. Strengthening of steel structures with fiber-reinforced polymer composites. *J. Constr. Steel Res.* **2012**, *78*, 131–143. [CrossRef]
112. Al-Saidy, A.H. *Structural Behavior of Composite Steel Beams Strengthened/Repaired with Carbon Fiber Reinforced Polymer Plates*; Iowa State University: Ames, IA, USA, 2001.
113. Tabrizi, S.; Kazem, H.; Rizkalla, S.; Kobayashi, A. New small-diameter CFRP material for flexural strengthening of steel bridge girders. *Constr. Build. Mater.* **2015**, *95*, 748–756. [CrossRef]
114. Narmashiri, K.; Sulong, N.R.; Jumaat, M.Z. Failure analysis and structural behaviour of CFRP strengthened steel I-beams. *Constr. Build. Mater.* **2012**, *30*, 1–9. [CrossRef]
115. Karam, E.C. Retrofitting of Composite Steel Beams Pre-Damaged in Flexure using Fiber Reinforced Polymers. Master's Thesis, American University of Sharjah, Sharjah, United Arab Emirates, 2015.
116. Wang, H.-T.; Bian, Z.-N.; Chen, M.-S.; Hu, L.; Wu, Q. Flexural strengthening of damaged steel beams with prestressed CFRP plates using a novel prestressing system. *Eng. Struct.* **2023**, *284*, 115953. [CrossRef]
117. Yu, Q.-Q.; Wu, Y.-F. Fatigue durability of cracked steel beams retrofitted with high-strength materials. *Constr. Build. Mater.* **2017**, *155*, 1188–1197. [CrossRef]

118. Wu, G.; Wang, H.-T.; Wu, Z.-S.; Liu, H.-Y.; Ren, Y. Experimental study on the fatigue behavior of steel beams strengthened with different fiber-reinforced composite plates. *J. Compos. Constr.* **2012**, *16*, 127–137. [CrossRef]
119. Dawood, M.; Rizkalla, S.; Sumner, E. Fatigue and overloading behavior of steel–concrete composite flexural members strengthened with high modulus CFRP materials. *J. Compos. Constr.* **2007**, *11*, 659–669. [CrossRef]
120. Vatandoost, F. Fatigue Behaviour of Steel Girders Strengthened with Prestressed CFRP Strips. Master's Thesis, University of Waterloo, Waterloo, ON, Canada, 2010.
121. Nozaka, K.; Shield, C.K.; Hajjar, J.F. Effective bond length of carbon-fiber-reinforced polymer strips bonded to fatigued steel bridge I-girders. *J. Bridg. Eng.* **2005**, *10*, 195–205. [CrossRef]
122. Lenwari, A.; Thepchatri, T.; Albrecht, P. Debonding strength of steel beams strengthened with CFRP plates. *J. Compos. Constr.* **2006**, *10*, 69–78. [CrossRef]
123. Yu, Q.-Q.; Wu, Y.-F. Fatigue behaviour of cracked steel beams retrofitted with carbon fibre–reinforced polymer laminates. *Adv. Struct. Eng.* **2017**, *21*, 1148–1161. [CrossRef]
124. Li, J.; Zhu, M.; Deng, J. Flexural behaviour of notched steel beams strengthened with a prestressed CFRP plate subjected to fatigue damage and wetting/drying cycles. *Eng. Struct.* **2022**, *250*, 113430. [CrossRef]
125. Ghafoori, E.; Motavalli, M.; Nussbaumer, A.; Herwig, A.; Prinz, G.; Fontana, M. Determination of minimum CFRP pre-stress levels for fatigue crack prevention in retrofitted metallic beams. *Eng. Struct.* **2015**, *84*, 29–41. [CrossRef]
126. Linghoff, D.; Al-Emrani, M. Performance of steel beams strengthened with CFRP laminate—Part 2: FE analyses. *Compos. Part B Eng.* **2010**, *41*, 516–522. [CrossRef]
127. Deng, J.; Li, J.; Wang, Y.; Xie, W. Numerical study on notched steel beams strengthened by CFRP plates. *Constr. Build. Mater.* **2018**, *163*, 622–633. [CrossRef]
128. Wang, Z.; Xian, G.; Yue, Q. Finite element modeling of debonding failure in CFRP-strengthened steel beam using a ductile adhesive. *Compos. Struct.* **2023**, *311*, 116818. [CrossRef]

MDPI AG
Grosspeteranlage 5
4052 Basel
Switzerland
Tel.: +41 61 683 77 34

Applied Sciences Editorial Office
E-mail: applsci@mdpi.com
www.mdpi.com/journal/applsci